CURRENT TOPICS IN

DEVELOPMENTAL BIOLOGY

VOLUME 16

NEURAL DEVELOPMENT
PART II

Neural Development in Model Systems

CONTRIBUTORS

RUBEN ADLER

G. AUGUSTI-TOCCO

JOHN BARRETT

STANLEY M. CRAIN

ROBERT L. DEHAAN

S. DENIS-DONINI

L. C. DOERING

S. FEDOROFF

LAWRENCE D. GROUSE

JAN K. S. JANSEN

RANDAL N. JOHNSTON

NICOLE LE DOUARIN

CAROL H. LETENDRE

RAMON LIM

TERJE LØMO

PHILLIP G. NELSON

T. S. OKADA

BIRGIT ROSE

BRUCE K. SCHRIER

SILVIO VARON

NORMAN K. WESSELLS

CURRENT TOPICS IN
DEVELOPMENTAL BIOLOGY

EDITED BY

A. A. MOSCONA

DEPARTMENTS OF BIOLOGY AND PATHOLOGY
THE UNIVERSITY OF CHICAGO
CHICAGO, ILLINOIS

ALBERTO MONROY

STAZIONE ZOOLOGICA
NAPLES, ITALY

VOLUME 16

NEURAL DEVELOPMENT
PART II

Neural Development in Model Systems

VOLUME EDITOR

R. KEVIN HUNT

THOMAS C. JENKINS DEPARTMENT OF BIOPHYSICS
THE JOHNS HOPKINS UNIVERSITY
BALTIMORE, MARYLAND

1980

ACADEMIC PRESS

A Subsidiary of Harcourt Brace Jovanovich, Publishers

New York   London   Toronto   Sydney   San Francisco

ACADEMIC PRESS, INC.
111 Fifth Avenue, New York, New York 10003

United Kingdom Edition published by
ACADEMIC PRESS, INC. (LONDON) LTD.
24/28 Oval Road, London NW1 7DX

LIBRARY OF CONGRESS CATALOG CARD NUMBER: 66–28604

ISBN 0–12–153116–3

PRINTED IN THE UNITED STATES OF AMERICA

80 81 82 83   9 8 7 6 5 4 3 2 1

CONTENTS

CHAPTER 8.   Colony Culture of Neural Cells as a Method for the Study of
Cell Lineages in the Developing CNS:
The Astrocyte Cell Lineage
S. FEDOROFF AND L. C. DOERING

CHAPTER 9.   Glia Maturation Factor
RAMON LIM

CHAPTER 10.   Molecular and Lectin Probe Analyses of
Neuronal Differentiation
S. DENIS-DONINI AND G. AUGUSTI-TOCCO

LIST OF CONTRIBUTORS

Numbers in parentheses indicate the pages on which the authors' contributions begin.

RUBEN ADLER, *Department of Biology, School of Medicine, University of California, San Diego, La Jolla, California 92093* (207)

G. AUGUSTI-TOCCO, *Laboratory of Quantum Electronics, 50127 Florence, Italy* (323)

JOHN BARRETT, *Department of Physiology and Biophysics, University of Miami School of Medicine, Miami, Florida 33101* (1)

STANLEY M. CRAIN, *Departments of Neuroscience and Physiology, and the Rose F. Kennedy Center for Research in Mental Retardation and Human Development, Albert Einstein College of Medicine, Yeshiva University, Bronx, New York 10461* (87)

ROBERT L. DeHAAN, *Department of Anatomy, Emory University School of Medicine, Atlanta, Georgia 30322* (117)

S. DENIS-DONINI, *Laboratory of Molecular Embryology, 80072 Arco Felice, Naples, Italy* (323)

L. C. DOERING, *Department of Anatomy, University of Saskatchewan, Saskatoon, Saskatchewan S7N 0W0, Canada* (283)

S. FEDOROFF, *Department of Anatomy, University of Saskatchewan, Saskatoon, Saskatchewan S7N 0W0, Canada* (283)

LAWRENCE D. GROUSE,* *Neurobiology Section, Laboratory of Developmental Neurobiology, National Institute of Child Health and Human Development, National Institutes of Health, Bethesda, Maryland 20205* (381)

JAN K. S. JANSEN, *Institute of Physiology, University of Oslo, Oslo, Norway* (253)

RANDAL N. JOHNSTON, *Department of Biological Sciences, Stanford University, Stanford, California 94305* (165)

NICOLE LE DOUARIN, *Institut d'Embryologie du CNRS et du Collège de France, 94130 Nogent-sur-Marne, France* (31)

* Present address: Journal of the American Medical Association, Chicago, Illinois 60610.

CAROL H. LETENDRE, *Intermediary Metabolism Section, Laboratory of Developmental Neurobiology, National Institute of Child Health and Human Development, National Institutes of Health, Bethesda, Maryland 20205* (381)

RAMON LIM, *Brain Research Institute, The University of Chicago, Chicago, Illinois 60637* (305)

TERJE LØMO, *Institute of Neurophysiology, University of Oslo, Oslo, Norway* (253)

PHILLIP G. NELSON, *Laboratory of Developmental Neurobiology, National Institute of Child Health and Human Development, National Institutes of Health, Bethesda, Maryland 20205* (381)

T. S. OKADA, *Institute of Biophysics, Faculty of Science, University of Kyoto, Kyoto 606, Japan* (349)

BIRGIT ROSE, *Department of Physiology and Biophysics, University of Miami School of Medicine, Miami, Florida 33101* (1)

BRUCE K. SCHRIER, *Neurobiology Section, Laboratory of Developmental Neurobiology, National Institute of Child Health and Human Development, National Institutes of Health, Bethesda, Maryland 20205* (381)

SILVIO VARON, *Department of Biology, School of Medicine, University of California, San Diego, La Jolla, California 92093* (207)

NORMAN K. WESSELLS, *Department of Biological Sciences, Stanford University, Stanford, California 94305* (165)

PREFACE

Developmental neurobiology has made extraordinary progress in recent decades, increasing the need for dialog between neuroscientists and cell and developmental biologists. Both the progress and the dialog have been enhanced by the application of model systems to classical problems in neuroembryology. Cell, tissue, and organ culture methods, together with the use of defined cell lines from the nervous system, have made their mark on problems as diverse as neural cell migration, differentiation of excitable membrane, and synaptogenesis; and they have begun to draw the rich strategies of *in vitro* cell biology to bear on neurogenesis. Likewise, the developing membrane of muscle and heart cells, the migration of neural crest, and the application of electrophysiologic methods to embryonic cell communication all provide insights for the developmental neurobiologist. The present volume, "Neural Development in Model Systems," is the second in a three-volume collection of essays on Neural Development, to appear as Volumes 15, 16, and 17 of *Current Topics in Developmental Biology.*

Tissue culture was launched into prominence when a great embryologist used it to test one of the central issues in turn-of-the century neuroscience, the neuron doctrine. It was Ross Harrison whose tissue culture system allowed him to demonstrate that the axon originates as a growing protoplasmic extension of the individual neuron. Model systems continue at the heart of contemporary research on neural development.

R. Kevin Hunt

CHAPTER 1

INTRODUCTION TO TECHNIQUES
IN DEVELOPMENTAL ELECTROBIOLOGY

Birgit Rose and John Barrett

DEPARTMENT OF PHYSIOLOGY AND BIOPHYSICS
UNIVERSITY OF MIAMI SCHOOL OF MEDICINE
MIAMI, FLORIDA

1

I. Introduction

Our present level of understanding of cell physiology owes much to the investigative power of electrophysiological methods. The usefulness of electrophysiological approaches in the study of basic cell function has long been recognized. Biologists have closely watched the development of ever more sensitive instruments for the measurement of electrical parameters such as potential, current, resistance, and capacitance and have eagerly applied the new technology toward the study of electrical phenomena such as cell resting potential, electrical excitability of nerve and muscle cells, cell-to-cell coupling, receptor mechanisms, and nerve transmission, to name a few. To the developmental biologist, electrophysiological techniques may also have much to offer. Only recently have biologists begun to use electrical measurements in probing such questions as: When during embryonic development do the specialized membrane properties first arise? When are the sensory inputs into the nervous system specified? When are chemical and electrical synapses first established? This chapter is intended as a basic introduction to electrobiological techniques, outlining the general principles involved and pointing out limitations and possible pitfalls. Examples from developmental biology are used to illustrate the application of some of the methods.

II. Recording with Extracellular Electrodes

Extracellular recording is the oldest measuring technique in electrobiology—yet it remains a valuable tool. Basically, an electrode is placed close to the cell(s) studied and the potential difference between this electrode and a reference electrode located elsewhere on the preparation is measured. Since extracellular electrodes leave the cell membrane intact, they are less likely to impair or alter cell function and/or electrical activity than intracellular electrodes which necessarily puncture the membrane. The risk of changing the cell's "true" membrane permeability properties is greatly reduced by measuring the extracellular potential changes produced by the membrane currents. However, because of the high conductivity of the medium surrounding the cells, usually only a relatively high local current density (such as is generated by action potentials but not by small, steady electrical gradients) will result in voltage changes detectable by conventional extracellular electrodes.

A. ELECTRODES FOR RECORDING FAST TRANSIENT MEMBRANE CURRENTS

A fine metal wire insulated down to its tip may serve as an extracellular electrode. For best mechanical rigidity of the electrode, tungsten wire is used, which may be etched electrolytically to a fine (0.5–0.05 μm) yet sturdy tip. Except for the very tip, the wire is then insulated by repeated dipping into lacquer (Hubel, 1957) or by sealing it in a glass micropipet (Baldwin et al., 1965). Alternatively, glass micropipets with tip openings of 1 μm to several micrometers can be filled with an alloy of suitably low melting point (Gesteland et al., 1959). To further reduce the resistance of the metal electrodes, their tips may be platin black-plated by electrolytic deposition (Svaetichin, 1951; see also Gesteland et al., 1959). Electrolyte (NaCl)-filled glass micropipets are also employed for extracellular recording. A chlorided silver wire or Ag–AgCl pellet contacting the filling solution then serves as the electrical terminal. Metal electrodes have a much lower resistance (and thus a lower noise level) than electrolyte-filled glass electrodes of similar tip size. They act like high-pass filters, restricting their application to fast (high-frequency), transient events. The electrolyte-filled electrodes on the other hand respond well to dc potentials and low frequencies but become limiting in the resolution of very fast events. The major components of the extracellular signal from neuron action potentials fall in the frequency range of 0.5–5 kHz, well within the range of sensitivity of the metal electrode but close to the frequency response limit of the electrolyte electrode. The narrow frequency range of action potentials makes it possible to improve the signal-to-noise ratio by removing both the high- and the low-frequency (noise) components from the incoming signal with electrical filters.

Many important insights in sensory physiology and neurophysiology have been gained by extracellular recording. For instance, with relatively gross electrodes, the brain cortex was mapped according to the types of sensory input by recording the potentials evoked on the surface of the cortex following peripheral sensory stimulation (see e.g., Talbot and Marshall, 1941; Woolsey, 1958). The potentials recorded with such large-tipped electrodes represented the summed electrical activity of many neurons and were thus of poor spatial resolution. But once small-tipped electrodes were developed, it became possible to record action potentials from single neurons ("unit" action potentials) and to determine the receptive fields of individual neurons in the brain (Hubel and Wiesel, 1959).

In developmental biology, unit extracellular recording has been used to investigate development of the topographical organization of the nervous system. Gaze *et al.* (1972), recording unit action potentials from the optic tectum, mapped retinotectal projection during the development of *Xenopus* from the tadpole stage to metamorphosis. They found that the connections of given retinal ganglion cells with neurons in the optic tectum changed in an ordered fashion during development until metamorphosis, at which time they became stable. By similarly mapping the retinotectal projections after the metamorphosis of normal tadpoles and of those whose eyes had been surgically rotated at various embryonic stages, Jacobson (1968) showed that the (final) connectivity of retinal ganglion cells with neurons in the optic tectum was specific very early during the development of *Xenopus:* During the "critical period," a period actually anteceding outgrowth of the optic nerve fibers from the retina, the area on the tectum to which a retinal ganglion cell eventually will project becomes irreversibly specified according to the ganglion cell's relative position in the retina.

Extracellular recording techniques can also be used to determine when in development spontaneous action potential activity is present in various regions of the nervous system. For example, Ripley and Provine (1972) found spontaneous action potential activity as early as day 4 of incubation in the embryonic chick spinal cord. Their data show that the early movements of the chick embryo are of neural origin and raise the possibility that early spontaneous action potential activity itself has a role in development.

B. Vibrating Probe for Recording Small Steady Transcellular Currents

Until recently, the detection of transcellular steady currents was possible only in a few specialized epithelial tissues in which such (relatively large) currents are associated with the particular epithelium's transport function (e.g., frog skin and urinary bladder). But small, steady currents (electric fields), driven transcellularly, are also thought to be associated with cell differentiation, cell growth, and tissue regeneration (Lund, 1947; Woodruff and Telfer, 1974; Jaffe and Nuccitelli, 1977).

The detection of very small currents presents a technical problem. *Intra*cellular voltage differences arising from such currents are on the order of microvolts ($1 \ \mu V = 10^{-6} \ V$); such small potential differences are impossible to separate from artifacts associated with recording with high-resistance intracellular electrodes. The *extra*cellular voltage gradients produced are even more minute: They are on the order of nanovolts (1 nV = 10^{-9} V), which buries them in the noise level usually encountered with extracellular electrodes (microvolts), even under optimal shielding conditions.

Jaffe and Nuccitelli (1974) found a way to overcome this problem. They designed an ultrasensitive extracellular voltage probe capable of resolving steady potential differences of 1–2 nV between two locations 30 μm apart. With this probe the existence of tiny, steady, transcellular currents was demonstrated, and in fact the relative densities of current entering at one area of a cell and leaving at another could be mapped (Nuccitelli and Jaffe, 1976).

The probe design's main virtue lies in the fact that one and the same electrode is used to sample and compare the potential at two locations. Such a self-referencing system eliminates errors introduced by electrode drift, polarization, or changes in resistance. It also enables one to effectively reduce the electrical noise of the system. In principle, the probe consists of a voltage-sensing platin black tip which is vibrated back and forth between two points. The oscillations are produced by a piezoelectric crystal driven by a sinusoidal voltage generated by a "lock-in amplifier." Any steady voltage difference encountered by the oscillating tip is thereby converted into a sinusoidal signal of a frequency determined by the lock-in amplifier. The signal, whose peak-to-peak amplitude corresponds to the voltage difference between the two extreme tip locations, is fed back into the lock-in amplifier. The phase-sensitive detector of the amplifier converts only that part of the incoming signal that is synchronous (in phase) with the sinusoidal voltage driving the probe oscillations into a dc level proportional to the input signal's amplitude. The nonsynchronous portion of the input (electrical noise) remains as an ac signal and is removed by a low-pass filter (i.e., a filter with a long time constant). With a time constant of 10 seconds and a vibration amplitude of 30 μm, these authors reported a voltage resolution of 1–2 nV with this probe. It is clear that only relatively steady voltage differences can be detected by this method. Moreover, at least at present, the relatively large size of the probe (20–30 μm) limits the spatial resolution and therefore restricts its application to large cells.

Several exciting results have come from the laboratory that developed the probe. Nuccitelli and Jaffe (1976) found that embryos of the seaweed *Pelvetia* began to generate long-lasting current pulses along their prospective growth axis hours before actual growth began.

Based on the direction and relative current densities detected around the embryo, the largest intracellular current density is located just beneath the future growth tip. The currents thus herald not only the impending onset of growth but also its direction. This has led these authors to speculate that the direction of growth is determined by the current pattern. In a different study (referred to in Jaffe and Nuccitelli, 1977), pulses of current were detected entering the membrane at the future site of the cleavage furrow of a cell about to undergo division—the currents again heralding a cellular event and its location.

Because of its high voltage resolution this vibrating probe is a powerful and promising tool for investigations in developmental biology. It presently provides the most sensitive way to measure small electrical gradients such as are set up by cells and tissues during cell growth, differentiation, and tissue regeneration.

C. MEASUREMENT OF LOCAL MEMBRANE CURRENTS

Many cells have nonuniform membrane properties. For example, the membrane of a neuron dendrite usually has electrical properties different from those of the membrane of the neuron cell body. Measuring membrane currents locally is therefore essential in determining how these various membrane regions contribute to cell function. Several methods have been developed to this end; all employ extracellular electrodes for measuring the current through local patches of cell membrane. The simplest method is the focal extracellular recording technique used by del Castillo and Katz (1956) to record local synaptic currents at the neuromuscular junction. A glass micropipet (tip opening 1-2 μm) filled with Ringer's solution is placed against the membrane of the muscle cell near a neuromuscular junction. This pipet records the local extracellular voltage changes produced by synaptic currents and so gives a rough measure of the time course of the synaptic current. Neher and Sakmann (1976a) modified this technique to record currents from single acetylcholine channels in denervated frog muscle. They used electrodes with fire-polished tips (1- to 2-μm tip opening) and pretreated the cell membrane briefly with pronase in order to improve the seal between the electrode tip and the cell membrane. The micropipet was held at ground potential by a virtual ground circuit which also provided a measure of the current required to do so. This current is equal to the current flowing through the patch of membrane under the pipet tip opening, provided there is a good seal between the cell membrane and the micropipet tip. Neher and Lux (1969) used a similar method, but with a much larger pipet tip opening (100 μm), to record the currents through membrane patches of snail neurons.

Although the measurement of local membrane currents is still very difficult, it will certainly have an important role in future studies on cells with nonuniform membrane properties. Perhaps this method and the vibrating probe will have complementary roles in mapping the local membrane properties of developing cells: The vibrating probe can be used for measuring slow currents of very low amplitude, and the extracellular focal current recording method for obtaining greater spatial and temporal resolution of local membrane currents.

III. Recording with Intracellular Electrodes

Extracellular recording reveals the timing of transient membrane events such as action potentials but cannot inform us of the amplitude or time course of potential changes within the cells. Intracellular electrodes, on the other hand, directly measure cell potentials and, in combination with perfusion techniques, they can be used to identify many of the ion permeability mechanisms underlying these potentials. In addition, intracellular recording also provides a means for measuring steady-state cell parameters such as the resting membrane potential, input resistance, and capacitance.

The basic technique of intracellular recording is of simple design: A cell is impaled with the fine but open tip (0.05- to 1-μm diameter) of a microelectrode filled with an electrolyte solution. A silver wire (coated with AgCl) contacting the solution in the pipet leads the electrical signals from the pipet to a high-input impedance amplifier, and the voltage output of the amplifier is measured with respect to a reference electrode in the extracellular fluid (usually at ground potential).

A. MANUFACTURE OF INTRACELLULAR RECORDING ELECTRODES

One of the most critical steps in applying this technique to small cells, such as cells from embryonic tissue, is making the microelectrodes. The micropipets are made by locally heating glass capillary tubing (1- to 3-mm outside diameter) to the glass working-point temperature and then pulling the softened glass to a fine tip. Although originally undertaken by hand (Ling and Gerard, 1949), this is now usually done with a mechanical "electrode puller."

A variety of methods have been devised to fill the micropipets with electrolyte solutions for either intra- or extracellular recording. Until recently the most commonly used filling technique consisted of boiling the electrodes under an intermittent, partial vacuum in an electrolyte solution (or in methanol which later was exchanged for water and then for electrolyte solution) until the air inside the micropipets was displaced by fluid. This boiling procedure often blunts fine micropipet tips, making it difficult to impale and obtain good intracellular recordings from small cells. Many investigators now fill microelectrodes with methods that utilize capillary forces to draw solution into the tip of the electrodes. To this end, one or more fine glass fibers are inserted into the glass capillaries used in making the micropipets (Tasaki *et al.*, 1968). During the electrode-pulling procedure, the fine glass fiber is thinned together with the capillary wall to form a fine, open electrode tip. To fill these electrodes, solutions are injected into the shank of the micropipet with a fine hypodermic needle. The solution quickly follows the fine glass fiber by capillarity to the tip of the micropipet, filling the electrode within a few minutes. The major tedium with this technique lies in introducing the fibers into the glass capillaries. Capillaries are now available with built-in glass fibers (e.g., Omega Dot glass from Frederick Haer Company, Brunswick, Maine). "Theta" glass tubing (R. and D. Optical Systems, Spencerville, Maryland) also fills rapidly by capillary action, without even the need for glass fibers. (For a general reference concerning microelectrodes see Geddes, 1972, and Brown and Flaming, 1977).

B. Optimizing Electrode Performance

The performance of an electrode, i.e., its capability for following the time course of electrical events with high fidelity and a high signal-to-noise ratio, depends on its resistance and capacitance. Together they determine the noise level and time constant of the electrode and so influence its ability to record very small or very fast electrical events.

1. Ways to Lower Electrode Resistance

Electrode resistance is greatly influenced by the shape of the microelectrode. Electrodes with long, small-diameter shafts typically have very high resistances, while electrodes with a short, rapid taper to an equally fine tip possess considerably less resistance. In addition to choosing a rapid electrode taper (Brown and Flaming, 1977), the electrode resistance can also be reduced by grinding the electrode tip using microgrinding methods (Barrett and Graubard, 1970; Brown and Flaming, 1975; Ogden *et al.,* 1978) to obtain a beveled tip shape similar to that of a hypodermic needle. The larger tip opening produced in this way yields a lower electrode resistance without increasing the diameter of the electrode shaft. Since the point of a beveled electrode tip is very sharp, these electrodes penetrate cell membranes and tissue with greater ease and with reduced risk of tip breaking or clogging (Barrett and Barrett, 1976). However, the larger tip openings at the same time increase the rate of solute diffusion from the electrode into the cell. For many applications electrodes should therefore be filled with a lower-concentration electrolyte solution. On the other hand, the facilitated diffusion from the electrode tip makes beveled electrodes especially useful for intracellular injection of ions, dyes, and other substances.

Since electrode resistance is dependent on the inner diameter of the micropipet tip, another way to reduce electrode resistance without increasing the outside diameter of the micropipet tip is to use thin-walled glass such as "Ultra-Tip" glass (Frederik Haer Company, Brunswick, Maine).

2. Compensation for Electrode Capacitance

The ability of a microelectrode to record fast, transient events is limited in part by the capacitance of the electrode. It is possible to compensate for some of the electrode capacitance by using a preamplifier with capacitative feedback (Amatneik, 1958). Partial capacitative compensation can improve the response time of a microelectrode by as much as 10 times. Since further compensation produces a tremendous increase in electrical noise, perfect compensation for all of the electrode capacitance cannot be obtained (see Kootsey and Johnson, 1973).

C. CELL IMPALEMENT TECHNIQUES

The quality of intracellular recordings depends not only on the properties of the microelectrode and preamplifier (see below) but also on the manner in which the cell membrane is penetrated by the electrode. If an electrode is moved slowly toward the membrane of a cell, it indents the membrane considerably before eventually penetrating it. The deep penetrations produced in this way usually result in abnormally low resting potentials and input resistances. This problem can be avoided by using electrical or mechanical pulses to enable the microelectrode to penetrate the cell membrane before substantial indentation has occurred. An intense, brief (50–100 nA, 0.1–10 msec) negative current pulse applied through the microelectrode is especially effective in achieving membrane penetration with minimal cell damage. Deliberately induced amplifier oscillation as well as mechanical pulses are also effective. All electrophysiologists are familiar with techniques of tapping the electrode micromanipulator in order to "jar" an electrode into a cell. Tomita (1965) added control to this technique by using a speaker coil located underneath the preparation chamber to produce the required brief mechanical pulse. Alternatively, a small speaker placed on the electrode micromanipulator and oriented so as to produce movement along the axis of the electrode also provides controlled mechanical pulses for good membrane penetration. Alternatively, stepping motor systems are available which will advance a microelectrode in pulses.

IV. Problems in Measurements of Intracellular Potentials

Nearly all cells maintain a potential difference across their membrane, with the cell interior negative with respect to the extracellular fluid. The magnitude of this potential difference, generally referred to as resting potential or simply membrane potential, varies with the cell type and perhaps also with the stage in the cell cycle (Cone and Tongier, 1973). To measure the resting potential, an electrode is inserted into the cell and the potential difference between it and a reference electrode (Ag–AgCl$_2$ wire in an agar–salt bridge) in the bathing fluid is amplified and displayed on an oscilloscope. The evaluation of this recorded potential as a faithful representation of the actual resting potential requires caution, however. Two main sources of error in determination of the resting potential are the electrode potential and the leakage currents introduced by the microelectrode. Furthermore, diffusion of ions from the electrode itself may result in cell potential changes.

A. Electrode Potential

With only the microelectrode and the reference electrode in the bath, a small potential (electrode potential) often is observed. This potential is generated both by fixed charges of the glass itself and by the liquid junction potential that develops when two solutions of different ionic composition come into contact. The contribution of the fixed charge may be reduced by the addition of small amounts of ThCl to the filling solution (Agin and Holtzman, 1966). With the electrode in the bathing solution, the electrode potential is easily determined. However, once the microelectrode is inside the cell (which has an ion composition different from that of the extracellular solution), the electrode potential and thus its contribution to the recorded resting potential are uncertain. Adrian (1956) demonstrated that the liquid junction potential of a 3 M KCl microelectrode differed by about 12.5 mV when in a solution of 125 mM NaCl as compared to 125 mM KCl. This value thus contributes to the resting potential recorded by an electrode upon passing from a 125 mM NaCl bath through the membrane into a cell containing 125 mM KCl (in this example, 12.5 mV should be added to the recorded potential). Usually, only microelectrodes with a small electrode potential (< 5 mV) are chosen by the experimenter.

B. LEAKAGE CURRENTS

Particularly in the case of cells with very high input resistance, even small leakage currents caused by the recording microelectrode will result in large changes in membrane potential. For example, a current as small as 0.1 nA (10^{-10} A) would produce a 50-mV change in the membrane potential of a cell with an input resistance of 500 MΩ. Such leakage currents may be introduced either by the recording preamplifier or by ionic leakage around the electrode shaft at the membrane penetration site. Leakage current from the preamplifier can be reduced to low levels by using a preamplifier with a very high input resistance (10^{11}–10^{13} Ω) and of at most 10^{-11} A leakage current. But even with a preamplifier having these characteristics, the measured resting membrane potential and input resistance for small cells are often much less than expected from theory. Often this is due to current leakage around the shaft of the microelectrode at the location where the electrode penetrates the cell membrane. A gap of molecular dimensions (5 Å) encircling the shaft of a microelectrode 0.5 μm in diameter will constitute a resistance of about 11 MΩ. Since this resistance is smaller than most cells' input resistance, current will "leak" through this resistance (hence the terms "leakage resistance" and "leakage current") and significantly affect the cell membrane potential. In the above example the leakage resistance was calculated assuming a resistivity of 50 ohm-cm for both the extra- and intracellular solutions (this is the resistivity of a mammalian balanced salt solution at 37°C) and a membrane thickness of 100 Å. The resistance of the cylindrical gap itself around the electrode is calculated to be 6 MΩ, and the convergence resistances at both sides of the membrane are 2.5 MΩ each, giving a total resistance of 11 MΩ for these resistances in series. Such a leakage resistance would drastically affect a cell with an input resistance of 500 MΩ. The resting potential would be reduced to less than one-tenth its normal value, and the measured input resistance would be slightly less than 11 MΩ. A case in point may be the discrepant results obtained from unfertilized sea urchin eggs. Many different authors have reported membrane potential values for echinoderm eggs between +5 and −20 mV. Occasionally, however, investigators encountered potentials as high as −60 to −80 mV (a value found in the majority of animal cell types), especially in media with high Ca^{2+} concentrations (Miyazaki et al., 1974). The latter finding seems to point to a leak conductance introduced by the electrode insertion—a leak whose repair is aided by the increased Ca^{2+} concentration (see, e.g., Oliveira-Castro and Loewenstein, 1971, for a membrane repair function of Ca^{2+} in a different cell system). With a fast electrode impalement technique, minimizing membrane damage, resting potentials of −60 to −70 mV could be routinely measured in sea urchin eggs at normal external Ca^{2+} concentrations (Jaffe and Robinson, 1978; Chambers and de Armendi, 1979).

This caveat in regard to error in the measured resting potential due to leakage resistance applies in particular to small, single cells, since input resistance is a function of membrane surface area. Thus, even at similar values of specific membrane resistance, cells of large surface area (e.g., nerve) or cells interconnected by low-resistance membrane junctions, have comparatively low input resistance. The effect of leakage current on the resting potential of those cells is therefore greatly reduced, and the recorded values may be considered more accurate.

Since the leakage pathway (resistance) produced by the electrode results in a flux of ions through the cell membrane, it thereby can change the internal ionic composition of the cell so much as to alter some cells' properties. For example, an increase in internal Ca^{2+} produced by such a leak can be sufficient to change the membrane permeability (Kostyuk and Krishtal, 1977). It is thus important to reduce membrane damage around the microelectrode. Two approaches to this problem are possible: (1) reducing the outer diameter of the electrode tip so as to reduce the area of membrane affected, and (2) improving the seal between the microelectrode glass and the cell membrane. Since the interior of the cell membrane is composed largely of hydrophobic fatty acid chains, a better seal might be achieved by coating the electrode shaft with a hydrophobic substance. In some studies, the electrode glass is treated with silicone compounds (Desicote or Silicad; Maloff et al., 1978) to make it more hydrophobic. Electrodes with very small tip diameters (0.05–0.2 μm) produce less membrane damage and thus less leakage current than larger electrodes. On the other hand, the resistance of microelectrodes increases enormously (approximately with the reciprocal of the square of the tip inner diameter) as the electrode tip diameter decreases. High electrode resistance is often a disadvantage, since the long passive time constant of such electrodes makes it difficult to measure fast, transient potential changes accurately.

C. DIFFUSION OF ELECTROLYTES FROM MICROELECTRODES

In order to reduce the resistance of fine-tipped electrodes, the micropipets are frequently filled with concentrated electrolyte solutions (3 M KCl, 4 M potassium acetate, 3 M potassium citrate, or 2 M K_2SO_4). Such highly concentrated electrolyte solutions reduce the resistance of the microelectrodes, but they also introduce a new, potential source of damage to small cells. Diffusion of ions from such electrodes can be sufficiently large to produce damaging intracellular ion concentrations. Thus it is often best to use a more physiological electrolyte concentration (e.g., 0.15 M K_2SO_4), especially if the events being studied have a relatively slow time course and can therefore be resolved with a high-resistance electrode.

D. EXAMPLES OF RESTING POTENTIAL MEASUREMENTS IN EMBRYONIC CELLS

Intracellular potential measurements in embryos are fraught with difficulties because of the small cell size and high input resistance. Nonetheless, valuable information about membrane events during development has been obtained from intracellular potential measurements at various stages of development, from the unfertilized egg to differentiated neurons. A few examples are pointed out in the following:

1. Recording potentials of sea urchin eggs, Steinhardt et al. (1971) have shown that an early effect of fertilization is depolarization of the egg membrane. Jaffe (1976) has demonstrated that it is this depolari-

zation that is crucial in preventing polyspermy—an important factor for normal development.

2. Recording resting potentials and their dependence on external K^+ concentration during cell division cycles of early embryos, Slack and Warner (1973) found periodic changes in the resting potential. These changes were consistent with the idea that newly formed membrane inserted into the existing (original egg) membrane during cleavage has a high K^+ permeability as compared to the original egg cell membrane. Input resistance measurements made by DeLaat and Bluemink (1974) on cleaving cells support this interpretation.

3. An increase in the resting potential of presumptive neural cells during the passage of axolotl embryos from the early to the midneural plate stage was observed by Blackshaw and Warner (1976a). This "hyperpolarization" was interpreted as the result of activation of the sodium pump at this stage of development—an interpretation supported by the results of several pharmacological experiments (ibid). The change in resting potential as an early consequence of neural induction thus marks the onset of neural differentiation in these cells.

V. Methods for Measuring Electrical Membrane Properties

Many experimental paradigms require both intracellular current injection and simultaneous measurement of the cell membrane potential. For example, measurements of membrane resistance, membrane capacitance, membrane conductance changes, and electrical coupling between cells all require intracellular potential measurements during current injection.

A. Current Injection Methods

In principle, a single intracellular microelectrode can be used for both current injection and simultaneous voltage measurements. However, the voltage recorded from such an electrode during current injection will be the sum of the voltage drop across the electrode resistance and of the true intracellular voltage. Determining the true intracellular potential during current injection thus requires correction for the voltage drop across the electrode resistance. This is usually done with a wheatstone bridge which, when correctly balanced, subtracts a voltage equal to the current-induced voltage drop across the electrode resistance from the total voltage recorded by the microelectrode. Such bridges are usually balanced by the experimenter, assuming that the cell's transient response to the current pulse is slow compared to the time constant of the electrode. For spherical cells with membrane time constants greater than 10 times the effective electrode time constant, this is a reasonable assumption. However, the voltage response of a nonisopotential cell (e.g., a neuron with processes) to a current pulse includes charge redistribution components considerably faster than the passive time constant of the cell membrane (Rall, 1969). Thus bridge balancing provides at best an approximate correction when applied to neurons or other cells with complex geometry. Another problem arises at high currents where the electrode resistance becomes both nonlinear and time-dependent, making proper balancing of the bridge impossible for all practical purposes.

A relatively new method of current injection bypasses the current-induced voltage drop across the electrode resistance by using trains of brief, high-frequency pulses for injecting current (Brennecke and Lindemann, 1971) and by sampling the intracellular voltage during the brief pulse intervals when current no longer flows through the electrode, hence when no voltage drop occurs across its resistance. This method may be applied when the time constant of the microelectrode recording system is much shorter than the time constant of the cell. However, in cells with complex geometry (nonspherical, nonisopotential cells) the voltage response to current pulses contains fast, transient components (Rall, 1969) which can approach the response time of the microelectrode and thus would go undetected.

These methods are also subject to error when high currents are injected through the electrode. High current density produces a redistribution of the ion concentrations within the microelectrode tip. At the end of an intense current pulse the ion concentrations slowly relax back to the original gradient but in the process generate a voltage drop across the electrode tip. This "battery-like" potential due to the redistribution of ions with different mobilities has a time constant much longer than either the time constant of the microelectrode or the cell. For example, a transient potential (10–20 mV peak) lasting hundreds of milliseconds is recorded following an intense current pulse (50 nA, 10 msec) through an electrode (100 MΩ) with its tip immersed in a 150-mM NaCl or KCl solution (Barrett, unpublished observations). At intense currents such an ion redistribution potential will be present during the sampling time, thus adding to the true membrane potential.

The problems generated by ion redistribution potentials and by the nonlinearities of electrode resistance are avoided by using two separate intracellular electrodes: one electrode to inject current and a second electrode to record the intracellular voltage. In many preparations the separate current and voltage electrodes can be placed in the same cell by using direct visual control while positioning the micro-electrodes. Even when direct visualization of the cell is not feasible, it is still possible in some cases to place two electrodes in the same cell. For example, cat motoneurons have been impaled with two electrodes either by gluing the current and voltage electrodes together (Araki and Terzuolo, 1962) or by using a micro guide system to direct the electrodes (Barrett and Crill, 1980). However, even under visual control it is difficult to place these electrodes in a small cell without severely damaging the cell membrane.

Electrodes with two or more barrels side by side can be made from special glass tubing containing multiple chambers (for a review see Brown and Flaming, 1977). Between the barrels of such electrodes there is a coupling resistance due to a shared "convergence" resistance in the solution just outside the electrode tips. Current passed through one barrel of the electrode produces a voltage change across this coupling resistance, and this change is seen by the voltage-recording barrel. Thus the voltage drop across the coupling resistance must be balanced (compensated) using a bridge circuit similar to that for single-barrel electrodes. Fortunately this coupling resistance is usually considerable less than the resistance of the individual electrode barrels and is less susceptible to polarization and nonlinear changes with increasing current intensity. Capacitative currents between the adjacent electrode barrels can be reduced by the presence of an intermediate barrel driven by the output of a unity-gain recording preamplifier to the same voltage as that of the voltage barrel (Barrett and Barrett, 1976). Since the voltage barrel and this intermediate "shield" barrel are thereby always at the same potential, no capacitative current will flow between them.

B. Cell Input Resistance and the Identification of Ion-Selective Permeability Changes

Membrane permeability changes are reflected in changes in cell input resistance. Measurements of input resistance can thus be used to detect changes in membrane permeability and to determine the direction of these changes. An increase in membrane permeability to any ion decreases input resistance, whereas a decrease in membrane permeability increases input resistance. The cell input resistance R_n is determined by injecting a small current pulse into the cell and measuring the resulting voltage change. The quantity R_n is defined as this steady-state voltage change V divided by the magnitude of the applied current I: $R_n = V/I$. Small currents are used in making this measurement, so that the voltage change produced will not affect voltage-dependent membrane permeability systems. Averaging the voltage responses to repeated small current pulses greatly increases the accuracy of voltage change measurements and makes it possible to inject very small current pulses (Smith *et al.*, 1967; Nelson and Lux, 1970; Burke and ten Bruggencate, 1971). When it is necessary to resolve the time course of fast, transient membrane permeability changes, very brief current pulses are used. Rather than the steady-state voltage change, the area delineated by the voltage response (the integral of the voltage response curve) is then used as a measure proportional to the input resistance (Barrett and Barrett, 1976), because the time required to completely charge the membrane to a steady-state voltage may be longer than the duration of the permeability change.

Baccaglini and Spitzer (1977) used measurements of cell input resistance to help determine the nature of the mechanism underlying long-duration action potentials found in *Xenopus* tadpole Rohon Beard cells during early stages of development (stages 21–25). Such long-duration action potentials could conceivably be generated by an electrogenic pump, by an increase in membrane permeability selective for an ion with an equilibrium potential positive to the resting potential, or by a decrease in membrane permeability to an ion with an equilibrium potential negative to the resting potential. By measuring the voltage response to current pulses applied during the long action potential, Baccaglini and Spitzer found that the cell input resistance decreased during the action potential. This result indicated that the action potential was due to an increase in membrane permeability to some ion(s) with an equilibrium potential(s) positive to the peak of the action potential.

To determine the specific ion(s) involved in this permeability change Baccaglini and Spitzer (1977) altered the ionic composition of the extracellular fluid surrounding the Rohon Beard cells and observed the effects of different ion concentrations on the action potential. The action potential was not altered by changing the sodium concentration in the extracellular fluid but was sensitive to the concentration of Ca^{2+} ions and was blocked by ions such as Co^{2+} and Mn^{2+} which are known to block voltage-dependent Ca^{2+} permeability systems in other cells (Hagiwara, 1973). Hence Baccaglini and Spitzer concluded that the early long-duration action potentials of Rohon Beard cells were due to a voltage-sensitive increase in permeability of the membrane to Ca^{2+}. Since the calculated flux of Ca^{2+} into the cells during one of these action potentials would be sufficient to increase the concentration of internal calcium by 100 times, Baccaglini and Spitzer proposed that this increase in internal calcium could influence the development of other properties of the Rohon Beard cells. Indeed, at about the same time, Walicke *et al.* (1977) found that calcium (but not sodium) entering during action potentials, evoked by electrical stimulation of rat sympathetic neurons grown in tissue culture, could alter the type of neurotransmitter synthesized by these neurons. These *in vitro* experiments suggest that membrane permeability changes such as those that occur during action potentials may have a role in directing development.

C. Voltage Clamp for the Measurement of Membrane Currents

Many of the ion permeability systems that contribute to the generation of action potentials are voltage-dependent; that is, their magnitude varies with the membrane voltage. It is often essential to control the membrane potential when studying such systems. This control can be attained with voltage clamp techniques which use a feedback circuit for injecting the amount of current required to hold the cell membrane potential at a desired value. For isopotential cells this "clamp current" is equal to the current flowing through the cell membrane. The clamp current following a step change in membrane potential contains a fast, transient component due to the capacitance of the cell membrane, and slower components which depend on the ionic permeability systems of the cell membrane. The voltage clamp technique usually requires separate current and voltage electrodes, especially if the currents studied are large. But even with separate current and voltage electrodes, it is difficult to obtain control over the membrane potential in all regions of cells with complex shapes. For example, if both current and voltage electrodes are placed in the soma of a neuron, the soma voltage will be under control, but distant portions of the cell's processes (dendrites and axon) may maintain membrane potentials different from the desired clamp potential. Currents from these nonisopotential regions of a cell contribute to the total current required to clamp the cell soma and so can easily complicate analysis of the currents recorded at the well-clamped soma. This source of error can be avoided by measuring membrane currents from only that portion of the cell membrane under good voltage clamp control. A method for measuring local membrane currents has been developed that does exactly this (see Section II,C), but it is difficult to apply to small cells.

In view of the technical difficulties, it is not surprising that voltage clamp techniques have not yet been used to study the membrane properties of developing neurons. However, in combination with noise analysis and tissue culture preparations, the voltage clamp technique could be a very precise tool for studying the development of voltage-sensitive membrane properties.

The voltage clamp has been used very successfully to study membrane currents in unfertilized egg cells. These studies have revealed that both mammalian (Okamoto et al., 1977) and starfish (Hagiwara et al., 1975) egg cells have voltage-dependent calcium permeability systems that may play a role in events immediately following fertilization, such as the cortical granule reaction that forms a protective sheath around the egg cell (Vacquier, 1975).

D. Noise Analysis for the Measurement of Single-Channel Conductance Characteristics

Many ion-specific membrane permeability systems appear to be mediated by discrete "channels" which fluctuate between an open and a closed state. The fluctuation with time in the number of channels in the open state generates fluctuations in membrane conductance and membrane current, hence electrical noise. Both the voltage-dependent sodium channel (Conti et al., 1975) and the chemically activated acetylcholine channel (Katz and Miledi, 1972; Anderson and Stevens, 1973) have been shown to generate electrical noise by fluctuations in the number of open channels. Since the noise pattern reflects the properties of the individual channels producing the noise, its analysis can reveal some of the basic characteristics of the individual channels, such as individual channel conductance, mean channel lifetime, and the average number of channels open under different conditions.

Channel noise is best measured as current noise under voltage clamp, so that the passive properties of the cell will not affect the characteristics of the noise. In order to reduce other sources of membrane noise an attempt is usually made to selectively activate only one type of channel and to keep the other channels in the "off" or closed state using, for example, selective blocking agents. The characteristics of current noise from a single channel type can yield information about the average "open" time of individual channels and the conductance of a single open channel. This information is extracted by analysis of the frequency components comprising the noise (using power spectrum or autocorrelation analysis, see Anderson and Stevens, 1973; Neher and Sakmann, 1976b). Usually a number of assumptions underlie many of the conclusions drawn from this analysis. For example, the unit channel event is often assumed to be a simple on-off pulse like that found by Neher and Sakmann (1976a) using a microspot, focal extracellular recording technique to record the currents produced by individual channels in muscle membrane when activated by acetylcholine or cholinergic agonists.

Noise analysis has been useful in helping to identify neurotransmitters, since the current noise produced by iontophoresis of a putative transmitter must display channel lifetime characteristics consistent with the time course of the synaptic currents produced by physiological release of the real transmitter. Noise analysis can also be used to determine whether the prolonged time course of synaptic currents at early stages of development is due to characteristics of the receptor, to properties of the presynaptic transmitter release mechanism, and/or to the mechanisms by which the active transmitter is removed from the synaptic cleft (see e.g., Fischbach and Laas, 1978).

E. MEASUREMENT OF ELECTRICAL CELL-TO-CELL
    COUPLING VIA INTERCELLULAR MEMBRANE
    JUNCTIONS AND OF JUNCTIONAL MEMBRANE
    CONDUCTANCE

The cells of many tissues are connected to each other by permeable junctions. These juctions provide a direct pathway from cell to cell for inorganic ions, as well as for larger molecules (Loewenstein, 1966; Furshpan and Potter, 1968). The junctional pathways are thought to be well-insulated, aqueous channels spanning the width of both of the apposed membranes of adjacent cells (Loewenstein, 1966; Gilula, 1974). The transfer of ions from cell to cell via these junctional membrane channels underlies the so-called electrical or electrotonic coupling between cells. Measurement of electrical coupling therefore can reveal the presence or absence of such permeable intercellular junctions (channels) between cells in a tissue.

In many embryonic systems, all cells are electrically coupled to each other at early stages of development (e.g., amphibia: Ito and Hori, 1966; squid: Potter et al., 1966; Chick: Sheridan, 1968). In adult, differentiated organisms, on the other hand, only cells within an organ or tissue are coupled to each other. Cells therefore can and do selectively abolish junctional coupling with particular neighboring cells during development. Since intercellular junctions were found permeable to larger molecules as well as to small ions (see Loewenstein, 1975, for a list of junction-permeant molecules in various tissues and organisms), the hypothesis was put forward (Furshpan and Potter, 1968; Loewenstein, 1968) that intercellular junctions may play an important role during development by providing a direct pathway from cell to cell for the diffusion of substances controlling cell growth and differentiation.

Intracellular recording and dye injection techniques with microelectrodes are useful tools for unraveling the properties and the role of such junctions. For a detailed description and a discussion of various methods employed for their investigation, see Socolar and Loewenstein (1979).

Where feasible, three intracellular microelectrodes are used to measure electrical coupling between two cells. With one electrode, square pulses of current I are passed between the inside of a cell (cell 1) and the external medium. A second electrode in the same cell measures the steady-state voltage displacement (V_1) produced by the current. The third electrode records the voltage in the second cell (cell 2). In cells coupled via permeable junctions, part of the current injected into cell 1 (causing the voltage change V_1) will flow into cell 2, altering its potential as well. The ratio of the voltage displacements in the two cells, V_2/V_1, is called the coupling coefficient. The coupling coefficient is a function of the relative size of both junctional and nonjunctional membrane conductance, and it also depends on cell size and cell topography. Therefore it should not be used as a comparative index of junctional conductance in different tissues unless corrections are made for differences in cell size, cell topography, and nonjunctional membrane resistance. (See Socolar, 1977, for an analysis of V_2/V_1 as an index for junctional conductance in cell systems of various topographies.)

When it is not feasible to insert two electrodes into the same cell because cells are simply too small or cannot be visualized, a bridge circuit is employed, which permits the simultaneous injection of current and recording of voltage with one and the same electrode. Another electrode is then used for recording voltage in the second cell, often with a second bridge circuit permitting current injection from either side of a cell junction.

The bridge circuit may not be an adequate tool where relatively large currents are required to produce detectable voltage changes—

such as in cells with a low nonjunctional membrane resistance or in extensively coupled cells (both cases resulting in low input resistance). The nonlinear voltage response of the electrode due to changes in its resistance at higher currents introduces errors in the voltage record of cell 1 (V_1) and thus in the coupling coefficient. Double-barreled electrodes may be used, where one barrel serves for current injection and the other for voltage recording. When the current source and voltage sampling site are very close to each other (as is the case with most double-barreled and certainly with single-barreled electrodes), the recorded voltage may not be a true representation of the voltage seen by most of the cell membrane (see Eisenberg and Johnson, 1970, for point source effects near the current electrode; see also Engel *et al.*, 1972). In cases where the exact voltage change produced in the first cell cannot be assessed, the ratio of the voltage change in the second cell (V_2) to the magnitude of the current injected into the first cell (I_1) is used as a rough index of cell coupling ("transfer resistance" V_2/I_1).

Although these electrical coupling measurements do not permit a quantitative comparison of junctional conductances of cells in different tissues, they do provide a sensitive means for investigating the existence and the spatial extent of coupling between cells. Electrical coupling has been shown to exist between the cells of very early embryos of many species (see Sheridan, 1976, for a review of cell coupling in embryonic systems).

A number of investigations have been directed at the question of when the selective uncoupling takes place between cells differentiating into various tissues. Blackshaw and Warner (1976b), for example, investigated the spatial extent of electrical coupling between cells in the mesoderm before and during the development of trunk muscles in amphibia. While differentiated skeletal muscle cells lack coupling, the cells in the unsegmented region of the mesoderm are found to be electrically coupled. At the time of muscle somite formation, coupling is lost between cells of the unsegmented mesoderm and the cells destined to form the next somite. The uncoupling precedes the actual segmentation and thus marks the future intersomite border, signaling the impending differentiation event. In *Xenopus* and *Bombina*, the somite muscle cells thereafter reestablish electrical coupling to each other once somite formation is complete. Kalderon *et al.* (1977) have demonstrated that prefusion chick myoblasts also interact via ionic and metabolic coupling, while the communication via junctions is no longer expressed between mature muscle fibers. In an earlier study, Warner (1973) showed that electrical coupling between neural ectoderm and presumptive neural cells was lost at the time of closure of the neural tube, while coupling persisted between neural plate cells, as well as between ectoderm cells. Here then, too, the selective uncoupling may be viewed as a differentiation step, although it is not the first detectable one: The membrane potential of the presumptive nerve cells rises above that of the surrounding ectoderm before uncoupling is observed. Similarly, Furshpan and Potter (1968) have found that cells that already show visible signs of differentiation are still coupled to cells of other tissues; e.g., the elongate cells of the neural retina in the squid embryo are still coupled to the yolk sac via many intervening cells of different type at a time when the masking pigment and the rudiments of a lens have appeared.

Several caveats must be borne in mind when interpreting the results of electrical coupling measurements between cells of a tissue. (1) The experimenter must be aware of and control for the possibility of current spread from cell to cell via *non*junctional membrane. In intact blastulas, for example, current passed between a blastomere and the bath spreads from cell to cell via *non*junctional membrane because of the existence of a high-permeability barrier between the blastocoel and the medium bathing the embryo (Ito and Loewenstein, 1969; Bennett and Trinkaus, 1970). (2) Since cells uncouple in response to cell injury (Loewenstein and Penn, 1967) or to an elevation of their cytoplasmic free Ca^{2+} concentration (Rose and Loewenstein, 1976), the finding of a lack of electrical coupling must be interpreted with great caution. (3) The size and polarity of the current used in the coupling measurements may influence the junctional permeability, as it does for instance in the rectifying electrotonic synapse (Furshpan and Potter, 1959; Auerbach and Bennett, 1969; Baylor and Nicholls, 1969; Nicholls and Purves, 1970), in the case of uncoupling by large depolarizing currents in the insect salivary gland (Socolar and Politoff, 1971), and in the voltage-dependent junctions found in amphibian embryos (Spray *et al.*, 1979). Other possible alterations induced by the measurement process must also be considered. For example, in cells that possess voltage-dependent Ca^{2+} channels, depolarization might lead to an increase in intracellular Ca^{2+}, which in turn could affect junctional conductance.

To resolve the mechanisms controlling the permeability and formation of intercellular junctions, a quantitative measure of junctional conductance is desirable. With the use of a preparation of only two coupled cells, it becomes possible to calculate junctional conductance from parameters obtained by electrical coupling measurements. Ito *et al.* (1974) studied the development of junctional conductance during *de novo* formation of a junction between two embryonic cells. In an extension of this work, Loewenstein *et al.* (1978) showed that junction formation proceeds by quantal increments of junctional conductance, i.e., by the establishment of unitary cell-to-cell channels. In order to resolve the minute voltage changes produced by the current flowing through a single, newly established cell-to-cell channel, they refined the standard method for electrical coupling measurement by the use of sinusoidal current pulses and lock-in amplifiers (see Section II,B).

For direct measurement of the current flowing through a junction between a coupled cell pair, a double voltage clamp has been devised (Spray *et al.*, 1979). Here both cells are clamped independently at their respective resting potentials with separate clamp circuits. A voltage step is then applied to one cell to induce a junctional current. The clamp circuit of the second cell, in order to maintain that cell's resting potential, injects a current equal and opposite this junctional current. The magnitude of the *second* cell's clamp current divided by the transjunctional potential difference produced by the voltage step applied to the *first* cell is a direct measure of the junctional conductance.

VI. Iontophoresis of Pharmacological or Marker Substances

Charged molecules can be ejected from a micropipet by applying a voltage of the same sign as the charge on the molecule. Nastuk (1953) found that this method, referred to as iontophoresis, could be used as a means for applying controlled amounts of substances (e.g., neurotransmitters) from a micropipet to a locally restricted membrane region (e.g., a synapse).

For iontophoresis, a micropipet is filled with a solution containing the test molecule to be applied. Usually the concentration of other ions is kept to a minimum, so that a large portion of the total current will be carried by the desired molecule. If the net charge of the test molecule is pH-dependent, the filling solution is buffered to avoid pH changes within the micropipet during iontophoresis. To prevent leakage of the test substance from the electrode, an "offset" or "bucking" voltage of sign opposite that of the molecule's charge is applied to the electrode barrel; while for ejection of the test substance, a voltage of the same sign as that of the molecule's charge is applied. The current driven through the micropipet by the applied voltage provides an estimate of the amount of substance being iontophoresed; but since other ions will always be present (notably H^+), only a fraction of the total current is actually carried by the test molecule. This fraction is called the transfer number of the particular molecule. Although hard to quantify absolutely, the method permits locally restricted application in a relatively controlled fashion. Extracellular iontophoresis has been extremely valuable in studies of the development of receptors for synaptic transmitters (see Fambrough and Rash, 1971; Fischbach and Cohen, 1973; Kuffler and Yoshikami, 1975).

Intracellular iontophoresis of marker substances is used extensively as a means for identifying the particular cells being recorded from in electrophysiological experiments, or for determining neuronal geometry. For this purpose, stains are chosen that will bind to cytoplasm and survive fixation and histological preparation procedures (Prussian Blue, Niagara Sky Blue, or Chicago Blue, for example). Fluorescent dyes are particularly effective because of their excellent detectability (Procion Yellow M4RS: Kravitz et al., 1968; Lucifer Yellow CH: Stewart, 1978). A concentrated solution of the dye is mixed with the electrolyte solution filling the microelectrode used for intracellular recording. Upon termination of the electrophysiological experiment, the dye is iontophoresed into the cell, often for long periods, and the tissue is then processed for histology (Kater and Nicholson, 1973).

Another useful marker which can be iontophoresed into cells is the enzyme, horseradish peroxidase (Graybiel and Devon, 1974; Brown et al., 1977). When reacted with the substrates diaminobenzidine and hydrogen peroxide, this enzyme forms a reaction product that is so dense that even the fine processes of an injected neuron can be identified using light or electron microscopy.

When marking cells by dye microinjection, the possibility of intercellular dye transfer via permeable junctions or as a fixation artifact must be taken into account. Even if the dyes bind to cytoplasm and thus are restricted in their diffusibility, junctional transfer may occur. For example, Procion Yellow M4RS, a fluorescent tracer frequently used in the identification and mapping of neuronal dendrite arborizations, traverses electrotonic synapses (Payton et al., 1969), as well as junctions of other cell types (see e.g., Rose, 1971), and so does Lucifer Yellow CH (Stewart, 1978; Bennett et al., 1978).

Iontophoretic injection of fluorescent tracer molecules has in fact been an important tool in studies probing the permeability of cell-to-cell junctions in adult as well as in embryonic cells (see Loewenstein, 1975, for a review of junction-permeant molecules in a variety of tissues).

VII. Ion-Selective Electrodes for the Measurement of Intracellular Ion Activities

Intracellular microelectrodes are now frequently used to determine intracellular ion activities. The most commonly measured activities are those of Na^+, K^+, H^+, Cl^-, and Ca^{2+}. For manufacturing techniques, theory, and a comparison of various types of ion-selective electrodes, see Koryta (1975) and Thomas (1978).

Ion-selective electrodes can be made from special ion-selective (ion-sensitive) glass or by filling ordinary glass micropipets with ion-selective solvents. Some of the most reliable measurements of intracellular ion activities using ion-sensitive glass microelectrodes have been made with Thomas-type electrodes (Thomas, 1972, 1974). In this type of electrode, the ion-sensitive glass micropipet is enclosed within and sealed to an insulating (not ion-selective) glass microcapillary, which prevents access to the ion-sensitive glass except for a small portion inside the open tip of the insulating glass. Thus the closed tip of the ion-sensitive glass lies recessed within the open tip of the insulating glass.

Ion-sensitive glass or ion selective solvent-filled microelectrodes are calibrated in standard solutions of the respective ion. The theoretical (Nernst) response of either type of electrode is a 58-mV slope for a 10-fold change in activity of the particular ion if monovalent, and a 29-mV slope if divalent. Provided there is no damage to the insulating glass, there is no contribution of the extracellular fluid to the measured ion activity once intracellular location of the electrode tip is established. But then, in addition to the particular ion's activity, the electrode also senses the cell's membrane potential. To obtain the voltage change due to only the ion activity, the membrane potential must be measured with an additional voltage recording electrode in the same cell and subtracted from the total voltage recorded by the ion-selective electrode. Intracellular location of the ion-selective electrode can be ascertained by injecting a square current pulse into the cell: If both electrodes are in the same cell, the amplitude of the resulting voltage deflections seen by the ion-selective and by the standard electrode should be equal.

Recently, double-barreled electrodes have been used where one barrel tip is sealed off with an ion-sensitive glass membrane (Coles et al., 1979) or is filled with an ion-selective solvent (Walker, 1973), while the other barrel serves as reference, measuring the membrane potential. The proximity of the reference electrode is an advantage, but care must be taken to eliminate common mode artifacts between the barrels.

In general, the time resolution of ion-selective electrodes is rather poor (hundreds of milliseconds to many seconds) compared with some of the fast permeability changes of cell membranes. Hence they serve mainly for recording steady-state ion activities or slow changes thereof. For example, Shen and Steinhardt (1978) determined intracellular pH (pH_i) of sea urchin eggs during fertilization. After sperm penetration, the pH_i increased from 6.8 to 7.2. This increase in pH_i is thought to be instrumental in the derepression of protein and DNA synthesis in these cells (Johnson et al., 1976; Shen and Steinhardt, 1978; Grainger et al., 1979), and thus of their development. Investigating intracellular Na^+ activity, Slack et al. (1973) found that, while the total sodium content in the egg (Xenopus) was about 100 mM, the intracellular Na^+ activity was only 14 mM. Moreover, the intracellular Na^+ activity remained constant, while the total cell sodium content decreased significantly during the development of the egg to the blastula stage. Cytoplasmic Ca^{2+} activity was measured with Ca^{2+} sensitive microelectrodes during cell division in Xenopus embryos (Rink et al., 1980).

VIII. Conclusion

The sensitive monitoring power of electrobiological techniques promises continuing new insights into the development of organisms in general and of the nervous system in particular. Since the nervous system utilizes electrical changes as the basis for its major functional activity, electrical recording methods are uniquely advantageous in monitoring its functional development. In addition, intracellular electrical recordings, in particular the more refined approaches such as voltage clamping and membrane noise analysis, provides a means for investigating the basic ionic mechanisms underlying neuron or cell membrane function as well as for studying the development of the various membrane properties during differentiation.

REFERENCES

Adrian, R. H. (1956). J. Physiol. (London) 133, 631–658.
Agin, D., and Holtzman, D. (1966). Nature (London) 211, 1194–1195.
Amatneik, E. (1958). IRE Trans. Med. Electron. PGME-10, 3–14.
Anderson, C. R., and Stevens, C. F. (1973). J. Physiol. (London) 235, 655–691.
Araki, T., and Terzuolo, C. A. (1962). J. Neurophysiol. 25, 772–789.
Auerbach, A. A., and Bennett, M. V. L. (1969). J. Gen. Physiol. 53, 211–237.
Baccaglini, P. I., and Spitzer, N. C. (1977). J. Physiol. (London) 271, 93–117.
Baldwin, H. A., Frenk, S., and Lettvin, J. Y. (1965). Science 148, 1462–1464.
Barrett, E. F., and Barrett, J. N. (1976). J. Physiol. (London) 255, 737–774.
Barrett, J. N., and Crill, W. E. (1974). J. Physiol. (London) 239, 301–324.

Barrett, J. N., and Crill, W. E. (1980). *J. Physiol. (London)* (in press).
Barrett, J. N., and Graubard, K. (1970). *Brain Res.* 18, 565–568.
Baylor, D. A., and Nicholls, J. G. (1969). *J. Physiol. (London)* 203, 591–609.
Begenisich, T., and Stevens, C. F. (1975). *Biophys. J.* 15, 843–846.
Bennett, M. V. L., and Trinkaus, J. P. (1970). *J. Cell Biol.* 44, 592–610.
Bennett, M. V. L., Spira, M. E., and Spray, D. L. (1978). *Dev. Biol.* 65, 114–125.
Blackshaw, S., and Warner, A. E. (1976a). *J. Physiol. (London)* 255, 231–247.
Blackshaw, S., and Warner, A. E. (1976b). *J. Physiol. (London)* 255, 209–230.
Brennecke, R., and Lindemann, B. (1971). *J. Life Sci.* 1, 53–58.
Brown, K. T., and Flaming, D. G. (1975). *Brain Res.* 86, 172–180.
Brown, K. T., and Flaming, D. G. (1977). *Neuroscience* 2, 813–827.
Brown, A. G., Rose, P. K., and Snow, P. J. (1977). *J. Physiol. (London)* 270, 747–764.
Burke, R. E., and ten Bruggencate, G. (1971). *J. Physiol. (London)* 212, 1–10.
Chambers, E. L., and de Armendi, J. (1979). *Exp. Cell Res.* 122, 203–218.
Coles, J., Levy, S., and Mauro, A. (1979). *Biophys. J.* 25, 267a.
Cone, C. D., and Tongier, M. (1973). *J. Cell Physiol.* 82, 373–386.
Conti, F., DeFelice, L. J., and Wanke E. (1975). *J. Physiol. (London)* 248, 45–82.
Coombs, J. S., Eccles, Y. C., and Fatt, P. (1955). *J. Physiol. (London)* 130, 291.
DeLaat, S. W., and Bluemink, J. G. (1974). *J. Cell Biol.* 60, 529–540.
del Castillo, J., and Katz, B. (1956). *J. Physiol. (London)* 124, 630–649.
Eisenberg, R. S., and Johnson, E. A. (1970). *Prog. Biophys. Molec. Biol.* 20, 1–65.
Engel, E., Barcilon, V., and Eisenberg, R. S. (1972). *Biophys. J.* 12, 384–403.
Fambrough, D., and Rash, J. E. (1971). *Dev. Biol.* 26, 55–68.
Fischbach, G. D., and Cohen, S. A. (1973). *Dev. Biol.* 31, 147–162.
Fischbach, G. D., and Laas, Y. (1978). *J. Physiol. (London)* 280, 515–526.
Furshpan, E. J., and Potter, D. D. (1959). *J. Physiol. (London)* 145, 289–325.
Furshpan, E. J., and Potter, D. D. (1968). *Curr. Top. Dev. Biol.* 3, 95–127.
Gaze, R. M., Chung, S. H., and Keating, M. J. (1972). *Nature (London)* 236, 133–135.
Geddes, L. A. (1972). "Electrodes and the Measurement of Bioelectric Events." Wiley, New York.
Gesteland, R. C., Howland, B., Lettvin, J. Y., and Pitts, W. H. (1959). *Proc. IRE, Nov.* pp. 1856–1862.
Gilula, N. B. (1974). *In* "Cell Communication" (R. P. Cox, ed.), pp. 1–29. Wiley, New York.
Grainger, J. L., Winkler, M. M., Shen, S. S., and Steinhardt, R. A. (1979). *Dev. Biol.* 68, 396–406.
Graybiel, A. M., and Devor, M. (1974). *Brain Res.* 68, 167–173.
Grundfest, H., Sengstaken, R. W., Oettinger, W. H., and Gurry, R. W. (1950). *Rev. Sci. Instr.* 21, 360–361.
Hagiwara, S. (1973). *Adv. Biophys.* 4, 71–102.
Hagiwara, S., Ozawa, S., and Sand, O. (1975). *J. Gen. Physiol.* 65, 617–644.
Higashi, A., and Kaneko, H. (1971). *Annot. Zool. Jpn.* 44, 65–75.
Hubel, D. H. (1957). *Science* 125, 549–550.
Hubel, D. H., and Wiesel, T. N. (1959). *J. Physiol. (London)* 148, 574–591.
Hubel, D. H., and Wiesel, T. N. (1963). *J. Neurophysiol.* 26, 994–1002.
Ito, S., and Hori, N. (1966). *J. Gen. Physiol.* 49, 1019–1027.
Ito, S., and Loewenstein, W. R. (1969). *Dev. Biol.* 19, 228–243.
Ito, S., Sato, E., and Loewenstein, W. R. (1974). *J. Membr. Biol.* 19, 339–355.
Jacobson, M. (1968). *Dev. Biol.* 17, 202–218.
Jaffe, L. A. (1976). *Nature (London)* 261, 68–71.
Jaffe, L. F., and Nuccitelli, R. (1974). *J. Cell Biol.* 63, 614–628.

Jaffe, L. F., and Nuccitelli, R. (1977). *Annu. Rev. Biophys. Bioeng.* **6**, 445–476.
Jaffe, L. A., and Robinson, K. R. (1978). *Dev. Biol.* **62**, 215–228.
Johnson, J. D., Epel, D., and Paul, M. (1976). *Nature (London)* **262**, 661–664.
Kalderon, N., Epstein, M. L., and Gilula, N. B. (1977). *J. Cell Biol.* **75**, 788–806.
Kater, S. B., and Nicholson, C., eds. (1973). "Intracellular Staining in Neurobiology." Springer-Verlag, Berlin and New York.
Katz, B., and Miledi, R. (1972). *J. Physiol (London)* **224**, 665–699.
Kootsey, J. M., and Johnson, E. A. (1973). *IEEE Trans. Biomed. Eng.* pp. 389–391.
Koryta, J. (1975). "Ion Selective Electrodes." Cambridge Univ. Press, London and New York.
Kostyuk, P. G., and Krishtal, O. A. (1977). *J. Physiol. (London)* **270**, 545–568.
Kravitz, E. A., Stretton, A. O. W., Alvarez, J., and Furshpan, E. J. (1968). *Fed. Proc.* **27**, 749.
Kuffler, S., and Yoshikami, D. (1975). *J. Physiol. (London)* **251**, 465–482.
Ling, J., and Gerard, R. W. (1949). *J. Cell. Comp. Physiol.* **34**, 383–390.
Loewenstein, W. R. (1966). *Ann. N.Y. Acad. Sci.* **137**, 441–472.
Loewenstein, W. R. (1968). *Dev. Biol. Suppl.* **2**, 151–183.
Loewenstein, W. R. (1975). *Cold Spring Harbor Symp. Quant. Biol.* **40**, 49–63.
Loewenstein, W. R., and Kanno, Y. (1964). *Cell Biol.* **22**, 565–598.
Lund, E. J. (1947). "Bioelectric Fields and Growth." Univ. of Texas Press, Austin.
Maloff, B. L., Scordilis, S. P., and Tedeschi, H. (1978). *Science* **199**, 568–569.
Miyazaki, S., Takahashi, K., and Tsuda, K. (1974). *J. Physiol.* **238**, 37–54.
Nastuk, W. L. (1953). *Fed. Proc. Fed. Am. Soc. Exp. Biol.* **12**, 102–128.
Neher, E., and Lux, H. D. (1969). *Pflügers Arch.* **311**, 272–285.
Neher, E., and Sakmann, B. (1976a). *Nature (London)* **260**, 779–802.
Neher, E., and Sakmann, B. (1976b). *J. Physiol. (London)* **258**, 705–729.
Nelson, P. G., and Frank, K. (1967). *J. Neurophysiol.* **30**, 1097–1113.
Nelson, P. G., and Lux, H. D. (1970). *Biophysical J.* **10**, 55–73.
Nicholls, J. G., and Purves, D. (1970). *J. Physiol. (London)* **209**, 647–667.
Nuccitelli, R., and Jaffe, L. F. (1974). *Proc. Natl. Acad. Sci. U.S.A.* **71**, 4855–4859.
Nuccitelli, R., and Jaffe, L. F. (1976). *Dev. Biol* **49**, 518–531.
Ogden, T. E., Citron, M. C., and Pierantoni, R. (1978). *Science* **201**, 469–470.
Ohmori, H., and Sasaki, S. (1977). *J. Physiol. (London)* **269**, 221–254.
Okamoto, H., Takahashi, K., and Yamashita, N. (1977). *J. Physiol. (London)* **267**, 465–495.
Oliveira-Castro, G., and Loewenstein, W. R. (1971). *J. Membrane Biol.* **5**, 51–77.
Potter, D. D., Furshpan, E. T., and Lennox, E. J. (1966). *Proc. Natl. Acad. Sci. U.S.A.* **55**, 328–335.
Rall, W. (1969). *Biophys. J.* **9**, 1483–1508.
Ransom, B. R., and Holz, R. W. (1977). *Brain Res.* **136**, 445–453.
Rink, T. J., Tsien, R. Y., and Warner, A. E. (1980). *Nature (London)* **283**, 658–660.
Ripley, K. L., and Provine, R. R. (1972). *Brain Res.* **45**, 127–134.
Rose, B. (1971). *J. Membrane Biol.* **5**, 1–19.
Rose, B., and Loewenstein, W. R. (1976). *J. Membrane Biol.* **28**, 87–119.
Schuetze, S. M., and Fischbach, G. D. (1978). *Neuroscience Abstr.* **4**(1195), 374.
Shen, S. S., and Steinhardt, R. A. (1978). *Nature (London)* **272**, 253–254.
Sheridan, J. D. (1968). *J. Cell Biol.* **37**, 650–659.
Sheridan, J. D. (1976). *In* "The Cell Surface in Animal Embryogenesis and Development" (G. Poste and G. L. Nicolson, eds.). Elsevier, Amsterdam.
Slack, C., and Warner, A. E. (1973). *J. Physiol. (London)* **232**, 313–330.
Slack, C., Warner, A. E., and Warren, R. L. (1973). *J. Physiol. (London)* **232**, 297–312.

Smith, T. G., Wuerker, R. B., and Frank, K. (1967). *J. Neurophysiol.* **30**, 1072-1096.
Snow, P. J., Rose, P. K., and Brown, A. G. (1976). *Science* **191**, 312-313.
Socolar, S. J. (1977). *J. Membr. Biol.* **34**, 29-37.
Socolar, S. J. and Loewenstein, W. R. (1979). *In* "Methods in Membrane Biology" (E. D. Korn, ed.), Vol. 10, pp. 121-177. Plenum, New York.
Socolar, S. J., and Politoff, A. L. (1970). *Science* **172**, 492-494.
Spray, D. C., Harris, A. L., and Bennett, M. V. L. (1979). *Science* **204**, 432-434.
Steinhardt, R. A., Lundin, L., and Mazia, D. (1971). *Proc. Natl. Acad. Sci. U.S.A.* **68**, 2426-2430.
Stewart, W. W. (1978). *Cell* **14**, 741-759.
Svaetichin, G. (1951). *Acta Physiol. Scand.* **24**, Suppl. (86), 5-13.
Talbot, S. A., and Marshall, W. H. (1941). *Am. J. Ophthalmol.* **24**, 1255-1264.
Tasaki, K., Tsukahara, Y., Ito, S., Wayner, M. J., and Yu, W. Y. (1968). *Physiol. Behav.* **3**(6), 1009-1010.
Thomas, R. C. (1972). *J. Physiol. (London)* **220**, 55-71.
Thomas, R. C. (1974). *J. Physiol. (London)* **238**, 159-180.
Thomas, R. C. (1978). "Ion-Sensitive Intracellular Microelectrodes." Academic Press, New York.
Tomita, T. (1965). *Cold Spring Harb. Symp. Quant. Biol.* **30**, 559-566.
Vacquier, V. D. (1975). *Dev. Biol.* **43**, 62-74.
Van den Berg, R., Siebenga, E., and DeBruin, G. (1977). *Nature (London)* **265**, 177-179.
Walicke, P. A., Campenot, R. B., and Patterson, P. H. (1977). *Proc. Natl. Acad. Sci. U.S.A.* **74**, 5767-5771.
Walker, J. L. (1973). *In* "Ion Specific Microelectrodes" (N. C. Hebert and R. N. Khuri, eds.). Dekker, New York.
Warner, A. E. (1973). *J. Physiol. (London)* **235**, 267-286.
Woodruff, R. I., and Telfer, W. H. (1974). *Ann. N. Y. Acad. Sci.* **223**, 408-419.
Woolsey, C. N. (1958). *In* "Biological and Chemical Bases of Behavior" (H. F. Harlow and C. N. Woolsey, eds.). Univ. of Wisconsin Press, Madison, Wisconsin.

CHAPTER 2

MIGRATION AND DIFFERENTIATION OF NEURAL CREST CELLS

Nicole Le Douarin

INSTITUT D'EMBRYOLOGIE DU CNRS
ET DU COLLÈGE DE FRANCE
NOGENT-SUR-MARNE, FRANCE

The evolution of the part of the neural primordium of the vertebrate embryo designated the neural crest has recently become a subject of active investigation in several laboratories. The interest of this structure lies mainly in the fact that it can provide a model for investigating not only cell differentiation but also other developmental mechanisms, such as cell migration and cell–cell recognition and interactions, which have long been recognized to play a major role in embryogenesis. The neural crest is a transitory structure whose disappearance is the result of the spreading of crest cells throughout the embryo soon after fusion of the lateral ridges of the neural plate. The migration of crest cells in the developing body is a phenomenon of considerable precision, which may in some cases last several days and lead them far from their source. In view of the intricate and orderly manner in which crest cell derivatives become localized, it is not surprising that the nature of the pathway they follow, the forces that in-

31

itiate and maintain their movements, and the means by which they
are informed about their final destination are considered to be among
the most intriguing problems posed by morphogenetic processes in
developmental biology.

Some advances in our knowledge about the neural crest have
emerged recently and have been reported in previous reviews by
Weston (1970), Le Douarin (1974, 1976), and Noden (1978a). The present chapter will only briefly mention certain of the aspects extensively
treated in the above-mentioned papers, while it will concentrate on
subjects still under active investigation in various laboratories. Thus,
after reviewing briefly the state of our knowledge concerning the cell
types derived from the neural crest, recent advances concerning the
migratory process will be discussed. Finally I shall devote a large part
of this chapter to the differentiation of the peripheral nervous system.

I. The Neural Crest Derivatives

Soon after they have left the neural primordium, the neural crest
cells become indistinguishable from the tissues through which they
move. Various experimental methods have therefore had to be devised
to follow both their migration pathways and their developmental fate.
Early ablation of the presumptive moving cells, explantations either
in vitro or in heterotopic grafting, and cell marking *in situ* or in
association with isotopic grafting have been the general procedures
employed in studying these problems. Ablation experiments, although useful, yield uncertain results because of the regulatory
capacities of the early embryonic territories. As a consequence, the extent of the deficiencies consecutive to the extirpation of a given area
of the crest is not necessarily an exact reflection of its normal
presumptive fate. In fact, the remaining anterior and posterior regions
left *in situ* tend to replace the ablated area as a result of the extensive
ability of neural crest cells to migrate and proliferate. In addition, the
absence of a particular structure following extirpation cannot be readily interpreted to mean that the structure is derived from the crest.
This can be established only after it has been demonstrated that the
defect is not due to the lack of an inductive interaction between the
crest cells and another embryonic tissue rather than to the absence of
the prospective material.

Explantation experiments either *in vitro* or in heterotopic grafting
[for instance in the coelomic cavity or on the chorioallantoic membrane (CAM)] have long been felt to provide only an indication of the

developmental capabilities of the neural crest. This is, in fact, a common embryological practice by which a morphogenetic field is operationally defined by its ability to form specific, organized structures in isolation. However, in such an artificial situation, it is obvious that some, but not all, of the crest derivatives will develop. The failure of inductive or stimulating signals from the proper surrounding nonneural tissues can account for this fact in both *in vivo* heterotopic grafting and *in vitro* culture. In the latter situation, the imperfect ability of the medium to fulfill all nutritional requirements and the failure of the tissue to establish the appropriate tridimensional arrangements are only two causes of the only partial adequacy of the technique for the problems under investigation.

The use of cell markers under conditions that minimize the disturbance of normal development is evidently the most convenient method of revealing the ultimate fate of crest cells, as well as of following the pattern of their migration. Several types of markers have been successfully used in amphibians, such as cytoplasmic inclusions (yolk granules and pigment) and differences in nuclear size or staining properties (Raven, 1937; Triplett, 1958). They are not conspicuous enough to enable an individual cell isolated in a tissue to be recognized with certainty and, as a consequence, can only be of limited use. A better marker is that provided by isotopic labeling of the nucleus by tritiated thymidine. It was applied by Weston (1963, 1967) and by Johnston (1966) and Noden (1973, 1975) to the migration of neural crest cells of the chick embryo, and by Chibon (1966, 1967) in studying the developmental capabilities of the neural fold of Pleurodeles. This method, accurate and precise, is more useful in following the early steps of crest cell migration than in obtaining information about their ultimate destination and fate. The isotopic labeling becomes diluted through the rapid proliferation of neural crest cells, which precedes the expression of their final phenotype.

A noticeable advance in our knowledge in the field of neural crest ontogeny has recently been made possible by the use of quail–chick chimeras by several groups of workers. The principle of this technique, which has already been described (Le Douarin, 1969, 1971, 1973a), is based on differences in nuclear structure in the quail (*Coturnix coturnix japonica*) and chick (*Gallus gallus*) species. Whereas in chick cell nuclei heterochromatin is distributed in several small chromocenters, it is, in contrast, highly condensed in a large mass associated with the nucleolus in the quail. This difference, particularly evident after DNA staining by the Feulgen–Rossenbeck technique, is encountered in practically all embryonic and adult cell types. Electron microscopic study of the nucleus reveals that spatial and quantitative relationships between RNP and DNP vary according to the cell category considered in the quail. Three types of nucleoli can be distinguished in quail cells according to the relative disposition of RNP and heterochromatin (Le Douarin, 1973b), but in any case they clearly differ from their counterparts in the corresponding chick cell types. As a result, the labeling provided by quail and chick cell associations has the advantage over the isotopic method of being stable and totally devoid of the deleterious effects that characterize most artificial markers (cf. Weston, 1967, for a review on cell marker systems).

Cell Marker Analysis of Neural Crest Derivatives

The quail–chick system was applied in studying the derivatives of the neural crest in avian embryos in two ways. One consisted of carrying out isotopic and isochronic grafts of fragments of the total neural primordium (i.e., neural tube plus neural fold) between quail and chick embryos (see Le Douarin, 1976, for a review). In the other, the substitution was restricted to small areas of the neural fold and was performed only at the cephalic level where the neural crest is more developed than in the trunk (Johnston et al., 1974, 1979; Noden, 1973, 1975, 1978b,c; Narayanan and Narayanan, 1978a).

Labeling experiments, involving either [³H]thymidine or the quail–chick markers, were applied to the avian embryo to study a number of problems. The principal results can be summarized as follows (Table I).

1. The level of origin of various crest-derived cell types was determined with precision. Such is the case for the neurons of the autonomic nervous system (Le Douarin and Teillet, 1973, 1974), for the sympathetic chains and the enteric ganglia (see Section III; Narayanan and Narayanan, 1978b), for the ciliary ganglion), and for the cells of various cranial nerve ganglia (Noden, 1975, 1978c).

2. The migratory pathways followed by the cells and, in some cases, their behavior during migration were described (Weston, 1970; Le Douarin and Teillet, 1974; Johnston, 1966; Noden, 1975).

3. Derivatives of the neural crest so far unknown were identified at the cephalic level of the neural axis. The calcitonin-producing cells that develop in the ultimobranchial body, and the secretory and accessory (respectively, type I and type II) cells of the carotid body,

TABLE I

DERIVATIVES OF THE NEURAL CREST

Neuronal cells

Peripheral nervous system

Sensory ganglia
Some neurons of trigeminal (V), facial root ganglia (VII), superior (IX), and
jugulare (X) ganglia[a]
All the neurons of the spinal dorsal root ganglia
Autonomic ganglia

Supportive cells of the nervous tissue

Schwann and sheath cells
Supporting cells (glia and satellite cells) of dorsal root ganglia and autonomic ganglia
Supporting cells but not neurons of geniculate (VII), acoustic (VIII), petrosal (XI),
and nodose (X) ganglia[b]

Pigment cells

Melanocytes of dermis, mesenteries, internal organs, epidermis, etc., and melano-
phores of the iris

Endocrine and paraendocrine cells

Adrenomedullary cells and other adrenergic paraganglia
Calcitonin-producing cells in the ultimobranchial bodies
Type I and type II cells of the carotid body

Mesectoderm

Bones and cartilages of the facial and visceral skeleton
Dermis of the face and the ventral part of the neck
Connective tissue (except the endothelium of the blood vessels) of the buccal and
pharyngeal glands (salivary glands, thyroid, and parathyroid)
Fibroblast and "endothelium" of the cornea
Connective component (except endothelium of the blood vessels) of the thymus
Musculoconnective wall of the large arteries derived from the aortic arches
Ciliary muscles
Some participation to striated muscles in the facial and visceral regions

[a] The other neurons of these ganglia are of placodal origin.
[b] All the neurons of these ganglia are derived from placodes.

were shown to be derivatives of the rhombencephalic neural primordium (Le Douarin and Le Lièvre, 1970, 1971, 1976; Le Douarin et al.,
1972, 1974; Fontaine, 1973; Pearse et al., 1973; Polak et al., 1974; Le
Douarin, 1976).

4. On the other hand, the extent of the contribution to head and
neck morphogenesis of the neural crest-derived mesenchyme (i.e., the
mesectoderm, according to Platt, 1894, 1898) was shown to be much
more diversified as far as cell type was concerned and quantitatively

more significant than had been previously believed. In fact, experiments performed mainly on amphibians had shown that the cephalic neural crest gave rise to mesenchymal cells that participated in formation of the facial and visceral skeleton (see Hörstadius, 1950, for
references), but very little was known about the precise extent of this
ectodermal contribution to the head mesenchyme, especially in higher
vertebrates. The only information available had been provided by extirpation experiments performed on the chick embryo by Hammond
and Yntema (1953, 1964) who showed that various visceral and cranial
cartilages were made up of mesectodermal cells.

The massive lateroventral migration of crest cells at the cephalic
level was first clearly demonstrated by Johnston (1966) who implanted tritium-labeled neural folds into unlabeled embryos. However,
these studies did not fully reveal the developmental capabilities of the
crest cells, because of the transient nature of the label.

Based on studies using the quail–chick marker system, the facial
part of the head skull appeared to be derived from the neural crest,
while the vault and the dorsal part were shown to be mesodermal. Cartilages and bones found in the intermediary region are of mixed origin
(Fig. 1) (Le Lièvre, 1978).

These structures are derived from crest cells originating from
various levels of the encephalon as revealed by heterospecific implan-
tations of definite segments of the neural primordium between quail
and chick embryos (Fig. 2). From the prosencephalon, neural crest
cells migrate into the frontal nasal process and mix with mesencepha-
lon-derived cells in the lateral nasal processes around the optic cup
and beneath the diencephalon. The main fate of mesencephalic crest
cells is to form the mesenchyme of the maxillary processes and man-
dibular arch. The rhombencephalic neural crest cells migrate more
posteriorly in the branchial arches.

Mesectodermal structures other than skeletal have a large
distribution in the face and the ventral aspect of the neck where they
form the dermis (including the smooth arrector feather muscles and
the adipose layer), the connective tissue of the tongue and of the
glands developed as appendages of the buccal and pharyngeal regions
(salivary glands, thyroid, and parathyroids) (Fig. 3) (Le Lièvre and Le
Douarin, 1975; Le Lièvre, 1978).

The mesenchymal cells that partly comprise the thymic tissue
were shown to be derived from the rhombencephalic crest and to form
a thin layer of perivascular tissue in both the cortex and the medulla
(Le Lièvre and Le Douarin, 1975; Le Douarin and Jotereau, 1975).

A contribution of neural crest mesenchyme was found in the striated muscles, which, however, appeared to be mainly of mesodermal origin.

In all the tissues of the head and neck that are of mesectodermal origin, the endothelial wall of the blood vessels was uniformly found to be of the host type and therefore mesoderm-derived.

In the large blood vessels arising from the heart, the segments corresponding to the aortic arches are composite in nature in the chimeric embryos: The endothelium is of mesodermal origin, while the musculoconnective wall is made up of mesectodermal cells (Fig. 4) (Le Lièvre and Le Douarin, 1975).

5. The problem of the origin of the cells of the diffuse endocrine system was investigated. The diffuse endocrine system, also called the *diffuse endokrine epitheliale Organe* was first defined by Feyrter (1938) who grouped under this term a number of cells characterized by a faintly stainable cytoplasm (*helle Zellen*) mainly located in the gut and its appendages. He later considered that the clear cells were not actually endocrine, but rather paracrine in nature, that is, they acted at short range on their immediate neighbors. As to their origin, he assumed that they were derived from enterocytes of the gastrointestinal tract or from the endodermal epithelium lining the ducts of foregut origin.

The development of cytochemical and ultrastructural studies on the endocrine cells associated with the endodermal structures, along with the recent advances in our knowledge of the peptides they produce (see Pearse, 1976), have led Pearse to develop the concept of the APUD series (Pearse, 1969). APUD is an acronym for one of the major cytochemical characteristics shared by the system of clear cells, i.e., the ability to take up from the blood the precursors of fluorogenic monoamines and to decarboxylate them by means of an amino acid decarboxylase (*a*mine *p*recursor *u*ptake and *d*ecarboxylation). Besides this property, APUD cells possess common ultrastructural features and are responsible for the secretion of low-molecular-weight polypeptide hormones. In addition, neuron-specific enolase, a molecular marker for nerve cells shown to be homologous with the 14–3–2 protein isolated from bovine brain (Moore, 1973), has recently been found in a variety of cells of the APUD series (Schmechel *et al.*, 1978).

The APUD series includes not only the endocrine and enterochromaffin cells of the gut epithelium but also the pancreatic islet cells, the adrenomedulla, the calcitonin-producing cells, the carotid body

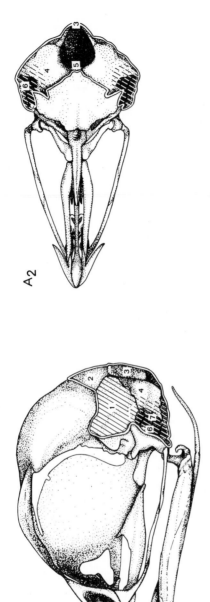

Fig. 1. Diagrams of the skull of a 14-day chick embryo showing the localization of the cartilages and bones partly (light shading) or totally (dark shading) derived from the mesectoderm. Double outline indicates the boundary between the regions of the skull considered. Regions: Occipital (A); orbital (B); maxillary and palatal (C); nasal (D); mandibulary and hyoid (E). 1, Squamosal; 2, parietal; 3, supraoccipital; 4, exoccipital; 5, basioccipital; 6, otic capsule; 7, columella auris; 8, basisphenoid; 9, alisphenoid; 10, orbitosphenoid; 11, rostrum of parasphenoid; 12, interorbital, internasal septum; 13, supraorbital cartilage; 14, anteorbital cartilage; 15, frontal; 16, lacrymal; 17, sclerotic cartilage (not been represented in this figure); 18, palatine; 19, pterygoid; 20, maxillary; 21, jugal; 22, quadratojugal; 23, quadrate; 24, premaxillary; 25, nasal; 26, concha nasalis; 27, articular; 28, angular; 29, supraangular; 30, dentary; 31, splenial; 32, mentomandibular; 33, Meckel's cartilage; 34, entoglossum; 35, basihyal; 36, ceratobranchial; 37, epibranchial; 38, basibranchial. (From Le Lièvre, 1976.)

38

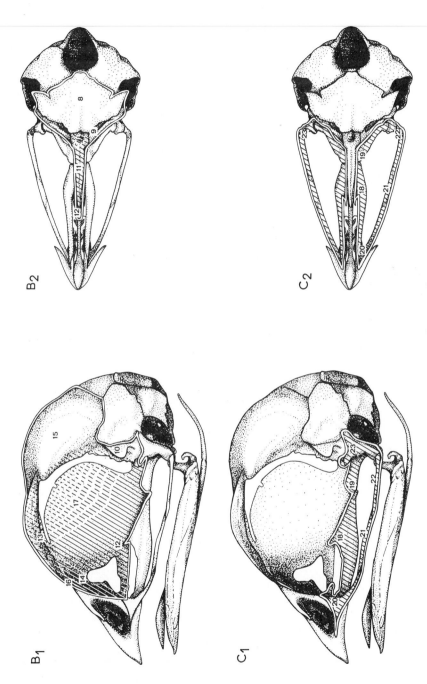

B_2

C_2

B_1

C_1

Fig. 1 (*Continued*)

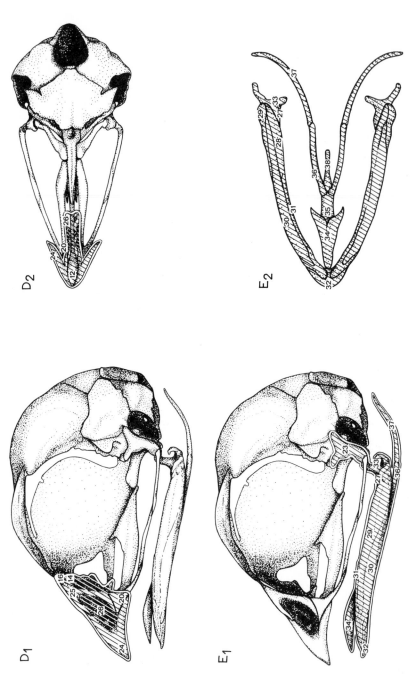

FIG. 1 (Continued)

40

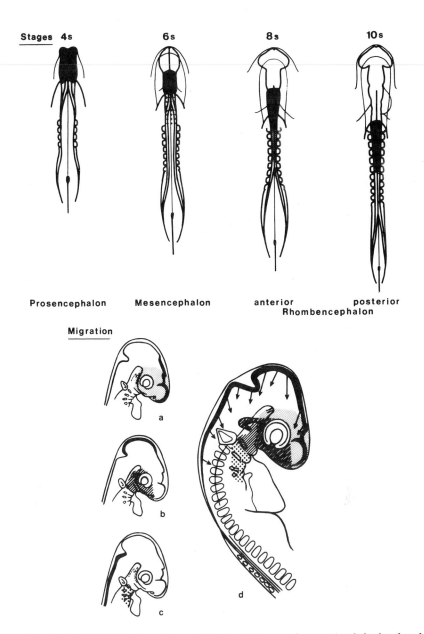

FIG. 2. (Top) Participation of neural crest cells in morphogenesis of the head and pharynx. Levels of the graft and stages of the operated embryos (4- to 10-somite stages) in the various experimental series: grafts of the prosencephalon, mesencephalon, and anterior and posterior rhombencephalon. (Bottom) Localization of neural crest-derived cells in the head and branchial regions of a stage-20 (Hamburger and Hamilton, 1951) chick embryo, according to their level of origin: (a) Prosencephalon; (b) mesencephalon; (c) rhombencephalon; (d) derivatives of the whole cephalic neural crest.

41

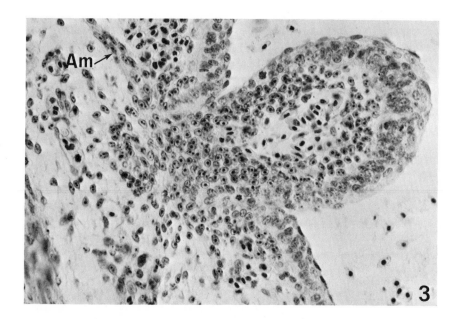

Fig. 3. Feather germ in the face of a 12-day chick embryo which received at the 9-somite stage the isotopic and isochronic graft of a quail rhombencephalon. The epidermis is of chick origin, while the dermis originates from the graft. Am, Arrector feather muscle of mesectodermal origin. Precise examination of the sections reveals that the endothelial cells of the blood vessels are of chick type and therefore of mesodermal origin. Feulgen–Rossenbeck staining. × 250.

type I cells and some of the anterior pituitary cells [ACTH and melanocyte-stimulating hormone (MSH)-secreting cells] (for the complete list of the APUD series see Pearse and Polak, 1978).

In terms of pathology, the APUD concept has the advantage of providing an explanation for the relationships of a number of endocrine disorders and syndromes such as the so-called pluriglandular syndrome and multiple endocrine tumors (Pearse, 1969). According to Pearse (1969) the cytochemical and ultrastructural similarities of APUD cells are related to their origin from a common embryological ancestor. Throughout their particular differentiating pathways, the APUD cells retained from their progenitors a common set of functions and characteristics. The neural crest (Pearse, 1966, 1969), and more recently the neurectoderm (Takor Takor and Pearse, 1975), have been proposed as antecedents of the APUD cells.

Pearse's hypothesis prompted us to undertake a series of embryological studies on the origin of various cell types of the APUD system. As mentioned above, we demonstrated the derivation from the rhombencephalon primordium of the calcitonin-producing cells and of the type I cells of the carotid body of the avian embryo (see a review of this work in Le Douarin, 1974).

The possibility that enterochromaffin cells originate from the neural crest is an interesting hypothesis because of the chromaffinity and argentaffinity common to enterochromaffin cells on the one hand, and to chromaffin cells of the adrenal medulla on the other. A number of authors have at various times considered whether or not there may be a relationship in terms of embryological origin between enterochromaffin and adrenomedullary cells.

The first experiment designed to test this hypothesis was done by Simard and Van Campenhout (1932). They based their work on the fact that no "nervous elements" are present in the intestine of chick embryos prior to 92 hours of incubation; in grafts on the CAM of gut from 82-hour embryos, enterochromaffin cells differentiated. The authors concluded that they were not of neural origin. However, since the migration of neural crest cells starts before the gut is removed from the donor embryo, the experiments of Simard and Van Campenhout are not conclusive. Much later, Andrew (1963, 1974) reinvestigated this question by transplanting on the CAM pieces of chick blastoderm from primitive streak to 25-somite stages. Definitive and potential neural crest and neural plate were excluded from experimental explants and included in controls. In addition, an experimental series was devised where the endoderm and adherent mesoderm alone were transplanted. Enterochromaffin cells occurred in similar numbers in control and experimental grafts, and it was concluded that they were not of neural origin. Although these results were very suggestive, further evidence was needed to draw a definitive conclusion. It could be objected that morphogenetic processes are significantly disturbed in the graft, and a possibility remained that the migration of the neural crest could likewise be modified. For this reason, we reinvestigated the problem of the origin of the enterochromaffin and endocrine cells of the gut epithelium by using the quail–chick marker system.

Isotopic and isochronic grafts of quail neural primordium (and vice versa) were made along the whole neural axis, and the gut and pancreas of the host were later removed for neural crest cell migration analysis (Le Douarin and Teillet, 1974). Although intramural ganglia of the gut were made up of grafted nerve cells when the graft had been performed at the appropriate level of the embryo, neural crest cell migration was never observed in the intestinal epithelium. In all

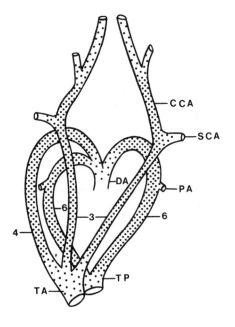

4A

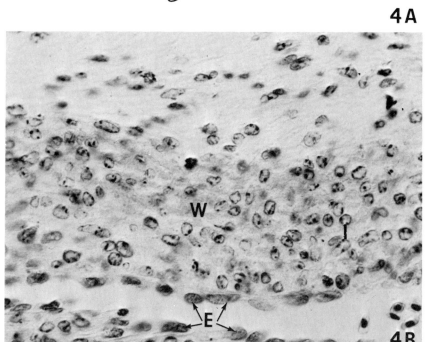

4B

cases, however, enterochromaffin and other endocrine cells normally developed in the gut but were always of the host type. No derivation from the neural crest could therefore be attributed to these elements. The experiments described above, however, did not rule out a possible migration of cells from the neurectoderm at a developmental stage preceding neural crest formation. To investigate this possibility, interspecific combinations were made between the endomesoderm of chick and the ectoderm (comprising the presumptive neural primordium area) of quail embryos at stages ranging from head process to two-somite. Similar associations of quail endomesoderm with chick ectoderm were also made. The chimeric embryos resulting from this association were grown on the CAM and examined histologically after Feulgen–Rossenbeck staining when organogenesis was completed. Anarchic morphogenesis occurs in such conditions; however, neural tissue and gut structures developed in most explants. Moreover, neural crest cell migration can occur, as indicated by formation of the myenteric plexuses from cells originating from the labeled neurectoderm. In no case were cells of ectodermal origin ever seen in the endoderm; the latter, as in the previous experiment, contained cells with APUD characteristics (Fontaine and Le Douarin, 1977a) (Fig. 5). It can therefore be concluded that the enterochromaffin cells and the endocrine cells of the gut epithelium do not originate in the neurectoderm.

The problem of the origin of pancreatic islet cells stimulated a certain number of embryological observations which gave rise to similar conclusions. Pictet et al. (1976) showed that, in early rat embryos deprived of the ectodermal precursor of the neural primordium and cultured in vitro, the pancreas developed and insulin production by normal B cells could be evidenced. Andrew (1976) and Fontaine and Le Douarin (1977b) applied the quail–chick system to this problem. By

---

FIG. 4. (A) In aortic arch-derived arteries (light shading) the vessel wall is entirely made up of mesectodermal cells originating from mesencephalic and rhombencephalic neural crest, except for the endothelium which is derived from the mesoderm. In darkly shaded areas mesectodermal and mesodermal cells are mixed in the musculoconnective wall of the vessel. The latter, in the dorsal aorta (DA), is entirely mesodermal in origin. 3–6, Arteries deriving from the third, fourth, and sixth aortic arches; CCA, common carotid artery; PA, pulmonary artery; SCA, subclavian artery; TA, aortic trunk; TP, pulmonary trunk. (B) Longitudinal section of the wall of the common carotid artery of a chimeric embryo (chick embryo into which the rhombencephalon of a quail was grafted at the 9-somite stage). Age of the host at the time of sacrifice was 12 days. E, Endothelial cells of chick host type; W, wall of the blood vessel made up of quail cells. Feulgen–Rossenbeck staining. × 180.

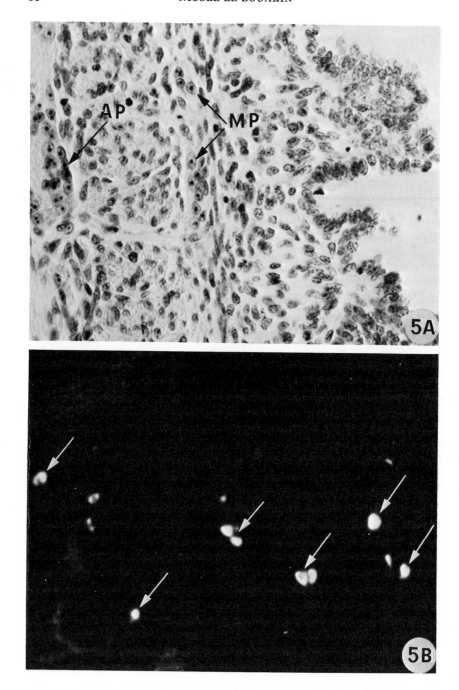

means of quail grafts into chick embryos (or vice versa), we showed that neural crest cell migration to the pancreas took place when the neural primordium was implanted at the vagal level of the neural axis (level of somites 1-7). They formed small parasympathetic ganglia that developed apart from the endocrine structures. Particular attention was focused on the somatostatin-producing cells, which could be identified by immunocytochemistry. No coincidence was ever found between the somatostatin cells and the quail cells present in the pancreas following the graft of a vagal neural primordium of quail into a chick embryo (Fontaine, Le Lièvre, and Dubois, unpublished).

It was clear that the possibilities offered by the avian embryo, along with the use of the quail–chick marker system, have encouraged the investigations on neural crest development preferentially in this class of vertebrates during the last 10 years. It was, however, necessary to see whether similar patterns of differentiation took place in mammals. Fontaine (1979), aiming to follow the evolution of calcitonin cell precursors during ontogeny of the mouse pharynx, demonstrated that they reached their final localization in the thyroid gland after a multistep migration. By isolating selectively either the fourth branchial pouch endoderm with or without the associated mesenchyme at increasing developmental stages, she showed that C-cell precursors were primarily localized in the branchial arch mesectoderm (which is neural crest-derived), whereas they only reached the endoderm secondarily. C cells reach their final site and invade the developing thyroid glandular tissue only when the fourth branchial pouches join the medioventral thyroid bud (Pearse and Carvalheira, 1967).

II. Role of the Environment in Crest Cell Migration

The precise distribution of neural crest cells in the embryo raises a number of questions about which only little is known, although they have for a long time aroused the interest of many investigators (see Weston, 1970): What are the factors responsible for the onset of

FIG. 5. (A) Result of an experiment involving the association of chick endomesoderm with quail ectoderm at the head process stage. After 24 hours of *in vitro* organotypic culture, the explant was transferred to the CAM of a chick for 12 days. Organogenesis occurred in the explant, and intestinal structures developed. The myenteric plexuses were made up of quail cells [in ganglia of Auerbach's (AP) and Meissner's (MP) plexuses], while no quail cells were ever seen in the endodermal epithelium. × 610. (B) Same experiment. The explant was treated with the FIF technique according to Falck (1962). Enterochromaffin cells of chick origin normally develop in the grafted tissue (arrows). × 420.

migration? What is the driving force of the moving cells? What ensures the orientation of the movement? What factors control the localization and aggregation of crest cells?

A. EXTRACELLULAR MATERIAL AND MIGRATION ROUTES FOR NEURAL CREST CELLS

Interesting information has recently been obtained about the chemical and structural composition of the space through which crest cells migrate. At both the cephalic and trunk levels, the initial steps of crest cell migration take place in a cell-free (or virtually cell-free) space in which the extracellular matrix (ECM) has been shown to contain high levels of hyaluronic acid (HA) (Pratt et al., 1975; Derby, 1978). In addition, in this developmental system, as well as in some others, variations in relative concentrations of HA have been correlated with cellular migration (Toole and Trelstad, 1971; Toole, 1972; Meier and Hay, 1973; Trelstad et al., 1974; Pratt et al., 1975; Derby, 1978).

As originally postulated by Toole et al. (1972), the accumulation of HA expands the ECM, providing both spaces through which cells can move and a substrate propitious for migration. In contrast, subsequent decreases in HA concentration can, in some instances, be correlated with the condensation and stabilization of previously migrating cells. At the trunk level, for instance, Derby (1978) clearly demonstrated histochemically that crest cells initiated their migration in an ECM, rich in HA; somewhat later, the precursor cells of the dorsal root ganglion aggregate in a region where the ECM HA concentration has significantly decreased. The elevated levels of HA apparently result from the synthetic activity of several embryonic structures. In vitro synthetic studies and autoradiography have shown that the tissues adjacent to the crest cell migratory route (neural tube, ectoderm, and somites), as well as the migrating cells themselves, synthesize HA (Greenberg and Pratt, 1977; Pintar, 1978). Other kinds of glycosaminoglycans (GAGs) are also produced by the tissues lining the neural crest migration pathways. Following administration of [³H]glucosamine to early embryos, 65% of the label is incorporated into sulfated GAGs by somites (Pintar, 1978), while the neural tube secretes predominantly chondroitin sulfate and heparin sulfate and the notochord produces little, if any, nonsulfated GAG (Hay and Meier, 1974).

Recent studies have also focused on another constituent of the extracellular space through which the crest cells move. Fibronectin (or

LETS protein; see Yamada and Olden, 1978, for a review) is a cell-associated glycoprotein for which a role has been proposed in a variety of cellular phenomena such as migration and adhesion. In order to define the possible role of fibronectin in crest cell migratory behavior, its sites of synthesis and distribution have been studied, in our laboratory, in early chick and quail embryos by using both *in vitro* and *in vivo* procedures (Newgreen, Leben, and Thiery, unpublished). Fibronectin is produced by all the structures surrounding the migrating crest cells, namely, the ectoderm, the somites, the neural tube, and the notochord, whereas crest cells themselves do not synthesize detectable amounts of this substance before or during their migration and their differentiation into nervous tissue. *In vivo*, in the 2-day embryo, fibronectin is concentrated in basement membranes which in fact serve as the limiting boundaries of crest cell migration (Tosney, 1978) (Fig. 6). However, the three-dimensional matrix through which the cells move has a complex structural and chemical composition, and the respective role of each of its presently known components in the migration of crest cells is still poorly understood.

Whether the composition of the ECM in various tissues is related to the pattern of migration in normal development has not been established but is suggested by some observations. If fragments of the rectum and of a limb bud, both taken from 5-day chick embryos, are associated with a piece of the trunk neural primordium of a 2-day quail, the pattern of crest cell migration differs strikingly in each of the chick explants. The intestine is extensively invaded by quail cells which become distributed into well-organized myenteric plexuses. In contrast, crest cell penetration into the limb bud is restricted to some melanocytes and Schwann cells lining nerves arising from the quail explant. No ganglia of crest origin are seen in the limb bud mesenchyme. In fact, the distribution of crest cells under the graft conditions mimics perfectly that occurring in the corresponding structures of the body. The availability of numerous crest cells in close contact with each of the tissue explants does not significantly affect their distribution in the tissues (Teillet and Le Douarin, unpublished).

Another observation deserves to be reported in this context: Following intracoelomic grafting of trunk neural primordium from a 2-day quail embryo into a 3-day chick, crest cell behavior depends on the contacts established between grafted and host tissues. In most of the cases observed 2 days after the graft, the implant remains free in the coelomic cavity and is only attached to the host through a thin vascularized pedicle. The neural crest cell population has expanded significantly and appears as two bands of randomly arranged cells at-

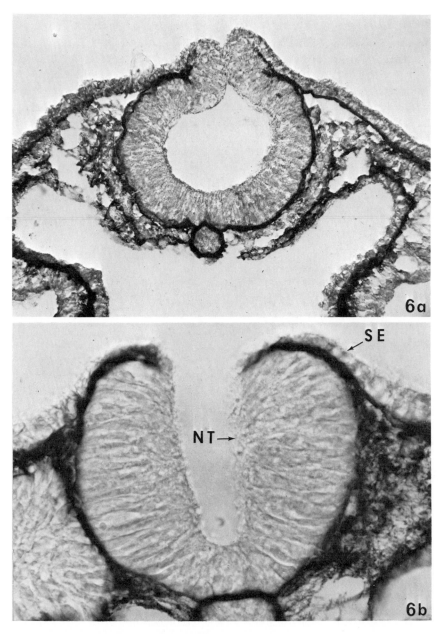

FIG. 6. Immunocytochemical localization of fibronectin in a transverse section of a chick embryo during individualization of the neural crest. The embryo (6–10 somites) was embedded in polyethylene glycol 1000 (Marzurkiewicz and Nakane, 1972). Sections were treated with anti-chick fibronectin antibodies (IgG fraction) prepared in rabbit and incubated with peroxidase-conjugated sheep antibodies (anti-rabbit IgG). Sections were stained with 3,3′-diaminobenzidine in the presence of H_2O_2. (a and b) Closure of

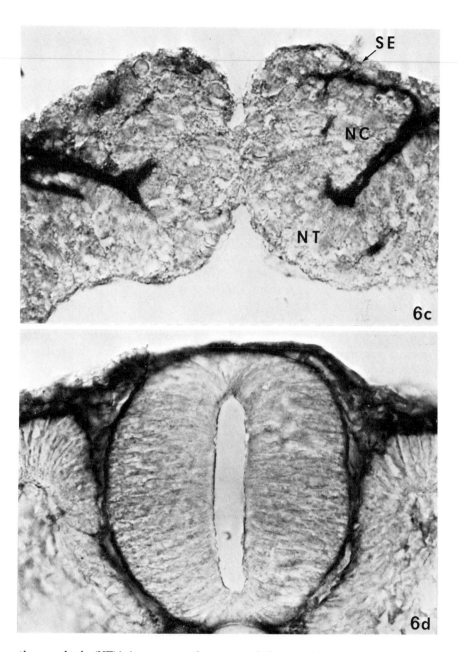

the neural tube (NT) is in process at the mesencephalic (a) and trunk level (b). (a) × 340. (b) × 810. (c) Individualization of the neural crest (NC) from the superficial ectoderm (SE) and the neural tube (NT) through synthesis of fibronectin-rich basement membranes. × 810. (d) Trunk level after closure of the neural tube. The area into which the crest cells will migrate is particularly rich in fibronectin (Thiery and Leben, unpublished). × 810.

tached laterally to the neural tube. If the duration of the graft is pro-
tracted, contacts are established between the explant and either the
splanchno- or the somatopleure. In the latter case, the neural crest
cells do not penetrate the body wall mesenchyme, except for melanocytes and Schwann cells (a result similar to that noted for the limb
bud in the previous experiment). In contrast, when the explant at-
taches to the splanchnopleure, crest cells invade the gut wall where
they differentiate into myenteric plexuses. When the graft lodges in
the umbilical cord, the crest cells migrate actively along the wall of
the umbilical artery where chains of ganglia develop (Lamers and Le
Douarin, unpublished).

Therefore, it seems clear that neither the limb bud nor the somatopleure provides a convenient substratum for crest cell migration,
whereas the gut mesenchyme (even if it is 2 or 3 days older than the
crest cells) is invaded by them. On the other hand, artery walls appear
to have suitable properties not only for crest cell migration but also
for their aggregation into ganglia. It is of interest to parallel this
observation with the fact that, at many sites in the body, adrenergic
ganglia are located along large arteries (e.g., in the trunk along the
dorsal aorta, where accessory adrenergic paraganglia are present).

Interesting information concerning the migratory behavior of crest
cells has been provided by scanning electron microscope (SEM) and
transmission electron microscope (TEM) observations of the crest during its dorsoventral progression in the space between the somites and
the neural tube. The extracellular environment into which the migration proceeds shows a network of fibrils of intricate microarchitecture,
with spherical bodies scattered among them. These bodies have a
dense, fibrillogranular appearance and are about 0.1–1.0 μm in
diameter (Low, 1970; Cohen and Hay, 1971; Tosney, 1978). Neural
crest cells show a wide range of variations in shape and orientation
during migration. In the axolotl embryo, they appear to initiate movement as a sheet when they leave their initial site, while at later stages
they proceed individually (Lofberg and Ahlfors, 1978). The fact that
numerous contacts with the matrix fibril network are established by
crest cell processes suggests that the latter could be the actual
material substrate for their progression in three-dimensional space.

In the chick, the progression of crest cells down the lateral aspects
of the neural tube has been carefully followed with a SEM by Tosney
(1978). When the neural crest is first distinguishable on the dorsolateral edges of the neural tube, large spaces are seen between the
cells, and the basal lamina overlying the crest appears discontinuous.
Cellular processes extend from crest cells into extracellular spaces.

After the initiation of crest cell movement, the cells take on an elongated form and become tangentially oriented with respect to the surface of the neural tube so that their long axis is perpendicular to the embryonic axis and parallel to the direction of their movement; long filopodia occur at the leading edge of the moving sheet of cells. A similar elongation is not observed in isolated cells which can be seen in areas where the population density is depleted. This occurs for instance during the late migration period in the regions near the dorsal neural tube. Similar observations can be made when cells happen to be insulated by a meshwork of fibrillar material; in this case, individual cells are not aligned and have no definable leading edge. These observations suggest that the behavior of migrating crest cells is mainly governed by the rules of contact inhibition of movement (CIM), which results in the alignment imposed upon cells by contact with their neighbors (Abercrombie and Heaysman, 1966; Abercrombie, 1970).

No directivity seems to exist in the orientation of the fibrillar meshwork into which the migration proceeds (Bancroft and Bellairs, 1976; Tosney, 1978). Since no experimental evidence has been reported showing that the direction of crest cell movement responds to environmental cues, CIM seems to be the simplest explanation accounting for the direction of their migration; crest cells disperse from zones of high cell density and progress into available cell-free space.

B. MEMBRANE PROPERTIES OF NEURAL CREST CELLS

Other points of interest raised by the problem of crest cell migration concern their membrane properties and the variations they may exhibit at various phases of their evolution, i.e., in the premigratory state, during the process of migration, and when they stop and differentiate.

Difficulties of various kinds are encountered in all meaningful approaches to this problem. These are due primarily to the small number of cells involved and the very close distribution in space and time of the various phases that have to be considered. One way, albeit imperfect, to overcome some of these problems consists in explanting the crest cells *in vitro* and looking at their evolution in terms of surface properties.

In vivo labeling of chick embryo with appropriate sugar nucleotides has revealed that neural crest cells are rich in glycosyltransferases—galactosyl-, *N*-acetylglucosaminyl-, fucosyl-, and sialyltransferase. Some of these enzymes seem to be developmentally regulated; for instance, fucosyltransferase activity present on the crest cells and neural tube in the 10-somite embryo is not detected at the 30-somite stage (Shur, 1977). For this reason, it has been proposed that embryonic cells migrate over cellular or noncellular substrates containing oligosaccharide chains, using their surface transferase to interact with the substrate. According to this hypothesis modifications in the migratory behavior of cells result from changes in cell surface enzymatic properties and from the constitution of the ECM (Roth et al., 1977).

Recently a cell surface glycoprotein, the cell adhesion molecule, was isolated from chick embryo neural tissue (Thiery et al., 1977). Monovalent antibodies directed against this glycoprotein inhibited adhesion among neural cells and disturbed the formation of neurite bundles (Rutishauser et al., 1978).

During the migration and further stabilization, changes in adhesive properties of crest cells obviously take place. It seemed, therefore, of interest to analyze the distribution of the cell adhesion molecule at different stages of migration.

Preliminary results indicate that, prior to or at the onset of migration in vivo, neural crest cells, which are still in clusters, carry large numbers of adhesion molecules at their surface, whereas lower numbers are detected during migration. In contrast, an increase in cell adhesion molecules occurs during the aggregation phase and can be easily seen, for instance, in the early stages of sensory ganglion formation. The same correlation also holds for crest cells in vitro, and the role of the cell adhesion molecule in the mechanisms controlling neural crest cell distribution in the body is now under investigation, as are its possible interactions with other ECM components (Thiery, Leben, and Newgreen, unpublished).

Lectins have recently been used as probes for detecting changes in surface properties of crest cells during their differentiation in culture (Sieber-Blum and Cohen, 1978a). The pattern and intensity of binding of several fluorescent lectins was found to change as the crest cells developed into melanocytes and adrenergic cells. Concanavalin A (Con A) and wheat germ agglutinin (WGA) bound to all kinds of unpigmented cells throughout the culture period, while melanocytes had much less affinity for lectins. Soybean agglutinin (SBA) only bound to unpigmented cells later in development, at the time when catecholamine could be detected histochemically. However, binding of SBA could be induced in unpigmented cells, but not in melanocytes, by pretreatment with neuraminidase. Cell processes resembling nerve fibers bound fluorescent lectins in high amounts. Similar results were recently found by another group (Rapin *et al.,* 1979) who also showed that undifferentiated crest cells bound *Ricinus communis* agglutinin 120 (RCA_{120}) but not RCA_{60}; in differentiating autonomic ganglion cells, the affinity for RCA_{60} appeared together with that for SBA.

The developmental significance of the changes observed in membrane properties of neural crest cells is not yet established.

III. Development of the Peripheral Nervous System

Cells originating from the neural crest differentiate into constituents of two major neuronal systems: the sensory and the autonomic. Considerable interest has been focused in recent years on the development of autonomic nerve cells, particularly on their chemical differentiation in terms of neurotransmitter synthesis. Two main lines of research, which can be regarded as complementary, have been followed: (1) The migratory behavior and differentiation capabilities of the presumptive autonomic neuroblasts have been studied by means of *in vivo* transplantation experiments; and (2) the stability of neurotransmitter synthesis in developing autonomic neurons has been examined in *in vitro* culture.

A. Origin of the Autonomic Ganglioblasts in the Normal Development of Birds

Although the ganglia of the autonomic nervous system and the adrenomedullary cells have long been recognized to be of neural crest origin, the migration pathways followed by their precursor cells has been a controversial matter. Some authors considered the vagal level of the neural primordium to be the only source of enteric ganglia (Yntema and Hammond, 1945, 1947, 1954, 1955), while others attributed a role in the constitution of these structures to the whole vagal plus trunk crest (Abel, 1909, 1912; Andrew, 1964, 1969, 1970, 1971; Van Campenhout, 1930, 1931, 1932; Kuntz, 1953; Uchida, 1927). As a prerequisite to the analysis we undertook of the mechanisms controlling autonomic cell differentiation, we reinvestigated the question of their origin in normal development by using the quail–chick chimera system (Le Douarin and Teillet, 1971, 1973). Grafts of small fragments (corresponding to a length of four to six somites) of quail neural primordium into chick embryos (and vice versa) were systematically made along the whole length of the neural axis. The developmental stages of host and donor embryos were identical and varied according to the level elected for the operation in order to ensure that crest cell migration had not started at the time of the intervention. The migrating neural crest cells were thereafter observed on serial sections of the trunk of the host (for the sympathetic chain and the adrenomedulla) and of its digestive tract (for the enteric and intravisceral ganglia). A correspondence was established between the level of the graft and the definitive location of the ganglion cells, as a result of the stability of the labeling provided by the quail–chick cell association. A schematic representation of the migrating process that gives rise to the autonomic system is shown in Fig. 7.

One of the striking observations made during this study was that, in the area of the trunk between somites 7 and 28, neural crest cell

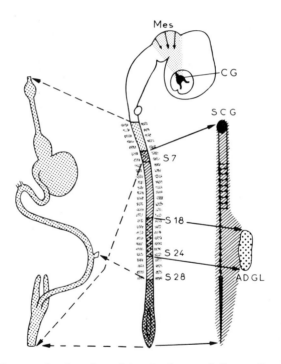

Fɪɢ. 7. Diagram showing the origin of adrenomedullary cells and autonomic ganglion cells. The spinal neural crest caudad to the level of the fifth somite gives rise to the ganglia of the orthosympathetic chain. The adrenomedullary cells originate from the spinal neural crest between the level of somites 18 and 24. The vagal neural crest (somites 1–7) gives rise to the parasympathetic enteric ganglia of the preumbilical region, the ganglia of the postumbilical gut originating from both the vagal and lumbosacral neural crest. The ganglion of Remak is derived from the lumbosacral neural crest (posterior to the somite-28 level). The ciliary ganglion (CG) is derived from the mesencephalic crest (Mes). ADGL, Adrenal gland.

migration was strictly confined to the dorsal mesenchymal region derived from the somites and the intermediate cell mass. Except for the Schwann cells that followed the nerve bundles to the periphery, neural crest derivative distribution was restricted to the sensory and sympathetic chain ganglia, the aortic and adrenal plexuses, and the adrenomedullary cords. No cells were ever found in the mesonephros or the gonads, but of more importance is the fact that they never penetrated the dorsal mesentery (Fig. 8). In contrast, orthotopic grafts carried out at the vagal and lumbosacral levels of the neural primordium resulted in colonization of both the dorsal mesenchyme and the splanchnopleure. The migration of the lumbosacral neural crest gives rise on the one hand to the caudal part of the sympathetic chain and to the coeliac and pelvic plexuses and, on the other hand, to the ganglion of Remak and some of the intramural neurons of the gut.

The largest contribution to the myenteric plexuses is made by the crest at the level of somites 1–7, whéreas that of the lumbosacral crest

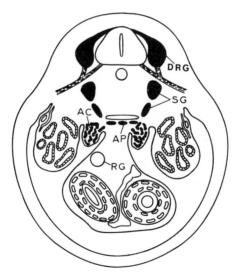

Fig. 8. Schematic transverse section of a 7-day embryo at the level of the suprarenal glands showing the extent of the migration of cells of the peripheral nervous system following orthotopic transplantation of the neural primordium between quail and chick embryos. The crest-derived ganglioblasts become localized in the dorsal root ganglia (DRG), the sympathetic ganglia (SG), the adrenomedullary cords (AC), and the aortic plexus (AP). No ganglion cells of the dorsal crest migrate in the dorsal mesentery. The cells of the ganglion of Remak (RG) and of the myenteric plexus at this level of the gut are, respectively, derived from the lumbosacral and the vagal neural crest.

located behind somite 28 is comparatively small. The cells migrate ventrally from the vagal region from the 7- to about the 14-somite stage. They become incorporated into the mesodermal wall of the foregut and thereafter undergo a long craniocaudal migration. During their progression in the splanchnopleural wall of the gut, they appear dispersed in the loose mesenchyme of the gut wall. They reach the level of the pancreatic ducts at about stage 20 of Hamburger and Hamilton (1951) and the umbilicus at about 5 days of incubation. The colorectum is not fully colonized before 8 days. When the muscular and connective structures of the gut are organized, the neural crest cells become distributed into ganglia located on each side of the circular muscle layer.

The contribution of the lumbosacral region of the crest to the intramural innervation of the gut is limited to the postumbilical intestine. Migration of neural crest cells into this region has been observed in the two kinds of grafts, i.e., quail neural primordium into chick embryo, and vice versa. The presumptive neuroblasts of the lumbosacral region do not invade the gut wall before 7–8 days of incubation (Le Douarin and Teillet, 1973).

The main parasympathetic structure arising from the lumbosacral crest is the ganglion of Remak. The ontogeny of the Remak ganglion has been the subject of a detailed study by Teillet (1978). It is a complex structure, peculiar to birds, which develops in the dorsal mesentery. The ganglioblasts arising from the neural crest posterior to the level of somite 28 accumulate first in the mesorectum at stage 24 of Hamburger and Hamilton in the chick and at stage 18 of Zaccei (Zaccei, 1961) in the quail. They subsequently migrate cranially, along the ileum and jejunum, to reach the level of the hepatic and pancreatic ducts. In addition to masses of ganglion cells distributed throughout its length, the ganglion of Remak appears to be the route for descending and ascending nerve fibers. At the level of the cloaca, it is, as mentioned by Browne (1953), in close relationship with the pelvic plexus; the large nerve network that develops in the vicinity of the cloaca and the bursa of Fabricius contains mostly adrenergic cells which extend to the posterior end of the ganglion of Remak itself (Bennett and Malmfors, 1970; Teillet, 1978). The part of the ganglion corresponding to the anterior two-thirds of the rectum does not show catecholamine-containing cells at any developmental stage, although numerous adrenergic fibers run along the whole length of the Remak ganglionated nerve.

The primordium of the ganglion of Remak was selectively labeled by means of a graft of quail neural primordium at the lumbosacral level of a chick embryo (Teillet, 1978). Subsequently, the complex consisting of colorectum plus mesorectum containing the labeled primordium of the Remak ganglion was taken from the chimeric host at 5 days of incubation and grafted onto the CAM of a chick host for 10 days. Passage of ganglioblasts from the ganglion of Remak to the gut was observed, showing that at least part of the lumbosacral ganglionic supply to the hindgut intramural innervation migrated through the ganglion of Remak, in which the crest cells probably stop for a while before undertaking the last part of their ventral progression.

B. Migratory Behavior of the Autonomic Ganglion Precursor Cells and Other Neural Crest Cells Studied in Heterotopic Transplantation *In Vivo*

The different migratory behavior exhibited by the autonomic ganglion cell precursors originating from the various levels of the neural axis is a very striking phenomenon. For example, cells arising from the trunk neural crest (somites 7–28) do not penetrate the mesentery, while those deriving from the part of the neural axis located behind the somite-28 level regularly do so.

A possible explanation of this fact is that a population of crest cells programmed to become cholinergic enteric ganglia is present in both the vagal and lumbosacral regions and does not exist in the cervicodorsal neural crest. Its migration would therefore be "motivated" toward the intestine through some kind of chemotactism. In a similar way, a population of predetermined adrenomedullary cells would be exclusively located at the level of somites 18–24 from which they migrate to colonize the adrenal gland.

The other alternative in explaining the behavior of neural crest cells is to consider that they migrate along pathways that are in fact organized by the morphogenesis of the surrounding tissues, such "routes" leading them to definite embryonic rudiments. The way in which they are led to stop at certain determined spots where they become arranged in well-defined patterns is another question to which no satisfactory answer has so far been given. In any case, if this second alternative were true, one would have to assume that the environment encountered by the crest cells plays a role in their differentiation. Various kinds of experiments have been designed with the purpose of clarifying this question.

Transplantations of segments of the neural primordium between quail and chick embryos were carried out as indicated in Fig. 9.

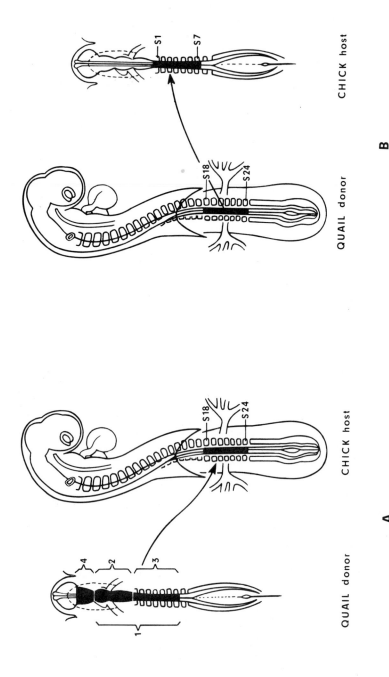

QUAIL donor   CHICK host   QUAIL donor   CHICK host

A   B

Fig. 9. Diagrammatic representation of the heterotopic and heterochronic transplantation between quail (donor) and chick (recipient) embryos of the cephalic neural primordium at the "adrenomedullary" level of the neural axis (A) and of the "adrenomedullary" neural primordium in the vagal region (B). (A) S18 and S24, Somites 18 and 24, Somites 1–4 indicate the various types of transplantations carried out. (B) S1 and S7, Somites 1 and 7.

The neural crest cells of the adrenomedullary level of the crest (which do not migrate into the gut in normal development) were shown to be able to colonize the gut and to give rise to functional cholinergic enteric ganglia when grafted at the vagal level of the neural axis.

In the same way, neural crest cells from the mesencephalon or rhombencephalon, when transplanted into the adrenomedullary region, populated the adrenal gland and differentiated into adrenomedullary cells (Le Douarin and Teillet, 1974; Le Douarin et al., 1975).

An interesting observation was made in this last series of transplantations. In the chick embryos that had received a graft of a quail neural primordium corresponding to somites 1–7 or to the mesencephalic or anterior rhombencephalic anlagen (see Fig. 9A), quail cells participated in formation of the enteric ganglia of the host at the level of the ileum and the large intestine.

Moreover, when the mesencephalic and anterior rhombencephalic primordia were grafted, mesenchymal cells originating from the graft were found in the host embryos in various locations: the dermis; the host vertebra, where they differentiated into cartilage; and the derivatives of the intermediate cell mass, i.e., the mesonephros and the wall of the Müllerian duct. In the mesonephros they were often found to differentiate into cartilage.

The migration pattern of the grafted mid- and hindbrain crest cells cannot be accounted for by a special migration route leading them to the gut, since the results of orthotopic grafting experiments have shown that crest cells in this region of the body never penetrate the splanchnopleure. It may be the ability of cephalic crest cells to migrate far from their source that must be invoked in this case. Another alternative explanation resides in the fact that, at the cephalic level, the crest cell population is much more numerous than in the trunk. Therefore, after having filled all the available sites of arrest in the dorsal mesenchyme during their dorsoventral migration, they continue to migrate further and colonize the intestine. At the present stage of our investigations no satisfactory answer can be given to this question.

The other point of interest concerning heterotopic cephalic transplants at the adrenomedullary level is the differentiation of mesenchymal derivatives of crest origin at this abnormal location. This result can be interpreted as meaning that the presumptive mesectodermal cell line is segregated early from the rest of the cephalic crest cell population, since it differentiates according to its normal fate in a foreign environment; the inductive cues leading to vertebral cartilage development (Strudel, 1955; Lash *et al.*, 1957) appear to be quite efficient in directing mesectodermal cell differentiation.

It is interesting to note that in amphibians the differentiation of cephalic crest cells seems to obey stricter requirements. Transplanted into the trunk, they did not develop into cartilage (Horstadius and Sellman, 1946) as they normally do in the head where, according to Holtfreter (1968), they receive appropriate differentiation signals from the pharyngeal endoderm.

Besides these points, the most important results of this series of experiments are the following. As far as crest cells colonization of the gut and the suprarenal glands are concerned, preferential pathways characterize the vagal and the adrenomedullary levels of the neural axis. They lead the cells to differentiate into enteric ganglia and into adrenomedullary cells, respectively. The phenotypic expression of crest cells appears therefore to be regulated by the environment they encounter after leaving the neural primordium. On the other hand, the developmental capabilities of crest cells were shown to be fundamentally identical in the cephalic and trunk regions, except for the mesectodermal precursor cells, the origin of which appears to be restricted to the head neural crest down to the level of the fourth somite (Le Lièvre and Le Douarin, 1975). The developmental capabilities of crest cells seem to be uniformly distributed along the neural axis. At least, the capacity to produce enteric ganglia, adrenergic neurons, and paraganglionic cells is not confined to the areas from which they originate during the normal process of embryogenesis but appears to be a property of all regions of the neural primordium so far tested.

Several questions are raised by these observations:

1. Do the environmental differentiating signals act on neural crest cells during their migration or when they are settled in their definitive location?
2. When do the developing autonomic neurons become irreversibly differentiated into adrenergic or cholinergic cells?
3. What are the factors in the external environment that regulate crest cell migration and localization?

C. DETERMINISM OF TRANSMITTER FUNCTION IN DEVELOPING NEUROBLASTS

Some recently accumulated data concerning these three questions will be reported.

1. Is the Chemical Differentiation of the Autonomic Neuron Precursor Determined during or after Its Migration?

Since—as described above—cholinergic differentiation of presumptive adrenergic neuroblasts occurs when the dorsal neural crest is transplanted into the vagal region, the question arose as to whether significant developmental changes occurred during neuroblast migration into the gut, rather than when they were settled in the gut itself. We therefore set up an experiment devised to suppress the extraintestinal migration phase and to see what kind of transmitter the neuron would synthesize in this case.

Since neural crest cells reach the hindgut at 7–8 days of incubation only, the colorectum remains totally aneural if it is removed from the embryo before this stage and subsequently cultured on the CAM; it can therefore be used as a culture medium for crest cells. When associated with the trunk neural crest (Fig. 10) and grafted for 10–12 days on the CAM, it becomes innervated according to a normal developmental pattern. We were able to show that Auerbach's plexus and

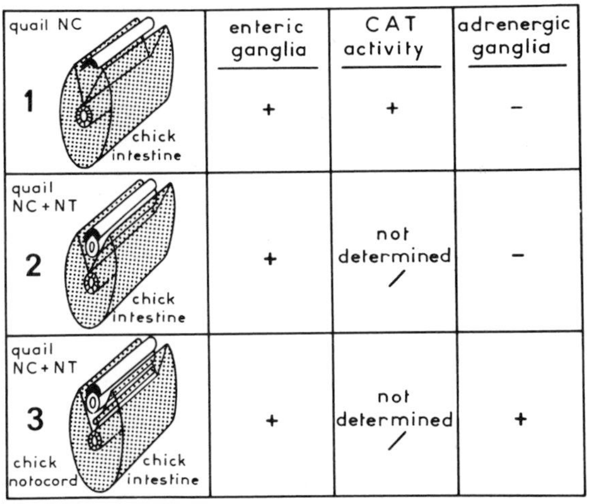

	enteric ganglia	CAT activity	adrenergic ganglia
1 quail NC / chick intestine	+	+	−
2 quail NC+NT / chick intestine	+	not determined /	−
3 quail NC+NT / chick notocord / chick intestine	+	not determined /	+

FIG. 10. Association of the aneural colorectum of a 5-day chick embryo with various dorsal trunk structures of 2-day embryos. (1) Chick intestine plus quail neural crest (NC); (2) chick intestine plus quail neural crest and neural tube (NT); (3) chick intestine plus quail neural crest, neural tube, and chick notochord. Cholinergic ganglia develop in all cases. Adrenergic ganglia differentiate only in explants containing the notochord.

Meissner's plexus developed apparently normally under these conditions. Catecholamine (CA)-containing cells were never observed (Smith *et al.*, 1977; Teillet *et al.*, 1978) (Fig. 11) but, in contrast, significant levels of choline acetyltransferase (CAT) and acetylcholinesterase (AChE) activity were present in the explants. This indicated that the migration phase of autonomic neuron precursor cells from the neural primordium to the intestine was not of importance in their orientation toward cholinergic metabolism. More likely is the alternative explanation that their localization in the gut is responsible for their cholinergic phenotype.

2. In Vivo *Studies of the Factors Responsible for Adrenergic Cell Differentiation*

Experimental studies on the factors involved in expression of the adrenergic phenotype suggest that, in this case, neural crest cells are probably subjected to differentiating cues during their dorsoventral migration.

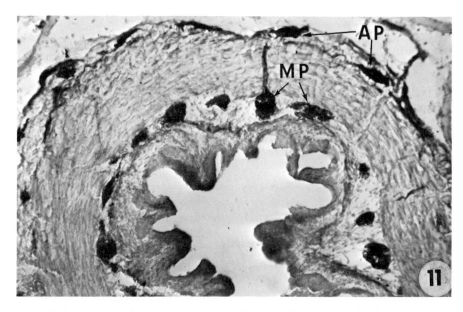

FIG. 11. Association of the quail neural crest with the aneural colorectum of a chick (see Fig. 10). Culture of the explant for 8 days on the CAM of a chick. Development of the enteric plexuses evidenced by the AChE reaction according to the technique of Karnovsky *et al.* (1964). MP, Meissner's plexus; AP, Auerbach's plexus. × 150.

In order to see whether adrenergic differentiation was determined through signals the migrating crest cells receive along their route or from the tissues in which they become localized, Cohen (1972) removed from embryonic axial trunks the future site of the primary sympathetic chain. The remaining tissues including the migrating crest cells were grown on the CAM of host chicks. Since CA-containing cells developed ectopically in the explants, Cohen concluded that crest cells gave rise to sympathetic neurons in response to the environmental signals encountered while they migrated ventrally. In tissue association experiments, he also showed that adrenergic differentiation required the presence of somitic mesoderm, since mesodermal tissues of other sources were unable to elicit sympathoblast differentiation.

The importance of the dorsal structure of the trunk, namely, the ventral neural tube, the notochord, and the somites, in the expression of adrenergic traits, was also demonstrated by Norr (1973). By comparing the effect of somites of various ages on adrenergic cell differentiation this author showed that somitic mesenchyme acquired its capacity to stimulate CA synthesis in neural crest cells through a previous inductive effect arising from the notochord and the neural tube.

In a recent series of experiments (Teillet et al., 1978), we have shown that, among the axial dorsal structures of the trunk, the one of decisive importance in promoting adrenergic differentiation in neural crest cells is, in fact, the notochord. If the aneural colorectum is associated with the trunk neural primordium or with the dissected neural crest, only cholinergic neurons develop in the explants. In contrast, when the notochord is added to the cultures, some groups of catecholaminergic cells develop in the gut wall itself (Fig. 10). The exclusive ability of the somitic mesenchyme to promote adrenergic cell differentiation, as suggested by previous authors (Cohen, 1972; Norr, 1973), is therefore not confirmed by these findings, whereas the fundamental role of the notochord is stressed. The neural tube by itself without the notochord was shown to be unable to promote adrenergic cell differentiation in the gut mesenchyme.

An interesting observation was made by Cochard et al. (1978) in the rat embryo. Both tyrosine hydroxylase (TH), revealed by immunofluorescence, and CA, evidenced by formaldehyde-induced fluorescence (FIF) (Falck, 1962) technique, were undetectable in the dorsal neural crest or in the ventrally migrating crest cells. They first appeared at 11.5 days of gestation (36- to 37-somite stage) in sympathetic ganglion primordia. In addition, TH and CA transiently appeared in scattered cells of the gut wall. The morphology of these cells with elongated, fluorescent cytoplasmic processes strongly suggests that they are enteric neuron precursors in the process of migration. The number of fluorescent cells progressively decreases so that, by 14.5 days they have practically disappeared. As proposed by these authors, CA differentiation in these cells may have been induced by interactions with the somite–notochord–neural tube environment during their dorsoventral migration toward the gut. A delayed appearance of the adrenergic phenotype takes place thereafter but cannot be maintained in the gut because of the lack of appropriate stimulation.

Although the experiments described earlier (Cohen, 1972; Norr, 1973; Teillet et al., 1978) demonstrate the influence of the dorsal trunk structures and, in particular, of the notochord in promoting the expression of adrenergic traits in neural crest cells, other tissues of the body have recently been shown to exert similar influences.

Fragments of the neural primordium from 2-day quail embryos were grafted into the coelomic cavity of 3-day chick hosts. Contacts were randomly established between the graft and the host tissues either at the level of the splanchnic mesoderm or with the body wall. In many instances, the graft was found to be attached to the umbilical cord. The migratory behavior of the neural crest cells differed considerably according to the kind of tissue they contacted. No infiltration of crest cells was ever found in the somatopleure, whereas extensive colonization of the gut occurred where the graft joined the splanchnopleure. As a result, enteric ganglia made up of quail cells were observed through an extended area of the host intestine.

Of particular interest was the umbilical cord tissue when the grafted neural tube was found attached to it. Neural crest cells migrated in this environment and aggregated to form ganglia scattered along the wall of the arteries. The FIF technique showed the CA content of the ganglionic cells (Lamers and Le Douarin, unpublished). The common characteristics provided by the notochordal surroundings and the arterial wall would be interesting to identify.

3. Lability of Transmitter Phenotype in Developing Autonomic Neurons

The possibility that the neurotransmitter phenotype expressed by neural crest cells is selected by the tissue environment to which they are subjected during the early phase of their development prompted us to try to change the fate of ganglion cells already in the course of differentiation (Le Douarin et al., 1978).

We took ganglia in the process of cholinergic differentiation and transplanted them into the dorsal trunk environment of a younger embryo at the precise stage when the neural crest cells receive the signals leading them to express the adrenergic phenotype. It has been shown that the first FIF-positive cells appear in the primary sympathetic chain of the chick embryo on the fourth day of incubation (Enemar et al., 1965). It is therefore between the stage at which neural crest cell migration starts and that at which the first sign of adrenergic differentiation appears (i.e., between 2 and 5 days of incubation) that the environmental signals are operative for orientation of the autonomic neuroblasts toward adrenergic metabolism. (The same or other types of signals are very likely necessary for maintenance of the differentiated state.)

Accordingly, ciliary ganglia or the part of the Remak ganglia previously shown to be cholinergic (Teillet, 1978) was removed from quail embryos and inserted into a slit made between the somite and the neural primordium of 2-day chick embryos at the level of somites 18–24 (i.e., the "adrenomedullary" level of the neural axis) (Fig. 12).

Normal development of the ciliary ganglion in the chick embryo has been extensively studied (Landmesser and Pilar, 1972, 1974a,b; Chiappinelli et al., 1976). It is a cholinergic parasympathetic ganglion in which fine, varicose, nonvascular adrenergic terminals are present, while adrenergic perikarya are absent in all the species studied so far (Ehinger, 1967; Cantino and Mugnaini, 1974). Both the ciliary and the Remak ganglia express the first signs of cholinergic differentiation at a very early stage. [^{3}H]choline can be converted to ACh in vitro as early as 96 hours of incubation in the ciliary ganglion and about 12 hours later in the Remak ganglion of the quail embryo. Approximately 20 fmoles of ACh are formed per ciliary ganglion during a 4-hour period of incorporation at 4 days, and the synthetic capacity is about 15-fold higher 2 days later (Le Douarin et al., 1978).

The age of the cholinergic ganglia grafted varied from 4.5 to 15 days for the ciliary ganglion and from 4.5 to 6 days for the Remak ganglion. Only parts of the ganglionic body were transplanted in the case of ganglia older than 6 days. The result of the experiment was observed in serial sections of the host stained with the Feulgen–Rossenbeck technique 6–8 days after the operation, i.e., when the host embryo was 8–10 days old. The anatomical integrity of the ganglion was not maintained in the graft situation irrespective of the age of the grafted ganglion. Quail cells were found dispersed in the host trunk structures, but their distribution was not random; they were always localized exclusively at normal sites of neural crest cell arrest (Fig.

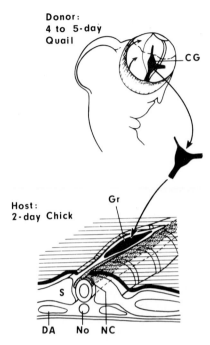

Fig. 12. Experimental design for back-transplantation of differentiating cholinergic ganglia from 4- to 15-day quail donor embryos into 2-day chick hosts. The ciliary ganglion (CG) is dissected and inserted either whole or in pieces into a slit made between the neural primordium and the somites (S). DA, Dorsal aorta; Gr, graft; NC, neural crest; No, notochord.

13). The extent of their migration was found to decrease with their age at transplantation, as shown in Fig. 14. It is interesting to note that colonization of gut ganglia only occurred with ciliary ganglion cells transplanted at the age of 4–5 days of incubation.

On the other hand, quail cells were found in the dorsal root ganglion (DRG) of the host in only about 3% of the transplantations of the ciliary and Remak ganglia. These cells were represented by a few scattered noneuronal (probably satellite) cell types. In most cases, the host tissues were treated by the FIF technique followed by Feulgen–Rossenbeck staining as previously described (Le Douarin *et al.*, 1978). It was then possible to observe that quail cells that had homed to the sympathetic chain ganglia, the aortic and adrenal plexuses, and the suprarenal gland exhibited a bright-green fluorescence. Electron microscopic observation of the host adrenal medulla showed the quail cells (characterized by their large DNA-rich nucleolus) to contain the secretory granules usually found in Ca-producing cells (Fig. 15).

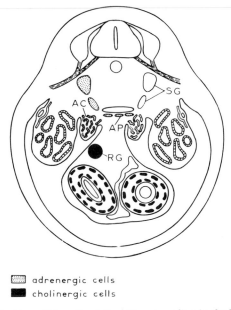

☒ adrenergic cells
■ cholinergic cells

FIG. 13. Localization of the cells of the ciliary ganglion in the host. They are found as Schwann cells in the rachidian nerve, the sympathetic chain ganglia (SG), the adrenomedullary cords (AC), the aortic plexus (AP), the ganglion of Remak (RG), and the myenteric plexuses. No grafted quail cells are found in the DRG. Most quail cells localized in the sympathetic chains and the adrenomedulla contain CA as evidenced by Falk's technique.

Although, in some cases, sympathetic chain ganglia of the host were found totally made up of quail cells, most often grafted cells and host neural crest-derived cells were intermixed. Therefore the possibility that the fluorescence observed in the grafted cells resulted from an uptake of CA synthesized and released by host cells had to be tested. For this purpose we designed the following experiment: The neural crest along with the dorsal quarter of the neural tube of the host embryo was removed on the side where the cholinergic ganglion was implanted at the level of somites 18–24. In order to prevent possible regulation by the anterior and posterior neural crest, the portion of the host embryo corresponding to somites 18–24 was cut off and cultured on the CAM. The implanted ganglion cells migrated in the chick tissues and became distributed in the same way as they did when grafted into the embryo *in ovo*. In the adrenal glands, the medullary cords were entirely of quail type on the operated side; only cells with the quail nuclear marker showed cytoplasmic fluorescence in these glands. This result excludes the possiblity that the ciliary and Remak ganglion cells that migrated into the adrenergic structures of the

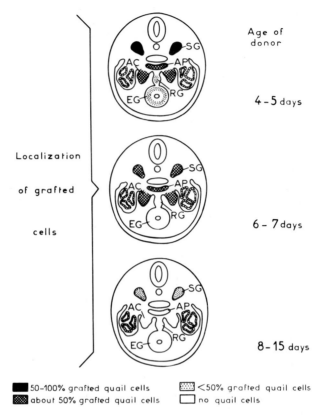

FIG. 14. The migratory capacity in the host embryo of the grafted ciliary ganglion cells decreases with the age of the donor embryo. However, the ability of the cholinergic neuroblasts to synthesize CA is maintained even in 15-day ganglion cells. AC, Adrenomedullary cords; AP, aortic plexus; EG, enteric ganglia; RG, Remak ganglion; SG, sensory ganglion.

chick host acquired their CA by means of uptake via the host adrenergic cells. Therefore, the fluorogenic amine content of the cells derived from the grafted ganglia certainly reflects their own ability to synthesize CA.

The cells of the ciliary and Remak ganglia that colonized the splanchnopleure and became localized in the host enteric and Remak ganglia did not exhibit any CA content after FIF treatment. It can therefore be assumed that they developed as cholinergic cells.

Besides the capacity of cells of the differentiating cholinergic ganglia to change their developmental fate when transplanted into the dorsal trunk structures of a young embryo, their ability to migrate in

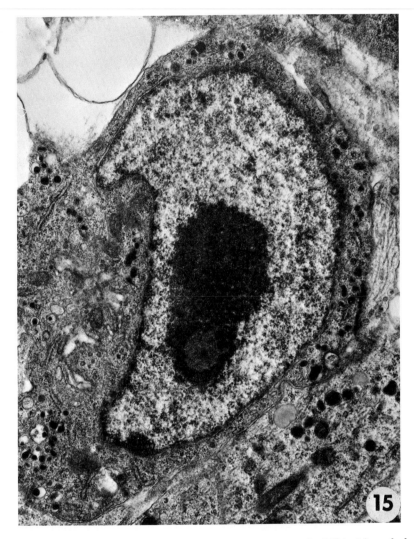

FIG. 15. Same experiment as in Fig. 12. Quail cell (note the DNA-rich nucleolus) with CA - secreting granules in the adrenomedullary cords of the chick host. × 18,900.

this new environment is a remarkable phenomenon. Our further investigations of this problem have consisted of following the progressive dissociation of the ganglion cells after the graft and their subsequent distribution in their definitive localizations; this was carried out following transplantation of 4.5- to 6-day ciliary ganglia. Quail cells become detached from the outer surface of the ganglion on the side facing the host neural tube. They are elongated and seem to move ventrally in the extracellular material surrounding the axial organs along with the host neural crest cells. Quail cells are also often found along the wall of the dorsal aorta, which may provide an appropriate substratum for their movements. During the few days following implantation, the size of the grafted ganglion diminishes and quail cells appear dispersed in the trunk structures mixed with somitic cells. At 4 days, numerous cells are present in the adrenal glands and in the dorsal mesentery. When the host reaches 6 days of development, the grafted ganglion, as such, has usually disappeared (unpublished observations).

The next step of our research on the unexpected behavior of ganglion cells transplanted into a younger embryo will consist of following, in the same way, the evolution of ciliary ganglion cells implanted at a later developmental stage (from 8 to 15 days of incubation). In particular, it would be of interest to see whether in such a case already differentiated neurons survive in the host. At the present stage of our studies, it is in fact impossible to tell which cells of the developing cholinergic ganglia are able to become adrenergic. Our experiments demonstrate that CA synthetic ability exists in certain cells of the ciliary and Remak ganglia. Such a capacity is not expressed in normal development but can be elicited by the dorsal trunk structures and the adrenal gland environment. Our results do not reveal, however, whether initially cholinergic cells become adrenergic or whether the appearance of CA-producing cells results from the stimulation of a pool of still undifferentiated or reversibly determined cholinergic neuroblasts.

Another point of interest is the ability of the grafted cells to proliferate in the younger host. From a mere consideration of the extent of quail nerve cell distribution in the 8-day host following the graft of a 4.5- to 6-day old ciliary ganglion it is clear that their number is higher than the total number of neurons of the mature ciliarly ganglion (the latter has been evaluated at 6500 in the chick at 9d–10d of incubation by Landmesser and Pilar, 1974b). It was, however, interesting to demonstrate directly the multiplication capacity of the grafted cells by [^{3}H]thymidine incorporation followed by histoautoradiography. Preliminary results obtained recently showed that quail cells of ciliary ganglion origin, which had colonized the suprarenal glands and formed adrenomedullary cords, incorporated [^{3}H]thymidine (unpublished data of Dupin, Ziller, and Le Douarin). Particular attention will be focused in this study on the fate of the morphologically differentiated postmitotic neurons present in the ciliary ganglia grafted relatively late in development. What is their migratory capacity? Is the environment of the young embryo capable of reinitiating cell division in the ciliary neuron population? Does CA synthesis appear in differentiated cholinergic neurons or in a population of undifferentiated neuroblasts, the former dying after their back-transplantation into a 2-day host?

Although it cannot be decided yet which of the two latter hypotheses is true under *in vivo* graft conditions, the ability of a developing autonomic neuron to change its transmitter metabolism in response to environmental influences has been undisputably demonstrated by a series of elegant experiments carried out *in vitro*. Manipulation of the fluid environment in which postmitotic dissociated neurons from newborn rat superior cervical ganglion (SCG) grow has been shown to influence the choice of transmitters and the type of synapses they make.

Culture conditions have been devised in which neuronal cells grow either in the absence of other cell types or with a variety of nonneuronal cells of known origin (Bray, 1970; Mains and Patterson, 1973a). When grown in the virtual absence of other cells, SCG neurons synthesize and accumulate norepinephrine (NE) from tyrosine (Mains and Patterson, 1973a,b) and take up, store, and release NE, as do adrenergic neurons *in vivo* (Claude, 1973; Rees and Bunge, 1974; O'Lague *et al.*, 1974; Burton and Bunge, 1975; Patterson *et al.*, 1975). These cells form synapses with each other that seem to be adrenergic (Rees and Bunge, 1974). In older cultures, synthesis of small amounts of ACh from [^{3}H]choline can be detected (Mains and Patterson, 1973a). Coculture of the same neurons with nonneuronal cells results in an increase in ACh synthesis by as much as 1000-fold. Cholinergic synapses can be established in the cultures between the neurons themselves (O'Lague *et al.*, 1974, 1975; Johnson *et al.*, 1976; Ko *et al.*, 1976) or between neurons and skeletal myotubes (Nurse and O'Lague, 1975) and on cardiac myocytes (Furshpan *et al.*, 1976).

The effect on transmitter synthesis in SCG neurons can be mediated through culture medium conditioned (CM) by appropriate nonneuronal cells. CAT activity in neuronal extracts, ACh synthesis from [^{3}H]choline by living cells, and cholinergic synapse formation between neurons are induced, while adrenergic properties markedly decrease (Patterson *et al.*, 1975; MacLeish, 1976; Landis *et al.*, 1976; Patterson and Chun, 1977). The extent of cholinergic differentiation varies with either the number of nonneuronal cells or the proportion of CM in the culture (Landis *et al.*, 1976; Patterson and Chun, 1977; MacLeish, 1976). It is also interesting to note that other components of the medium, such as buffers and sera, can influence ACh synthesis (Ross and Bunge, 1976; Patterson and Chun, 1977).

The proof that a change in transmitter synthesis in the cultures is not merely the result of selection due to the survival of only a proportion of neurons but reflects the action of the medium on most if not all the neuronal cells has been provided by single-cell cultures. It is possible to grow single neurons in microcultures containing various concentrations of CM or nonneuronal cells (Reichardt *et al.*, 1976). Under such conditions, 80–90% of the individual neurons grown on heart cells are cholinergic, while 0% are cholinergic under control conditions (Reichardt and Patterson, 1977; Nurse, 1977). In addition, a proportion of singly cultured neurons in 2-week cultures have been shown to be responsible for two simultaneous effects on heart cells: inhibition of heart myocyte contractions with an atropine-sensitive mechanism and speeding up of heartbeat with a propranolol-sensitive mechanism. These dual-function neurons that elicit both cholinergic and adrenergic responses in beating myocytes (Furshpan *et al.*, 1976), when examined with the electron microscope, are seen to contain dense-core vesicles combined with clear vesicles in varicosities and synapses (Landis, 1976). In older cultures (3–5 weeks), however, biochemical assays on transmitter production indicate that most of the neurons synthesize either NE or ACh according to culture conditions. Whether the dual-function neurons detected electrophysiologically correspond to a transient period leading to one of the two differentiated states detected in older cultures has not been established. It is interesting to report in this context data obtained with clonal cell lines of crest origin. For instance, rat pheochromocytoma PC12 (Greene and Tischler, 1976) produces both NE and ACh and responds to CM by an increase in CAT activity (Schubert *et al.*, 1977; Greene and Rein, 1977). Other neuronal clonal lines and glial-neuronal hybrids can synthesize more than one transmitter simultaneously (Prasad *et al.*, 1973; Schubert *et al.*, 1974; Hamprecht *et al.*, 1974). That this reflects the stabilization in transformed cells of a transient state in the normal history of neuronal differentiation can be suggested.

In looking at normal development, it is important to know from what stage onward the transmitter function becomes irreversibly determined in autonomic neurons. Various data concerning sympathetic SCG cells indicate that most, if not all, of them express adrenergic functions before birth and either continue their adrenergic differentiation or become cholinergic according to culture conditions if transplanted *in vitro* at birth. However, such a capacity exists only in explants taken from very young rats, whereas neurons develop exclusively adrenergic properties if they originate from older animals (Hill and Hendry, 1977; Ross *et al.*, 1977). An indication concerning the factors that can play a role in the stabilization of transmitter function has been provided by Walicke *et al.* (1977). SCG neurons cultured in the virtual absence of nonneuronal cells were treated with depolarizing agents: an elevated K^+ concentration, and addition of the drug veratridine which causes an influx of Na^+ into the neurons or electrical stimulation. Mass neuronal cultures, depolarized either in the presence of CM or for several days before the addition of CM, remained primarily adrenergic. The ratio of ACh synthesis to NE synthesis was depressed as much as 300-fold in depolarized cultures as compared with cultures that received only CM. In normal animals neuronal activity induced by the excitatory input from the central neurons can be considered the main stabilizing factor of sympathetic neuron differentiation. Artificial suppression of the spinal input to the sympathetic ganglia was actually shown to reduce the functional maturation of adrenergic neurons (Black and Geen, 1974).

One of the interesting points raised by these experiments concerns the cell types that are effective medium conditioners for ACh induction in sympathetic neurons. Among these are a glial cell line and nonneuronal cells of sympathetic ganglia (Patterson and Chun, 1974), as well as cells from a variety of newborn rat tissues (Ross and Bunge, 1976; Patterson and Chun, 1977). Targets that receive cholinergic or mixed innervation, such as skeletal and heart muscles, are good inducers, whereas tissues that receive only adrenergic innervation (brown fat and liver) are poor inducers.

However, these data reveal a paradoxical situation; although ganglionic nonneuronal cells induce ACh synthesis in adrenergically differentiating neurons *in vitro,* the latter are constantly subjected to the close contact of glial cells *in vivo* and yet mostly become adrenergic. A variety of influences must interact *in vivo* before the neurons reach a stabilized differentiated stage. The *in vitro* experiments have had the merit of providing clear evidence of a period during which the chemical differentiation of the neuron is sensitive to environmental cues. In addition, they may also make it possible to purify the factor which in CM induces ACh synthesis in sympathetic neurons.

4. In Vitro *Culture of Neural Crest as a Tool for Studying the Differentiating Capabilities of Autonomic Neuron Precursors*

In vitro culture of neural crest cells is presently under investigation in several laboratories. After the pioneering work of Dorris (1936), who showed that pigment cells could differentiate *in vitro* from chick embryo neural crest, Cohen and Konigsberg (1975) reported that explantation of a piece of quail neural primordium was followed by emigration of the crest cells around the neural tube. The removal of the initial explant after 48 hours leaves *in situ* the population of migrating cells adhering to the dish. These originate mainly from the crest, although a contribution of cells that have emigrated from the neural tube itself at the level of its anterior and posterior sections, or from areas damaged during its manipulation, cannot be excluded.

The outgrowth was isolated and dissociated with trypsin, and the cells plated at clonal density. Three types of colonies arose from the dispersed cells: some in which all cells were pigmented, some in which no cells were pigmented, and a third type in which pigmented and non-pigmented cells coexisted. The possiblity that the nonpigmented clones could give rise to nerve cells was suggested but not further investigated. Later, Cohen (1977) was able to demonstrate adrenergic cell differentiation in clusters of cells appearing in primary cultures of quail neural crest prepared in the same way.

More recently in a short report Sieber-Blum and Cohen (1978b) have indicated that, even at clonal density, trunk crest cells can differentiate into CA-containing cells provided they grow on an appropriate substrate.

In another study, CAT was shown to develop in cranial neural crest cells cultured in the presence of horse serum (HS) (Greenberg and Schrier, 1977). In this case, the culture resulted from the emigration of cells from chick encephalic vesicles explanted at stage 9 of Hamburger and Hamilton (1951). In the presence of HS, the cells formed aggregates from which prominent processes resembling fascicles of neurites connected the aggregates to one another. Similar morphological differentiation did not occur when the cells were cultured in fetal calf serum (FCS), which promoted essentially pigment cell differentiation.

The experiments reported above are open to criticism for two reasons: first, cells arising from the sectioned surfaces of the neural tube spread over the explant and mingle with the neural crest cells. It is far from certain that all—or any—of them are removed with the bulk of the tube after the first 48 hours in culture; second, the results of organotypic culture experiments have suggested that the neural tube has a positive influence on adrenergic cell differentiation (Norr, 1973). What is more, microexudates from the neural tube remain on the petri dish after the tube has been discarded; they may very well influence the further evolution of the cultured cells. Therefore, the introduction of neural tube explants in neural crest culture must be avoided if the ability of the crest to differentiate autonomously is to be investigated.

Attempting to test the abilities of neural crest cells from trunk and head regions to differentiate into autonomic neurons in culture, we have directed our efforts toward the isolation from the embryo of pure explants of neural crest. Two regions have been used for these experiments, the mesencephalic and the trunk crest taken, respectively, at the stages of 5–7 and 12–15 somites. At both levels, the neural fold is excised with the aid of steel needles sharpened to fine cutting edges. That the crest so removed is essentially free of neural tube fragments has been verified by implanting excised quail crest into the trunk of 2-day chick embryos according to a technique previously described by Teillet (1978). When a piece of neural tube was included in the implants, it developed as an epithelial vesicular structure, whereas the crest cells of the implant migrated away from the implantation site and gave rise to the expected crest derivatives. Epithelial vesicles were found in only 3 out of a total of 20 grafts, showing that the technique we used to remove the crest was reliable.

At the mesencephalic level, crest cells migrate massively away from the neural tube in a lateroventral direction. At stage 7–12 somites, they form a multilayered sheet of cells lying underneath the superficial ectoderm. They are, at this precise stage, easy to remove since they do not adhere to the sparse parachordal mesenchymal cells located anteriorly. The only possible contamination of the crest cells taken at this level of the embryo could be a few mesodermal cells of the parachordal mesenchyme. A detailed description of the culture method is given elsewhere (Ziller et al., 1979).

During the first few days in vitro, cells from both truncal and mesencephalic crest appeared microscopically as small, stellate cells migrating away from the explant. Subsequently, the histological aspect of the two types of cultures differed and depended largely on the type of serum used. In the presence of FCS, isolated melanocytes appeared in truncal crest cultures after 4–5 days, and numerous pigmented areas were macroscopically visible 2 or 3 days later. Under the same conditions, mesencephalic crest cultures contained only very few pigment cells, displaying instead large numbers of mesenchymal fibroblast-like cells which formed dense, multilayered sheets. Both glia-like and neuron-like cells were also present in the two types of culture, which could be maintained in vitro for several weeks.

In medium containing HS, both types of culture underwent rapid proliferation and differentiation. A number of melanocytes were also produced by the mesencephalic crest under these conditions. After 7–10 days in culture a great many of the cells were seen to contain vacuoles (lipid droplets?), and it became increasingly difficult to maintain the cultures beyond this period.

Autonomic neuron differentiation in culture can be monitored biochemically by examining the ability of the cells to synthesize and accumulate ACh and CA when provided with [³H]choline and [³H]tyrosine, respectively (Mains and Patterson, 1973a). We have routinely used this procedure to test neural crest cultures. In a number of cases, CAT activity was also assayed to provide an additional criterion of cholinergic differentiation.

A number of cultures have been examined for biochemical signs of differentiation after 7 days *in vitro* (Fauquet *et al.*, unpublished).

In medium supplemented with HS, significant CAT levels could be measured in cultures derived from both migrating and nonmigrating mesencephalic crest (no significant difference was observed between the two sources). When grown in FCS-containing medium, however, the CAT activity, whether expressed as specific activity or total activity per dish, was reduced by over 90%, thus confirming the findings of Greenberg and Schrier (1977).

The measurement of tritiated precursor incorporation into neurotransmitters provides a more sensitive assay for cholinergic or adrenergic differentiation than the determination of CAT or TH activity. A further, and important, advantage is that both ACh and CA synethesis can be monitored simultaneously in a culture. The results of numerous experiments of this sort are summarized briefly here.

All cultures of mesencephalic crest were found to produce ACh, and a certain number also synthesized CA. A marked serum effect was apparent here also: In the presence of HS, 40% of the cultures synthesized both ACh and CA, with a molar ratio of 34:1 in favor of the former. In FCS, 7 cultures out of 8 synthesized both types of neurotransmitter together and the mean ACh/CA ratio fell to 10, primarily because of the decreased ACh synthesis associated with FCS.

In the case of trunk crest cultures, once again all cultures were observed to synthesize ACh, albeit relatively feebly. On the other hand, adrenergic differentiation was much more difficult to demonstrate. None of the cultures grown in HS-supplemented medium synthesized CA, and only one out of seven grown in FCS did. However, when truncal crest cultures were derived from explantation of the total neural primordium (cf. Cohen, 1977), conversion of tyrosine to CA could be demonstrated in most of the cultures. One is tempted to conclude that this culture method facilitates subsequent adrenergic differentiation: the result observed may be an indirect effect of the temporary presence of the neural tube.

In conclusion, our results show that both cholinergic and adrenergic transmitters can develop *in vitro* from mesencephalic and trunk neural crest. Therefore the bipotentiality of each level of the crest observed in *in vivo* transplantation experiments is confirmed. In addition, the variations observed in CA and ACh synthesis under the various culture conditions described confirm also that the chemical differentiation of the autonomic neuroblasts is highly dependent upon the environment in which they grow. Transmitter synthesis in cultures of differentiating SCG cells have in the same way been shown to be influenced by various components of the culture medium such as buffers and sera (Ross and Bunge, 1976; Patterson and Chun, 1977).

IV. Conclusions

It appears from the data reported above that, among the fundamental problems raised by the ontogeny of the neural crest, the molecular basis of crest cell migration and localization still remains a poorly understood question. Since active investigation of this problem has been recently initiated in several laboratories, one can expect some progress to emerge in this field in the near future.

In regard to the other basic question concerning segregation of the various cell lines arising from the neural crest, interesting advances have been made.

The experimental analysis of autonomic nerve cell differentiation has shown that the choice of transmitter synthesis remains labile for a while during differentiation of the autonomic neuroblasts into fully functional adrenergic or cholinergic neurons. One of the most attractive hypotheses that could account for the experimental data is that the autonomic neuroblast normally goes through a state during which it is able to synthesize both transmitters (CA and ACh). Thereafter, environmental cues stimulate (or inhibit) selectively one or the other of these metabolic pathways until the stable state of chemical differentiation is finally reached.

This hypothesis suggests that the dual-function neurons observed in cultures of SCG cells subjected to the appropriate environment would parallel a normal developmental event: The switch of the metabolism from adrenergic to cholinergic, induced in this case by

CM, would result in a temporary reappearance of the bifunctional state the neurons had already gone through during their ontogeny. Testing this hypothesis in the early embryo is methodologically difficult; one of the possible means of doing so would be demonstration of the presence of the enzyme systems for both ACh and CA synthesis in a single cell of a developing adrenergic ganglion.

The fact that the autonomic ganglioblasts remain undetermined for so long, as far as their chemical differentiation is concerned, does not necessarily mean that the neural crest cell population is actually homogeneous and undetermined and that its differentiation into a variety of cell types depends entirely upon environmental signals. On the contrary, the early determination of certain cell lines has been established by clear experimental evidence. A good example is the mesectoderm, whose ability to differentiate autonomously according to its presumptive fate in heterotopic locations has been demonstrated in several instances. This determination does not exclude, however, the mesectodermal mesenchyme's requirement for differentiation signals from the environment. Cartilage differentiation, for instance, is expressed in this tissue if cephalic crest mesenchyme receives the proper extrinsic cues.

The developmental relationships between the different nerve cell lines, the supportive elements (satellite and Schwann cells) of the peripheral nervous system, and the melanocytes are also problems of great interest.

One question that requires investigation is whether the lability observed in neuroblasts with respect to their differentiation into sympathetic or parasympathetic neurons also exists for their evolution into sensory or autonomic nerve cells. In other words, does there exist a developmental state during which a neuronal precursor cell is determined as a "peripheral neuroblast" with the ability to choose between the sensory and the autonomic pathways?

On the other hand, the presence, in spinal ganglia of 4- to 7-day chick embryos, of cells with the capacity to differentiate along the melanocytic pathway has been reported by Nichols and Weston (1977) and Nichols et al. (1977). These authors suggest that prevention of the specific inductive action of neurons can allow melanocyte differentiation in cells that would normally become Schwann or satellite cells.

The existence of a certain degree of genetic determination, followed by a modulation of its expression through cell–cell interactions, seems to be an attractive hypothesis accounting for cell diversification during differentiation of the neural crest primordium.

ACKNOWLEDGMENTS

This work was supported by the CNRS DGRST and by NIH research grant RO1 DEO 4257 01CB4.

REFERENCES

Abel, W. (1909). *Proc. R. Soc. Edinburgh* **30**, 327–347.
Abel, W. (1912). *J. Anat. Physiol. (Paris)* **47**, 35–72.
Abercrombie, M. (1970). *In Vitro* **6**, 128–142.
Abercrombie, M., and Heaysman, J. E. M. (1966). *Ann. Med. Exp. Fenn.* **44**, 161–165.
Andrew, A. (1963). *J. Embryol. Exp. Morphol.* **11**, 307–324.
Andrew, A. (1964). *J. Anat.* **98**, 421–428.
Andrew, A. (1969). *J. Anat.* **105**, 89–101.
Andrew, A. (1970). *J. Anat.* **107**, 327–336.
Andrew, A. (1971). *J. Anat.* **108**, 169–184.
Andrew, A. (1974). *J. Embryol. Exp. Morphol.* **31**, 589–598.
Andrew, A. (1976). *J. Embryol. Exp. Morphol.* **35**, 577–593.
Bancroft, M., and Bellairs, R. (1976). *J. Embryol. Exp. Morphol.* **35**, 383–401.
Bennett, T., and Malmfors, T. (1970). *Z. Zellforsch. Mikrosk. Anat.* **106**, 22–50.
Black, I. B., and Geen, S. C. (1974). *J. Neurochem.* **22**, 301–306.
Bray, D. (1970). *Proc. Natl. Acad. Sci. U.S.A.* **65**, 905–910.
Browne, M. J. (1953). *Anat. Rec.* **116**, 189–203.
Burton, H., and Bunge, R. P. (1975). *Brain Res.* **97**, 157–162.
Campenhout, E. Van (1930). *C. R. Assoc. Anat., 25th Meet., Amsterdam* pp. 78–79.
Campenhout, E. Van (1931). *Arch. Biol.* **42**, 479–507.
Campenhout, E. Van (1932). *Physiol. Zool.* **5**, 333–353.
Cantino, D., and Mugnaini, E. (1974). *Science* **185**, 279–281.
Chiappinelli, V., Giacobini, E., Pilar, G., and Uchimura, H. (1976). *J. Physiol. London* **257**, 749–766.
Chibon, P. (1966). *Mem. Soc. Fr. Zool.* **36**, 1–107.
Chibon, P. (1967). *J. Embryol. Exp. Morphol.* **18**, 343–358.
Claude, P. (1973). *J. Cell Biol.* **59**, 57a.
Cochard, P., Goldstein, M., and Black, I. B. (1978). *Proc. Natl. Acad. Sci. U.S.A.* **75**, 2986–2990.
Cohen, A. M. (1972). *J. Exp. Zool.* **179**, 167–182.
Cohen, A. M. (1977). *Proc. Natl. Acad. Sci. U.S.A.* **74**, 2899–2903.
Cohen, A. M., and Hay, E. D. (1971) *Dev. Biol.* **26**, 578–605.
Cohen, A. M., and Konigsberg, I. R. (1975). *Dev. Biol.* **46**, 262–280.
Derby, M. A. (1978). *Dev. Biol.* **66**, 321–336.
Dorris, F. (1936). *Proc. Soc. Exp. Biol. Med.* **34**, 448–449.
Ehinger, B. (1967). *Z. Zellforsch. Mikrosk. Anat.* **82**, 577–588.
Enemar, A., Falck, B., and Hakanson, R. (1965). *Dev. Biol.* **11**, 268–283.
Falck, B. (1962). *Acta Physiol. Scand. Suppl.* **56**, 197, 1–25.
Feyrter, F. (1938). "Über diffuse endokrine epitheliale Organe." Barth, Leipzig.
Fontaine, J. (1973). *Arch. Anat. Micr. Morphol. Exp.* **62**, 89–100.

82   NICOLE LE DOUARIN

Fontaine, J. (1979). *Gen. Comp. Endocrinol.* **37**, 81–92.
Fontaine, J., and Le Douarin, N. M. (1977a). *J. Embryol. Exp. Morphol.* **41**, 209–222.
Fontaine, J., and Le Douarin, N. M. (1977b). *Gen. Comp. Endocrinol.* **33**, 394–404.
Furshpan, E. J., MacLeish, P. R., O'Lague, P. H., and Potter, D. D. (1976). *Proc. Natl. Acad. Sci. U.S.A.* **73**, 4225–4229.
Greenberg J. H., and Pratt, R. M. (1977). *Cell Differ.* **6**, 119–132.
Greenberg, J. H., and Schrier, B. K. (1977). *Dev. Biol.* **61**, 86–93.
Greene, L. A., and Rein, G. (1977). *Nature (London)* **268**, 349–351.
Greene, L. A., and Tischler, A. S. (1976). *Proc. Natl. Acad. Sci. U.S.A.* **73**, 2424–2428.
Hamburger, V., and Hamilton, H. L. (1951). *J. Morphol.* **88**, 49–92.
Hammond, W. S., and Yntema, C. L. (1953). *Anat. Rec.* **115**, 393.
Hammond, W. S., and Yntema, C. L. (1964). *Acta Anat.* **56**, 21–34.
Hamprecht, B., Traber, J., and Lamprecht, F. (1974). *FEBS Lett.* **42**, 221–226.
Hay, E. D., and Meier, S. (1974). *J. Cell Biol.* **62**, 889–898.
Hill, C. E., and Hendry, I. A. (1977). *Neuroscience* **2**, 741–750.
Hörstadius, S. (1950). "The Neural Crest: Its Properties and Derivatives in the Light of Experimental Research." Oxford Univ. Press, London and New York.
Hörstadius, S., and Sellman, S. (1946). *Nova Acta Soc. Scient. Uppsaliensis* **4**(13), 1–170.
Holtfreter, J. (1968). *In* "Epithelial-Mesenchymal Interactions" (R. Fleischmajer and R. E. Billingham, eds.), pp. 1–30. Williams & Wilkins, Baltimore, Maryland.
Johnson, M., Ross, D., Meyers, M., Rees, R., Bunge, R., Wakshull, E., and Burton, H. (1976). *Nature (London)* **262**, 308–310.
Johnston, M. C. (1966). *Anat. Rec.* **156**, 143–156.
Johnston, M. C., Bhakdinaronk, A., and Reid, Y. C. (1974). *In* "Oral Sensation and Perception: Development in the Fetus and Infant" (J. F. Bosma, ed.). U.S. Govt. Printing Office, Washington, D. C.
Johnston, M. C., Noden, D. M., Hazelton, R. D., Coulombre, J. L., and Coulombre, A. J. (1979). "Origins of Avian Ocular and Periocular Tissues." In press.
Karnovsky, M. J., and Roots, L. (1964). *J. Histochem. Cytochem* **12**, 219–221.
Ko, C. P., Burton, H., Johnson, M. I., and Bunge, R. P. (1976). *Brain Res.,* **117**, 461–485.
Kuntz, A. (1953). "The Autonomic Nervous System" pp. 117–134. Baillière, London.
Landis, S. C. (1976). *Proc. Natl. Acad. Sci. U.S.A.* **73**, 4220–4224.
Landis, S. C., Mac Leish, P. R., Potter, D. D., Furshpan, E. J., and Patterson, P. H. (1976). *Ann. Soc. Neurosci. Abstr.* **280**, 197.
Landmesser, L., and Pilar, G. (1972). *J. Physiol. (London)* **222**, 691–713.
Landmesser, L., and Pilar, G. (1974a). *J. Physiol. (London)* **241**, 715–736.
Landmesser, L., and Pilar, G. (1974b). *J. Physiol. (London)* **241**, 737–749.
Lash, J. W., Holtzer, S., and Holtzer, H. (1957). *Exp. Cell Res.* **13**, 292–303.
Le Douarin, N. (1969). *Bull. Biol. Fr. Belg.* **103**, 435–452.
Le Douarin, N. (1971). *Ann. Embryol. Morphol.* **4**, 125–135.
Le Douarin, N. (1973a). *Dev. Biol.* **30**, 217–222.
Le Douarin, N. (1973b). *Exp. Cell Res.* **77**, 459–468.
Le Douarin, N. M. (1974). *Med. Biol.* **52**, 281–319.
Le Douarin, N. (1976). *In* "Embryogenesis in Mammals," pp. 71–101. Ciba Foundation Symposium, Elsevier, Amsterdam.
Le Douarin, N., and Jotereau, F. (1975). *J. Exp. Med.* **142**, 17–40.
Le Douarin, N., and Le Lièvre, C. (1970). *C. R. Acad. Sci.* **270**, 2857–2860.
Le Douarin, N., and Le Lièvre, C. (1971). *C. R. Assoc. Anat.* **152**, 558–568.
Le Douarin, N., and Le Lièvre, C. (1976). *Congr. Natl. Soc. Savantes, 97th Nantes, 1972* Vol. III, pp. 405–412.

Le Douarin, N., and Teillet, M. A. (1971). *C. R. Acad. Sci.* **272**, 481–484.
Le Douarin, N., and Teillet, M. A. (1973). *J. Embryol. Exp. Morphol.* **30**, 31–48.
Le Douarin, N. M., and Teillet, M. A. (1974). *Dev. Biol.* **41**, 162–184.
Le Douarin, N., Le Lièvre, C., and Fontaine, J. (1972). *C. R. Acad. Sci.* **275**, 583–586.
Le Douarin, N., Fontaine, J., and Le Lièvre, C. (1974). *Histochemistry* **38**, 297–305.
Le Douarin, N. M., Renaud, D., Teillet, M. A., and Le Douarin, G. H. (1975). *Proc. Natl. Acad. Sci. U.S.A.* **72**, 728–732.
Le Douarin, N. M., Teillet, M. A., Ziller, C., and Smith, J. (1978). *Proc. Natl. Acad. Sci. U.S.A.* **75**, 2030–2034.
Le Lièvre, C. (1976). "Contribution des crêtes neurales à la genèse des structures céphaliques et cervicales chez les Oiseaux." Thése d'Etat, Nantes.
Le Lièvre, C. (1978). *J. Embryol. Exp. Morphol.* **47**, 17–37.
Le Lièvre, C., and Le Douarin, N. (1975). *J. Embryol. Exp. Morphol.* **34**, 124–154.
Löfberg, J., and Ahlfors, K. (1978). *Zoon* **6**, 87–101.
Low, F. N. (1970). *Am. J. Anat.* **128**, 45–56.
Mac Leish, P. R. (1976). "Synapse formation in cultures of dissociated rat sympathetic neurons grown on dissociated rat heart cells." Ph.D. Thesis. Harvard Univ., Cambridge, Massachusetts.
Mains, R. E., and Patterson, P. H. (1973a). *J. Cell Biol.* **59**, 329–345.
Mains, R. E., and Patterson, P. H. (1973b). *J. Cell Biol.* **59**, 361–366.
Mazurkiewicz, J. E., and Nakane, P. K. (1972). *J. Histochem. Cytochem.* **20**, 969–974.
Meier, S., and Hay, E. D. (1973). *Dev. Biol.* **35**, 318–331.
Moore, B. W. (1973). *In* "Proteins of the Nervous System" (D. J. Schneider, ed.). Raven, New York.
Narayanan, C. H., and Narayanan, Y. (1978a). *J. Embryol. Exp. Morphol.* **43**, 85–105.
Narayanan, C. H., and Narayanan, Y. (1978b). *J. Embryol. Exp. Morphol.* **47**, 137–148.
Nichols, D. H., and Weston, J. A. (1977). *Dev. Biol.* **60**, 217–225.
Nichols, D. H., Kaplan, R. A., and Weston, J. A. (1977). *Dev. Biol.* **60**, 226–237.
Noden, D. M. (1973). *In* "Oral Sensation and Perception: Development in the Fetus and Infant" (J. Bosma, ed.), pp. 9–36. U.S. Dept. H.E.W., Bethesda, Maryland.
Noden, D. M. (1975). *Dev. Biol.* **42**, 106–130.
Noden, D. M. (1978a). *In* "The Specificity of Embryological Interactions" (D. Garrod, ed.), pp. 4–49. Chapman Hall, London.
Noden, D. M. (1978b). *Dev. Biol.* **67**, 296–312.
Noden, D. M. (1978c). *Dev. Biol.* **67**, 313–329.
Norr, S. C. (1973). *Dev. Biol.* **34**, 16–38.
Nurse, C. A. (1977). "The formation of cholinergic synapses between dissociated rat sympathetic neurons and skeletal myotubes in cell culture." Ph.D. Thesis, Harvard Univ., Cambridge, Massachusetts.
Nurse, C. A., and O'Lague, P. H. (1975). *Proc. Natl. Acad. Sci. U.S.A.* **72**, 1955–1959.
O'Lague, P. H., Obata, K., Claude, P., Furshpan, E. J., and Potter, D. D. (1974). *Proc. Natl. Acad. Sci. U.S.A.* **71**, 3602–3606.
O'Lague, P. H., Mac Leish, P. R., Nurse, C. A., Claude, P., Furshpan, E. J., and Potter, D. D. (1975). *Cold Spring Harbor Symp. Quant. Biol.* **40**, 399–407.
Patterson, P. H., and Chun, L. L. Y. (1974). *Proc. Natl. Acad. Sci. U.S.A.* **71**, 3607–3610.
Patterson, P. H., and Chun, L. L. Y. (1977). *Dev. Biol.* **56**, 263–280.
Patterson, P. H., Reichardt, L. F., and Chun, L. L. Y. (1975). *Cold Spring Harbor Symp. Quant. Biol.* **40**, 389–397.
Pearse, A. G. E. (1966). *Vet. Rec.* **79**, 587–590.
Pearse, A. G. E. (1969). *J. Hist. Cytochem.* **17**, 303–313.
Pearse, A. G. E. (1976). *Nature (London)* **262**, 92–94.

Pearse, A. G. E., and Carvalheira, A. (1967). *Nature (London)* **214**, 929–930.

Pearse, A. G. E., and Polak, J. M. (1978). *In* "Gut Hormones" (S. R. Bloom, ed.), pp. 33–39. Churchill, Edinburgh.

Pearse, A. G. E., Polak, J. M. Rost, F. W. D., Fontaine, J., Le Lièvre, C., and Le Douarin, N. (1973). *Histochemistry* 191–203.

Pictet, R. L., Rall, L. B., Phelps, P., and Rutter, W. J. (1976). *Science* **191**, 191–192.

Pintar, J. E. (1978). *Dev. Biol.* **67**, 444–464.

Platt, J. B. (1894). *Arch. Mikrobiol. Anat.* **43**, 911–966.

Platt, J. B. (1898). *Morphol. Jahr.* **25**, 375–465.

Polak, J. M., Pearse, A. G. E., Le Lièvre, C., Fontaine, J., and Le Douarin, N. (1974). *Histochemistry* **40**, 209–214.

Prasad, K. N., Mandal, B., Waymire, J. C., Lees, G. J., Vernadakis, A., and Weiner, N. (1973). *Nature (London) New Biol.* **241**, 117–119.

Pratt, R. M., Larsen, M. A., and Johnston, M. C. (1975). *Dev. Biol.* **44**, 298–305.

Rapin, A. M. C., Berger, M. M., Ziller, C., and Le Douarin, N. M. (1979). *In* "Protides of the Biological Fluids" (H. Peeters, ed.), pp. 589–593. Pergamon, New York.

Raven, C. P. (1937). *J. Comp. Neurol.* **67**, 221.

Rees, R., and Bunge, R. P. (1974). *J. Comp. Neurol.* **157**, 1–11.

Reichardt, L. F., and Patterson, P. H. (1977). *Nature (London)* **270**, 147–151.

Reichardt, L. F., Patterson, P. H., and Chun, L. L. Y. (1976). *Ann. Soc. Neurosci. Abstr.* **327**, 197.

Ross, D., and Bunge, R. P. (1976). *Ann. Soc. Neurosci. Abstr.* **1094**, 769.

Ross, D., Johnson, M., and Bunge, R. (1977). *Nature (London)* **267**, 536–539.

Roth, S., Shur, B. D., and Durr, R. (1977). *In* "Cell and Tissue Interactions"(J. W. Lash and M. M. Berger, eds.), pp. 209–223. Raven, New York.

Rutishauser, U., Gall, W. E., and Edelman, G. M. (1978). *J. Cell Biol.* **79**, 382–393.

Schmechel, D., Marangos, P. J., and Brightman, M. (1978). *Nature (London)* **276**, 834–836.

Schubert, D., Heinemann, S., Carlisle, W., Tarikas, H., Kimes, B., Patrick, J., Steinbach, J. H., Culp, W., and Brandt, B. L. (1974). *Nature (London)* **249** 224–227.

Schubert, D., Heinemann, S., and Kidokoro, Y. (1977). *Proc. Nat. Acad. Sci. U.S.A.* **74**, 2579–2583.

Shur, B. D. (1977). *Dev. Biol.* **58**, 23–39, 40–55.

Sieber-Blum, M., and Cohen, A. M. (1978a). *J. Cell Biol.* **76**, 628–638.

Sieber-Blum, M., and Cohen, A. M. (1978b). *J. Cell Biol.* **79**, 31a, *Abstr.* CD 149.

Simard, L. C., and Van Campenhout, E. (1932). *Anat. Rec.* **53**, 141–159.

Smith, J., Cochard, P., and Le Douarin, N. M. (1977). *Cell Differ.* **6**, 199–216.

Strudel, G. (1955). *Arch. Anat. Microsc. Morphol. Exp.* **44**, 209–235.

Takor Takor, T., and Pearse, A. G. E. (1975). *J. Embryol. Exp. Morphol.* **34**, 311–325.

Teillet, M. A. (1978). *Roux's Arch. Dev. Biol.* **184**, 251–268.

Teillet, M. A., Cochard, P., and Le Douarin, N. M. (1978). *Zoon* **6**, 115–122.

Thiery, J. P., Brackenbury, R., Rutishauser, U., and Edelman, G. M. (1977). *J. Biol. Chem* **252**, 6841–6845.

Toole, B. P. (1972). *Dev. Biol.* **29**, 321–329.

Toole, B. P., and Trelstad, R. L. (1971). *Dev. Biol.* **26**, 28–35.

Toole, B. P., Jackson, G., and Gross, J. (1972). *Proc. Nat. Acad. Sci. U.S.A.* **69**, 1384–1386.

Tosney, K. W. (1978). *Dev. Biol.* **62**, 317–333.

Trelstad, R. L., Hayashi, K., and Toole, B. P. (1974). *J. Cell Biol.* **62**, 815–830.

Triplett, E. L. (1958). *J. Exp. Zool.* **138**, 283–312.

Uchida, S. (1927). *Acta School Med. Univ. Kyoto* **10**, 63–136.

Walicke, P. A., Campenot, R. B., and Patterson, P. H. (1977). *Proc. Natl. Acad. Sci. U.S.A.* **74**, 5767–5771.

Weston, J. A. (1963). *Dev. Biol.* **6**, 279–310.

Weston, J. A. (1967). *In* "Methods in Developmental Biology" (F. H. Wilt and N. K. Wessels, eds.), pp. 723–736. Growell, New York.

Weston, J. A. (1970). *In* "Advances in Morphogenesis" (M. Abercrombie, J. Brachet and T. J. King, eds.), Vol. 8, pp. 41–114. Academic Press, New York.

Yamada, K. M., and Olden, K. (1978). *Nature (London)* **275**, 179–184.

Yntema, C. L., and Hammond, W. S. (1945). *J. Exp. Zool.* **100**, 237–263.

Yntema, C. L., and Hammond, W. S. (1947). *Biol. Rev.* **22**, 344–357.

Yntema, C. L., and Hammond, W. S. (1954). *J. Comp. Neurol.* **101**, 515–542.

Yntema, C. L., and Hammond, W. S. (1955). *J. Exp. Zool.* **129**, 375–414.

Zacchei, A. M. (1961). *Arch. Anat.* **66**, 36–62.

Ziller, C., Smith, J., Fauquet, M., and Le Douarin, N. M. (1979). *Prog. Brain Res.* **51**, 59–74.

CHAPTER 3

DEVELOPMENT OF SPECIFIC SYNAPTIC NETWORKS IN ORGANOTYPIC CNS TISSUE CULTURES

Stanley M. Crain

DEPARTMENTS OF NEUROSCIENCE AND PHYSIOLOGY, AND
THE ROSE F. KENNEDY CENTER FOR RESEARCH IN
MENTAL RETARDATION AND HUMAN DEVELOPMENT
ALBERT EINSTEIN COLLEGE OF MEDICINE
YESHIVA UNIVERSITY
BRONX, NEW YORK

I. Introduction: Model Systems for Studies on Neuronal Specificity Mechanisms

Significant progress has been made during the past two decades in clarifying factors involved in the development of specific neuronal connections in the central nervous system (CNS). Many of these studies have focused on the retinotectal system where precise point-to-point topographic projections develop between two-dimensional arrays of retinal ganglion cells and their target tissues in the brain (see reviews in Sperry, 1965; Gaze, 1970; Jacobson, 1978). Attempts have been made to analyze some of the complex cellular mechanisms that underlie the remarkable positional and phenotypic specificity properties of these retinal ganglion cells as well as other types of central and peripheral neurons. Most of these studies have been carried out during CNS development in embryos and during CNS regeneration in adult animals. However, in view of the complexities involved in studies on the formation of specific patterned synaptic networks *in situ*, and the ambiguities and controversies engendered by these analyses (e.g.,

*CURRENT TOPICS IN
DEVELOPMENTAL BIOLOGY, Vol. 16*

Hunt and Jacobson, 1974; Meyer and Sperry, 1974), more direct studies with simpler cellular arrays, under more flexibly controlled conditions in culture, may provide valuable model systems for further analyses of basic neurospecificity mechanisms (see also Keating, 1976; Meyer and Sperry, 1976; Agranoff et al., 1976; Puro et al., 1977).

In the discussion following a paper by Crain et al. (1968b) on the formation of functional connections between explants of fetal spinal cord, brainstem, and cerebral tissues, Sperry (1968) commented:

> You mentioned something about the lack of specificity in your cultures, Dr. Crain. We should remember here that even the transplantation of a clump of nerve cells into a foreign part of the body may be sufficient to destroy the fine specificities that are involved in the functional hook-ups. If one wants to study specific connexions it might be better either to go back to something like Speidel's old method, utilizing the transparent tadpole tail, or to implant chambers in the body itself.

Crain (Crain et al., 1968b) responded:

> Although specificity has not yet been detected in our present studies with CNS explants, cultures of more favorable tissues, e.g., retina and optic tectum, may provide a basis for analysis of the minimal cellular organization prerequisite for development of the highly specific functional connexions characteristic of many parts of the CNS in situ.... With a judicious choice of embryonic CNS and associated tissues, for example from the visual system, neuronal arrays based on highly ordered selective connexions may nevertheless be feasible for direct study in culture as an extension of the elegant work, in situ, of Sperry, Gaze, and others.

The lack of progress in utilizing tissue cultures for studies on neuronal specificity was emphasized more recently in a review by James (1974). He noted that

> the explant situation makes it difficult to determine whether or not synaptic profiles develop in pre-programmed sites—indeed it has been suggested that in spinal-cord explants they do not, and that they may arise in relation to non-nervous elements (James and Tresman, 1969). There is no evidence that the pattern of synaptic distribution in vitro adheres to that of the in vivo material, and the absence of afferent fibres to the explant in fact makes it unlikely that this is the case.... The fact appears to be that tissue culture methods stand in need of further development before they can be used fruitfully to attack the enigma of specific connection.

That same year, however, we published our first report on a new spinal cord–sensory ganglion culture model which showed great potential for studies on the development of specific synaptic connections in vitro (Crain and Peterson, 1974). Our subsequent studies on patterned neuritic growth of dorsal root ganglion (DRG) cells in relation to synaptic network formation with regionally localized target neurons in explants of spinal cord and medulla (Crain and Peterson, 1975a,b, 1976; Crain, 1976) provide a prototype model system for further analyses in vitro of this fundamental problem in neurobiology.

Observations of "patterned migration *in vitro* of embryonic DRG perikarya attached to a spinal cord explant—forming organized dorsal roots—and outgrowth of orderly fascicles of ventral root fibers onto a homogeneous collagen substrate in the absence of target tissues" led Crain (1976a) to suggest

> that neurons in organotypic cultures can grow complex 3-dimensional neuritic arborizations with patterns that are determined by genetic programs which develop in relation to the position of the neuron perikaryon in "body-space" [see also Diamond *et al.*, 1976]. The concept of body-space in a culture may be useful in cases of organotypic CNS explants which include sufficient components to provide a coded representation of the original body axes in its geometrical array of neurons and glial cells. Perhaps *sets* of neurons in suitably prepared CNS explants *may* retain in culture complex properties which determine, not only the formation of connections between specific types of neurons (i.e., phenotypic specificity), but also the development of an organized spatial framework in which the neurons make synaptic connections at particular positions within the cell population, leading to functional patterns related to the 3-dimensional body axes, i.e., locus specificity (Hunt and Jacobson, 1974; see also Sperry, 1965). If this high degree of order can occur in culture it would permit experimental analyses of some of the basic principles which regulate development of the intricate spatial organization and precise regional localizations in the central nervous system. Analyses of these complex types of tissue cultures may provide clues to some of the fundamental aspects of spatial relationships between arrays of CNS neurons that are invariant to drastic environmental transformation.

The present chapter emphasizes recent studies with several types of cocultures which appear to be particularly useful for analyses of neuronal specificity mechanisms in the CNS. More general reviews of a wider variety of organotypic culture models of neurogenesis are already available (Crain, 1974a, 1976, 1978).

II. Formation of Specific Functional Synaptic Networks *in Vitro*

A. Peripheral Target Tissues (Muscle, Glands, and Ganglia)

The formation in culture of specific synaptic connections between spinal cord motoneurons and skeletal muscle fibers has been under study in many laboratories (see reviews by Fischbach, 1974; Nelson, 1975; Fischbach *et al.*, 1976; Obata, 1977). Evidence of neuron–muscle specificity *in vitro* has been obtained, but the results are somewhat ambiguous and further analyses are required. Selective growth of neurites from embryonic chick and rat autonomic and sensory ganglion cells to explants of peripheral target tissues, e.g., atrium and salivary gland, has been reported (Chamley *et al.*, 1973; Coughlin, 1975; Coughlin and Rathbone, 1977; Ebendal and Jacobson, 1977).

Adrenergic fibers from explants of neonatal rat sympathetic ganglion cells showed characteristic functional innervation of rat dilator pupillae muscle explants, whereas cholinergic fibers from these sympathetic ganglia appeared to form selective connections with sphincter pupillae muscle cells and not with dilator cells (Hill *et al.*, 1976). Neurons in fetal rat spinal cord explants formed abundant synapses with nearby sympathetic ganglion cells, as determined by electron microscopic analysis, whereas cerebral cortex neurons showed no evidence of innervating these cocultured autonomic cells (Olson and Bunge, 1973; see also Bunge *et al.*, 1974; Bunge, 1976a,b).

In a study on synapse formation between clonal neuroblastoma x glioma hybrid cells (NG108-14) and dissociated striated muscle cells in culture, Nelson *et al.* (1976) noted:

Establishment of synapses between nerve and muscle poses a paradox, for on the one hand, muscle movements are highly coordinated, which suggests that neuromuscular synapses and other synapses in the neural circuits are assembled with high precision, whereas the demonstrated ability of autonomic neurons of the vagus (Landmesser, 1971), sympathetic ganglion neurons (Nurse and O'Lague, 1975), and clonal (NG108-15) hybrid cells to synapse with striated muscle cells suggests that functional synapses can form that may not be dependent upon highly specific cell recognition molecules. . . . The molecular nature of the cell interactions that lead to synapse formation is not known. . . . The results suggest that synapse formation and the efficiency of transmission are regulated *in vitro*, apparently by independent processes. It may be possible to find conditions for the conversion of the early form of synapse to the late, mature form.

Puro and Nirenberg (1976) demonstrated, moreover, that

clonal NG108-15 hybrid cells form synapses with cells from different muscles and from different organisms such as chick, mouse and rat. . . . These results with clonal cells confirm and extend the findings of Crain *et al.* (1970) which show that explanted mouse spinal cord neurons form synapses with mouse, rat, or human muscle cells.

On the basis of these and related data, Puro and Nirenberg (1976; see also Changeux and Danchin, 1976) postulated

that much of the specificity of the normal neuromuscular synapse is acquired after the synapses form by a process of selection that reduces the number of synapses and that is dependent upon effective transmission across the synapse, rather than by a process of matching complementary molecules on neurons and muscle cells that code for different synaptic connections.

During a symposium in March 1977, Crain (see Crain, 1978) noted that

these hypotheses need to be tested by more systematic experiments with critical arrays of neurons and target cells in culture, especially in regard to their relevance for CNS synaptic network formation. It should be emphasized that the apparent lack of specificity of synapse formation in many of the types of cultures noted above may really be due to deficiencies in the *in vitro* environment which preclude phenotypic expression of a genotypic neuronal cell-recognition code. Judicious presentation of arrays of target and appropriate non-target cells to neurons growing under more *organotypic* culture conditions should help to clarify whether specific cell recognition molecules may indeed play a significant role in preferential formation of at least some types of specific synaptic connections.

Some aspects of this specificity problem were clarified in a subsequent study by Puro *et al.* (1977) utilizing cocultures of dissociated chick embryo retinal neurons and rat striated muscle cells. Abundant nonspecific synapses formed between the retinal neurons and the muscle cells within a few hours after incubation. All these synapses terminated, however, between the third and tenth days *in vitro*, concomitant with preferential aggregation of the retinal cells with one another. Ultrastructural studies showed (Puro *et al.*, 1977) that

synapses between retinal neurons become more abundant while neuron-muscle synapses are lost. These results suggest that synapses between neurons and muscle cells are terminated whereas some synapses between neurons are retained by a process of selection based on the preferential adhesiveness of retinal neurons for one another. We suggest that synapse turnover may be required for the assembly of certain neural circuits during embryonic development and perhaps also in the adult at some synapses with memory function.

B. CNS TARGET TISSUES (SPINAL CORD, MEDULLA, AND CEREBELLUM)

1. DRG–Spinal Cord

a. *Sensory-Evoked Dorsal Horn Responses.* Although formation *in vitro* of functional synapses within and between CNS explants has been well-documented (see reviews in Crain, 1976, 1978), no clear-cut evidence of synaptic connections between *specific* types of CNS neurons was obtained in the initial studies (Crain *et al.*, 1968b). Our recent demonstration of the formation of specific sensory-evoked synaptic networks in cultures of fetal mouse spinal cord and brainstem provides a valuable *in vitro* model system for analyses of neuronal specificity mechanisms in mammalian CNS. Focal stimuli to nerve growth factor (NGF)-enhanced DRGs (Fig. 1) evoked prominent negative slow-wave responses restricted to dorsal regions of spinal cord cross-sectional explants (Fig. 2), arising abruptly after latencies of 2–3 msec, with amplitudes up to 2 mV, and often lasting more than 500 msec (Crain and Peterson, 1974, 1975a). These potentials resemble primary afferent depolarization (PAD) and secondary sensory-evoked synaptic network responses in dorsal cord *in situ*. Simultaneous recordings in ventral cord regions generally showed small positive or

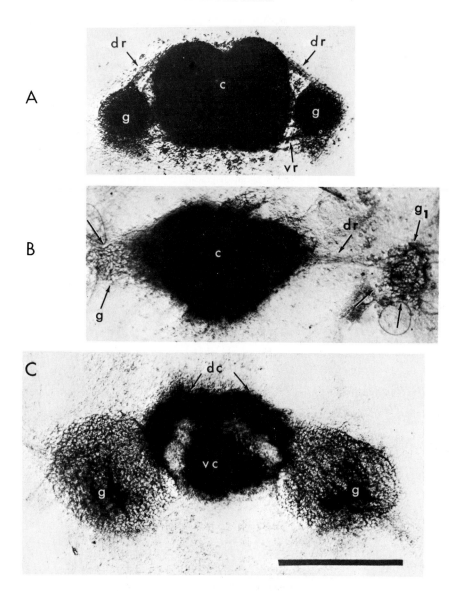

Fig. 1. Photomicrographs of 14-day fetal mouse spinal cord explants (cross sections) with attached DRGs; these are living, unstained cultures. Scale: 1 mm. (A) Shortly after explantation (1 day *in vitro*). Note size of DRGs (g) relative to cord tissue (c), also dorsal (dr) and ventral (vr) roots. (B) One month in normal culture medium. Many of the ganglion cells degenerated during the first few days *in vitro*, leaving a small, thinly spread array of DRG neurons (g) which have matured and retained characteristic (myelinated) dorsal root (dr) connections to the cord. Note that the DRGs are of similar size, although only one (g₁) shows the characteristic "migration" away from the

polyphasic slow-wave potentials and spike barrages after latencies of 5–10 msec. Microelectrode mapping of these cultures also indicated that the abundant NGF-induced growth of DRG neurites into the cord was restricted primarily to dorsal regions. Focal stimuli in dorsal cord regions evoked similar PAD-like responses as with DRG stimuli, whereas stimuli to nearby ventral cord regions (100–200 μm away) evoked only early-latency spikes in the dorsal cord, indicating that few stray collaterals of the primary afferent DRG input were present in the ventral cord (Figs. 4E and 8F). These electrophysiological analyses have now been extended with orthograde peroxidase labeling of the DRG neurites and their terminal arborizations with the spinal cord explants (Fig. 5; Smalheiser et al., 1978a,b, 1981). In all cases, DRG neurites were rarely found in ventral cord; rather, they showed a striking preference for dorsal cord regions, where their ramifications resembled those of cord in situ. The horseradish peroxidase (HRP) tracer studies in mature cocultures confirm our electrophysiological evidence of DRG–dorsal cord specificity relations in vitro and set further limits to the extent of aberrant growth and terminal arborization under organotypic culture conditions.

   b. *Selective γ-Aminobutyric Acid Enhancement of Dorsal Horn Responses.* Whereas strychnine showed relatively little effect on PAD potentials in dorsal cord [at concentrations (ca 10^{-5} M) that greatly enhanced complex long-latency spike barrage and slow-wave discharges in both dorsal and ventral cord], bicuculline and picrotoxin (10^{-5} M) produced marked attenuation of the PADs concomitantly with the appearance of convulsive discharges, especially in ventral cord (Fig. 2D). On the other hand, after the introduction of 10^{-3} M γ-aminobutyric acid (GABA) into the culture bath the PAD responses in dorsal cord were generally maintained or even augmented (Fig. 2E), in contrast to the rapid and sustained depression of almost all detectable synaptically mediated discharges in ventral cord regions as well as long-latency discharges in dorsal cord. Generation of the PADs by Ca^{2+}-dependent synaptic transmitter release is supported by the rapid

---

cord. Most of the other control cultures showed even lower survival of DRG neurons. (C) Another DRG–cord explant after 1 month in the same culture medium, but NGF was added at explantation (1000 BU/ml). Note remarkable enlargement of DRGs (g) relative to their initial size at explantation (A) and in contrast to the control culture (B). Many hundreds of ganglion cells form densely packed clusters close to the cord. (The major DRG volume increase was reached by the second week in vitro.) The relatively dense appearance of dorsal cord (dc) is due to large numbers of myelinated axons which represent central branches of DRG neurons. The dense region in ventromedial cord (vc) is due primarily to a "necrotic core" which generally develops in both treated and control explants. (From Crain and Peterson, 1974.)

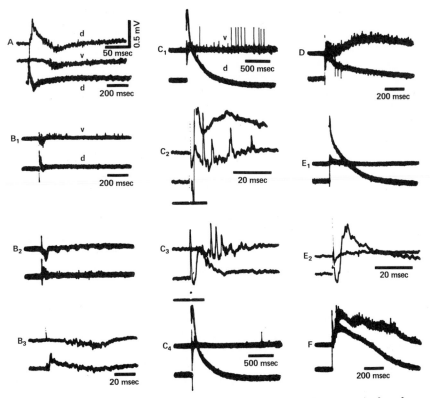

FIG. 2. Enhanced responses evoked in dorsal regions of fetal mouse spinal cord explants by stimuli appled to NGF-treated DRGs. (A) Control culture, 4 weeks *in vitro* (Fig. 1B). An early-latency negative slow-wave potential (resembling a PAD) is evoked in the dorsal cord (d) by a single DRG stimulus (via a focal 10-μm-tip electrode) and is followed by a positive slow wave concomitant with a high-frequency spike barrage (the dorsal cord response is shown at a slower sweep rate on the lowest record). The ventral cord response (v) begins after longer latency and involves primarily a positive slow wave and spike barrage. (B_1) A smaller PAD elicited in the dorsal cord (d) of another control explant (2 weeks *in vitro*); the ventral cord response (v) again consists of a primary positive slow wave and a repetitive spike barrage. (B_2 and B_3) After the introduction of strychnine (10^{-6} *M*), the ventral cord discharge becomes larger and more complex, but the PAD is relatively unchanged although it is now followed by a long spike barrage. A fast sweep (B_3) shows that the PAD begins shortly after DRG stimulus, whereas the ventral cord response occurs at longer latency (as in A). (C_1) A similar explant, 2 weeks *in vitro*, but NGF was added at explantation (1000 BU/ml). A PAD-like potential evoked in the dorsal cord (d) by a single stimulus to NGF-enhanced DRGs (Fig. 1C) is much larger in amplitude and longer in duration (cf. A and B_1), whereas the ventral cord response (v) is similar to the control pattern. (C_2) An early-latency, sharply rising phase and the complexity of the PAD response are seen at a faster sweep rate. (C_3) A tenfold reduction in the DRG stimulus intensity evokes a smaller but still prominent PAD, and the ventral cord discharge now begins after a

and complete block of PAD potentials after increasing the Mg^{2+} concentration from 1 to 10 mM, whereas spikes could still be directly evoked (Crain, 1974b, 1976). These and related pharmacological data suggest that the PADs in DRG–cord explants involve specific sensory-evoked synaptic circuits which may lead to depolarization via GABA-ergic interneurons at DRG terminals (Crain and Peterson, 1974), as in $situ$ (Barker and Nicoll, 1972, 1973; Benoist et $al.$, 1974; Davidoff, 1972; cf. Curtis et $al.$, 1971)—thereby mediating presynaptic inhibitory functions (Eccles, 1964; Wall, 1964).

    $c.$ $Selective$ $Opiate$ $Depression$ of $Dorsal$ $Horn$ $Responses.$ Major components of the sensory-evoked synaptic network responses in dorsal horn regions of mouse spinal cord explants are selectively depressed by acute exposure to low, analgesic concentrations of morphine and other opiates (Crain et $al.$, 1977). Introduction of morphine into the fluid bathing DRG–cord cultures at concentrations of 10^{-7}–10^{-6} M led to marked and sustained depression of major components of the DRG-evoked negative slow-wave responses in dorsal cord within 3–10 minutes, whereas ventral cord discharges were either unaltered or concomitantly enhanced. Furthermore, a series of endorphins and related synthetic opioid pentapeptides with a wide range of analgesic potencies produced selective depressant effects on the sensory-evoked dorsal horn network discharges of fetal mouse DRG–spinal cord explants at concentrations remarkably proportionate to their potency in the intact animal (Crain et $al.$, 1978).

    Introduction of the opiate antagonist naloxone or diprenorphine at

---

longer latency, during the falling phase of the PAD (as in A and B_3). (C_4) A larger DRG stimulus (as in C_1) again elicits a characteristic large PAD just before drug application. (D) The introduction of bicuculline (10^{-5} M) leads to a marked decrease in amplitude of the PAD concomitant with the onset of a convulsive negative slow wave and a repetitive spike discharge in the ventral cord (v). (E_1) After transfer to 1 mM GABA, the large PAD response is restored (cf. C_4), in contrast to an almost complete block of the ventral cord discharge. (E_2) A 10-fold reduction in DRG stimulus intensity still evokes a relatively large PAD (cf. C_3, in BSS). (F) A return to bicuculline (10^{-5} M) leads to partial depression of the PAD and the appearance of the secondary longer-lasting negative slow wave in the dorsal cord, concomitant with the onset of a huge negative slow wave and oscillatory discharge in the ventral cord. (From Crain and Peterson, 1974.)

    $Note:$ In this and all following figures, time and amplitude calibrations and specification of recording and stimulating sites apply to all succeeding records $until$ $otherwise$ $noted;$ upward deflection indicates negativity at an active recording electrode, and the onset of stimuli is indicated by the first sharp pulse or break in the baseline of each sweep. All recordings were made in BSS unless otherwise specified. All records were obtained extracellularly with Ag–AgCl electrodes via isotonic saline-filled micropipets (3- to 5-μm tips).

low concentrations (10^{-8}–10^{-6} M) generally restored opiate-blocked cord responses within minutes. Furthermore, exposure to naloxone (ca. 10^{-6} M) prevented development of the characteristic depression of sensory-evoked dorsal cord responses by morphine or etorphine. Naloxone often elicited a selective increase in amplitude and duration of the sensory-evoked negative slow-wave potentials in dorsal cord even when introduced without prior opiate exposure, suggesting that these dorsal horn networks may develop tonic opioid inhibitory control systems *in vitro* (Crain *et al.*, 1977), as occurs *in situ* (see review by Snyder and Simantov, 1977).

The development of specific opiate-sensitive functions in the dorsal horn regions of our DRG–cord explants provides another set of experimental parameters that can be utilized for studies on specificity mechanisms regulating the formation of organotypic CNS networks *in vitro* (Crain, 1980). Physiological analyses of these opioid networks have been correlated with opiate receptor binding assays showing that high levels of stereospecific opiate receptors develop in the neuritic outgrowth of isolated DRG cultures (Hiller *et al*, 1978a), as well as in the dorsal horn regions of DRG–cord explants (Hiller *et al.*, 1978b). These binding assays constitute strong evidence that opiate receptors are located on DRG nerve fibers destined to provide presynaptic afferent input into the spinal cord.

d. *Preferential Growth of DRG Neurites toward Dorsal Cord Targets.* Neurites from *isolated* DRGs can also grow across gaps of 0.5–1 mm on a collagen film substrate (in high NGF) and invade separate spinal cord explants. During the first weeks of coculture, abundant DRG neurites grew into slabs of *dorsal* cord tissue in contrast to the relatively sparse invasion of similarly apposed *ventral* cord slabs. Furthermore, when ventral cord was presented to DRG clusters, most of the DRG neuritic outgrowth actually appeared to be deflected from the CNS explant— more so than would be expected for a "neutral" nontarget tissue (Peterson and Crain, 1975, 1981). Large PAD responses were evoked by DRG stimuli in cocultures with dorsal, but not ventral, cord (Crain and Peterson, 1975b). Although longer-latency positive slow waves or spike barrages were occasionally evoked in the latter explants, these responses were rapidly blocked in 10^{-3} M GABA, whereas the DRG-evoked PADs in dorsal cord were unaffected or enhanced at this GABA concentration (Fig. 4).

In cultures where the cut end of longitudinal slabs of *whole* spinal cord was presented to DRG clusters, DRG neurites often formed prominent fascicles directly toward dorsal regions of the cord and appeared to avoid adjacent ventral regions at the same facing edge (Fig. 3; Peterson and Crain, 1980). For these experiments, a midline section

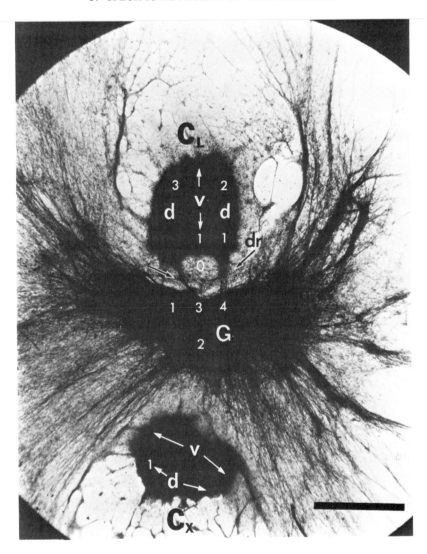

FIG. 3. Photomicrograph of coculture of fetal mouse DRG and deafferented fetal mouse spinal cord explants (3 weeks *in vitro,* silver impregnation). Note prominent fascicles of DRG neurites (dr) which formed *in vitro* in remarkably correct orientation with respect to the dorsal target regions (d) of the longitudinal (upper) spinal cord explant (C_L) (see also Fig. 4). The longitudinal strip of whole cord (C_L) was explanted in an "open book" orientation so that the dorsal regions (d) are clearly in lateral positions on both sides of the medially located ventral cord (v). In contrast, most DRG neurites appear to have diverged away from the ventral facing edge of the cross-sectional cord explant (C_x). The profuse outgrowth of DRG neurites in directions away from both cord explants consists of peripheral DRG fibers and probably some aberrant "central" neurites. Scale: 1 mm. (See Crain and Peterson, 1975b, 1981.)

of the cord is made from the central canal through the dorsal cord and meninges. The cord explant is then placed on the collagen substrate in an "open book" orientation so that the dorsal cord regions are clearly located in lateral positions on both sides of the medially located ventral cord (Fig. 3). Stimuli to these "*de novo* dorsal roots"—which can often be distinguished in living cultures from the less organized outgrowth of central neurites from the cord—evoked large PAD responses, whereas none were detected with stimuli to the adjacent neuritic growth zone bordering the ventral cord regions (Fig. 4; Crain and Peterson, 1975b, 1981). Mapping with stimulating microelectrodes revealed that the DRG fascicles forming these new functional afferent inputs were sharply demarcated and could often be traced from the DRG to the cord explants, even when outgrowing cord neurites tended to obscure DRG pathways in the living cultures. Our electrophysiological analyses of these types of cocultures have recently been confirmed by HRP labeling of DRG neurites (Fig. 5; Smalheiser *et al.*, 1978a,b). DRG fibers entered directly into dorsal regions and *not* via the adjacent ventral regions on the same facing edge; this was observed even within a day or two after the arrival of DRG neurites at the cord explant.

2. DRG-Cord-Medulla

Neurites from isolated DRGs can also innervate sensory target zones in separate explants of medulla cross-sections at the level of the cuneate and gracilis nuclei (Crain and Peterson, 1975b). Focal stimuli to DRG neurites located 1–2 mm from the medulla explant evoked characteristic PADs restricted to dorsal medulla target zones, whereas similar stimuli to nearby ventral regions of the medulla explant were generally ineffective. More organotypic explant arrays were prepared by positioning medulla cross sections near spinal cord cross sections with attached DRGs (Crain and Peterson, 1975a). A midline section of the cord fragment, from the central canal through the dorsal cord and meninges, ensured outgrowth of CNS neurites, glial cells, and DRG fibers comparable to dorsal column axons. Fetal mouse medulla explants were carefully positioned near the dorsal edge of the cord cross sections so as to be in the path of the outgrowing "dorsal column" neurites (Figs. 6 and 7). PADs similar to those evoked by DRGs in dorsal cord explants were detected in small regions of medulla explants connected to cord with NGF-enhanced DRGs (Fig. 6A). The medulla PADs evoked by single DRG stimuli ranged up to 1 mV and arose after longer (ca 3–10 msec) latencies. The large amplitude of these PADs indicates that relatively large numbers of

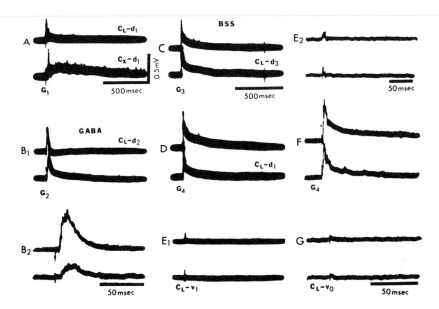

FIG. 4. Selective innervation of dorsal regions of deafferented longitudinal and cross-sectional spinal cord explants by isolated DRG (3 week coculture; see Fig. 3 for the location of recording and stimulating sites). (A) DRG stimulus (G_1) evokes PADs in dorsal cord regions of longitudinal cord (C_L-d_1) and cross-sectional cord (C_x-d_1). Note long-lasting spike barrage following PAD potential in the latter explant (lower sweep). (B_1) After the introduction of 1 mM GABA, PADS are more prominent in both explants and the spike barrage in the cord cross-section is blocked. The recording electrode in the longitudinal cord is now at a more distal site in the dorsal region (upper sweep, C_L-d_2). (The PAD at C_L-d_1 was even larger.) (B_2) Longer latency of PAD in the cord cross-section is more evident at a faster sweep, reflecting a longer circuitous pathway of DRG neurites to this explant, in contrast to the relatively straight DRG fascicle (dr) innervating the longitudinal cord. (No PADs were detected in the ventral regions of both cord explants.) (C) After a return to BSS, PADs are still quite large in both the "contralateral" and "ipsilateral" distal, dorsal regions of the longitudinal cord (C_L-d_2 and C_L-d_3, respectively). (D) PADs are stably maintained at distal and proximal (C_L-d_1) sites in the ipsilateral dorsal regions of the longitudinal cord in response to DRG stimulus (G_4). (E_1) In contrast, stimulus in the *ventral* region of the longitudinal cord (C_L-v_1) fails to elicit any response—except for early spikes (E_2)—even though the stimulus is applied much closer to the recording sites. (F) After returning the stimulating electrode to DRGs (G_4), large PADs are again evoked, whereas systematic series of stimuli throughout the central region between the "dorsal root" (dr) bridges (e.g., at C_L-V_0) failed to evoke any response at the dorsal cord recording sites (G). (See Crain and Peterson, 1975b; 1981.)

DRG terminals probably made synaptic connections with target neurons in the medulla explants. In DRG–cord cultures without added NGF, where only a few dozen DRG neurons may survive, dorsal cord PADs are often much smaller than medulla PADs in spite of the abun-

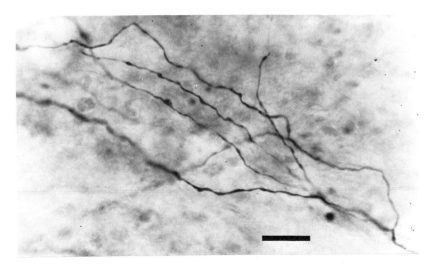

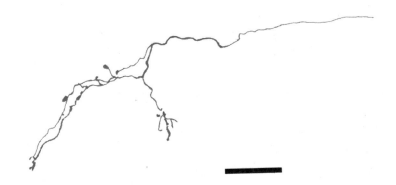

Fig. 5. Branching and arborization of DRG neurites within dorsal regions of cocultured fetal mouse spinal cord explants (whole mounts, 1 month *in vitro*). Selective orthograde Golgi-like labeling of these DRG fibers was produced by iontophoretic injection of HRP (via 8-μm pipets) into the DRG explant (located about 1 mm from the edge of the cord; see text). (Upper) DRG fibers ramify extensively upon entering the dorsal cord explant (from the the right edge of the field), as occurs in dorsal regions of whole-cord explants (photomicrograph). Scale: 25 μm. (Lower) A well-developed terminal arborization with many boutons in the dorsal region of a longitudinal cord strip (camera lucida, similar coculture as in Fig. 3). Scale: 25 μm. (See Smalheiser *et al.*, 1978a,b, 1981.)

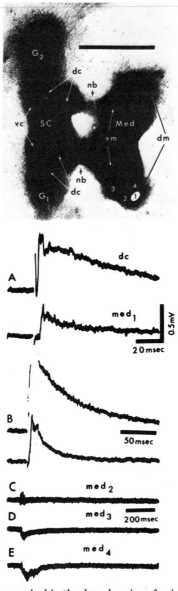

FIG. 6. PAD responses evoked in the dorsal region of spinal cord (SC) and medulla (med) explants (complete across sections) by DRG stimuli (14-day fetal mouse tissues, 14 days in culture). (Top) Medulla cross section is at the level of the cuneate and gracilis nuclei; dorsal closure has not yet occurred at this fetal stage, so that the dorsal medulla tissues (dm) are laterally displaced. Note the bridges (neurites and glia) that have formed between the dorsal edge of the cord (dc) and the ventral edge of the medulla

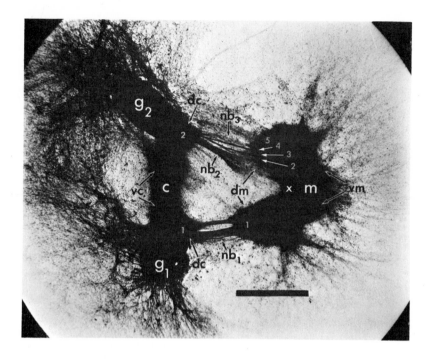

Fig. 7. Photomicrograph of a DRG–spinal cord explant cocultured with a medulla explant (as in Fig. 6 but separated by a larger gap) showing details of the neuritic bridges that formed between the two explants (14-day fetal mouse tissue). The medulla explant (m) consists of a cross section at the level of the cuneate and gracilis nuclei, and it is positioned so that the dorsal edge faces the dorsal edge of the cord explant (c) (cf. Fig. 6 where the ventral edge of the medulla faces the dorsal cord). The DRG–cord explant was added to a 1-week-old medulla culture, and the paired array was then maintained for two additional weeks *in vitro*. Note the prominent neuritic bridges (nb$_1$, nb$_2$, and nb$_3$) that formed between the dorsolateral regions of the cord (dc) and medulla (dm) explants (including "dorsal column" fibers—see Fig. 8). Scale: 1 mm. (From Crain, 1978; see also Crain, 1976.)

(vm) explants in two regions (nb); DRGs (G$_1$ and G$_2$) are located laterally in this explant, further away from the ventral cord (vc). Scale: 1 mm. (A) Simultaneous recordings of PADs in the dorsal cord (dc, lower left arrow in photomicrograph) and dorsal medulla (site 1) in response to a single DRG$_1$ stimulus. (B) After adding 1 mM GABA, PADs at this site in the medulla (med$_1$) and in the dorsal cord are augmented. (C) A large DRG$_1$ stimulus evokes only a spike burst at site 2 in the medulla (med$_2$) and small, positive, slow-wave responses at med$_3$ and med$_4$. (D and E) Systematic mapping of entire medulla explant showed no sign of PADs in response to DRG stimuli except in a small zone indicated in white around med$_1$ (ca. 0.1 × 0.2 mm), even during GABA exposure. (Mapping was not attempted in this culture with stimuli to DRG$_2$.) (From Crain and Peterson, 1975a.)

dant dorsal cord neurons available for establishing sensory synaptic networks with ingrowing DRG neurites (Crain and Peterson, 1974, 1975a). Introduction of 10^{-3} M GABA generally augmented the brainstem and dorsal cord PADs (Fig. 6B), whereas various cord-evoked brainstem network discharges were seriously depressed, as were ventral cord responses. Moreover, in cases where a midbrain explant was positioned between the cord and medulla, prominent DRG-evoked PADs were detected only in the latter explant, even when it was located more than 1 mm distal to the interposed midbrain tissue (Crain and Peterson, 1975a). Weak PADs, however, could be detected in some midbrain regions, especially when medulla target neurons were relatively distant or absent. Using this pharmacological marker technique, we have been able to map more than 20 DRG–cord–brainstem cultures, and in most cases PADs were sharply localized to one or two small zones (ca 100–300 μm) in each medulla explant. In 10 cultures where cross sections of the entire medulla at the level of the cuneate and gracilis nuclei were presented to the DRG–cord explant with controlled orientation, prominent PADs were evoked only in the *dorsal* medulla regions, precisely where dorsal column sensory fibers normally terminated and led to PADs *in situ* (Andersen et al., 1964; Eccles, 1964; Wall, 1964; Davidson and Southwick, 1971). Similar results were obtained in cultures where the medulla cross section was rotated 90° or 180° with respect to the axis of the cord explant (Figs. 7 and 8).

These *in vitro* experiments demonstrate that DRG neurites, after passing through spinal cord tissue, can grow across a homogeneous collagen substrate and, in mimicry of dorsal column fibers *in situ*, establish characteristic functional synaptic networks with programmed target neurons in brainstem explants, even in the presence of a variety of alternative CNS neurons with abundant synaptogenic receptor sites. Furthermore, although the initial neuritic and glial spinal cord outgrowth in relation to these nearby medulla explants was comparable to that extending toward nontarget CNS tissues, preliminary analyses in suitably arrayed cultures suggest that prominent fascicles of "dorsal column fibers" may become organized toward the target neuron zones in the medulla (see Fig. 7; Peterson and Crain, 1981). The remarkable degree of regional specificity of the sensory-evoked spinal cord and brainstem networks that can form under isolated conditions in culture provides the basis for direct analyses of cellular mechanisms regulating the formation and development of specific synaptic connections in the mammalian CNS.

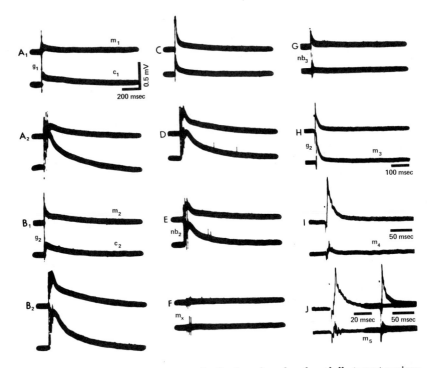

FIG. 8. DRG-evoked PAD responses in the dorsal cord and medulla target regions of the coculture illustrated in Fig. 7. (A_1) Simultaneous recordings of PADs evoked in the dorsal cord (dc, site 1, lower sweep) and dorsal medulla (dm, site 1) regions in response to a single DRG stimulus (at g_1) (DRG–cord explant, 2 weeks *in vitro;* medulla, 3 weeks *in vitro*). (A_2) A brief 100/second DRG volley (with smaller stimulus strength, at g_1) elicits much larger and longer-lasting PADs in both the cord and medulla. (B_1) Similar PADs were evoked on the other side of the dorsal cord and dorsal medulla (sites c_2 and m_2) by stimulus to a nearby DRG (g_2). Note that the response at this relatively distant (and initially separate) medulla site is as prominent as the PAD recorded in the proximal (initially attached) cord, both with single and 100/second DRG stimuli (B_2). (C) PADs evoked in the cord and medulla by a single DRG stimulus are enhanced in amplitude after the introduction of 1 mM GABA (cf. B_1). (D) After a return to BSS, cord and medulla responses to a 100/second DRG volley are still large (cf. B_2). (E) Similar responses are also evoked in the cord and medulla with a 100/second volley applied to the neuritic bridge (nb_2). (F) In contrast, no PADs are elicited with large stimuli to the medial region of medulla explant (m_x), nor in the ventral cord (vc) or ventral medulla (vm). (G) A single stimulus applied, on the other hand, to the diffuse neuritic bridge at nb_3 still evokes significant PADs at the same dorsal cord and medulla sites. (No PADs were elicited, however, with stimuli applied to the medulla growth zone, i.e., between nb_2 and the prominent bridges at nb_1.) (H) Prominent PADs are again evoked by a DRG stimulus (g_2) in two target regions of medulla, sites 2 and 3 (in 1 mM GABA). (I) The response to this DRG stimulus is much smaller, however, at nearby medulla site 4 (slightly lateral to m_3). (J) The response is even smaller in more lateral regions of the medulla, e.g., site 5 (in this record, responses to DRG stimuli are shown at a faster and then a slower sweep rate, partly superimposed). Systematic mapping of the entire medulla explant showed little or no sign of DRG-evoked PAD responses in medial (m_x) and ventral (vm) regions away from dorsal (dm) target zones.

3. Cerebellum

Electrophysiological studies on fetal mouse cerebellar explants have recently provided another demonstration of the development in culture of a regionally localized synaptic network—between Purkinje cells in cortical regions of these explants and neurons in characteristically clustered deep cerebellar nuclei (DN) (Wojtowicz et al., 1978; Hendelman et al., 1978). These analyses indicate that Purkinje cells make monosynaptic, GABA-mediated, inhibitory connections with DN neurons, as in situ (see also Seil and Leiman, 1977).

III. Formation of Functional Retinotectal Connections in Cocultures of Fetal Mouse Explants

Morphological studies on embryonic chick and rodent retinal explants have demonstrated that a remarkable degree of histological organization can develop under suitable conditions in culture (Strangeways and Fell, 1926; Lucas and Trowell, 1958; Sidman, 1961; Hild and Callas, 1967; Barr-Nea and Barishak, 1970; Kim, 1971), including characteristic synaptic ultrastructure (LaVail and Hild, 1971). Prior to our recent studies on fetal mouse retinotectal cocultures (Smalheiser et al., 1977), however, no electrophysiological analyses had been reported regarding the degree to which retinal explants could develop and maintain organotypic bioelectric activities in culture (see, however, studies on frog embryo retina in vitro by Hollyfield and Witkovsky, 1974); nor have there been any reports on cocultures of retina with target tissues, e.g., superior colliculus.

Explants of retina and superior colliculi (whole tecta) from 13- to 14-day mouse embryos have now been successfully cocultured on collagen-coated coverslips in Maximow slide chambers for up to 6 weeks (Smalheiser et al., 1977, 1981b; Smalheiser and Crain, 1978).

Tectal explants resemble typical cerebral explants in their general morphology and development of bioelectrical network activity (Crain, 1976). The tectal-evoked responses (recorded via 3- to 5-μm saline-filled micropipets) consisted of early-latency spikes followed by more complex spike and slow-wave potentials (Fig. 9A), and they were strongly enhanced by d-tubocurarine, strychnine, and bicuculline (10^{-6}–10^{-5} M) (Fig. 9D). Several aspects of their activity patterns suggest that they retain some regionally specific CNS functions in vitro, e.g., characteristic features and laminar localization of the evoked positive waves (Marchiafava and Pepeu, 1966), and marked excitatory effects of low concentrations of d-tubocurarine (Stevens, 1973).

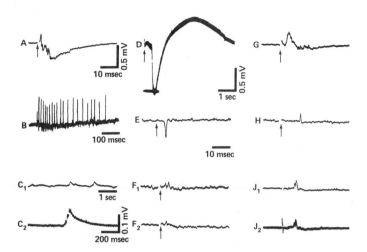

FIG. 9. Recordings from 14-day fetal mouse retinal and tectal explants, 2–4 weeks *in vitro*. (A) A typical tectal response evoked by a single stimulus (7 μA), showing an early compound action potential followed by a large, positive slow wave with concomitant spikes; the stimulating and recording electrodes were within the tectum. (Coculture, 1 month *in vitro*.) (B) Spontaneous ganglion cell spiking and concomitant slow wave in the retina (3 weeks *in vitro*). (C$_1$) Spontaneous, rhythmic, negative slow-wave complexes in the retina. Several distinct wave sizes are seen here. (In some cultures the rhythm is very regular; in others, they may occur in clusters and bursts.) (1 month *in vitro*.) (C$_2$) At a faster sweep and higher gain typical retinal negative-wave features can be seen. (D) Stereotyped convulsant response evoked in a tectal explant by a single stimulus (3 μA) after the introduction of *d*-tubocurarine (2 μg/ml) (cf. record A); concomitant spike bursting is not readily visible at this slow sweep rate. (1 month coculture.) (E) Synaptically mediated single-unit tectal spike in response to retinal stimulation. (F$_1$ and F$_2$) Two juxtaposed sweeps of recordings in the tectum following retinal stimuli (5 μA) at 1/second; the first spike is an early direct spike, while the second is inconstant even at 1/second (3 week coculture). (G) Array of spikes in the tectum evoked by a large retinal stimulus (30 μA). In this case, none followed 10/second stimuli (same culture as in record F). (H) A single early direct spike in the tectum, negative-positive in shape (the positive components were always small). The spike was evoked with a 9-msec latency by retinal stimulus with a 10-μA threshold and was seen at two sites (see map in Fig. 10) (3.5-week coculture). (J$_1$ and J$_2$) Three early direct spikes in the tectum (3- to 7-msec latencies). J$_1$ shows a single sweep at 1/second; J$_2$ consists of several superimposed sweeps at 10/second (retinal stimuli, 3 μA). (The small slightly later spikes were not included in the mapping analysis, but they added to the impression of multiple spikes evoked in a tectal target zone from a single retinal locus.) (From Smalheiser and Crain, 1978.)

Retinal explants reflect the distinctive morphology and development of this tissue *in situ* (Kim, 1971; LaVail and Hild, 1971). Sheets or "rosettes" of photoreceptors were surrounded by areas rich in interneurons and ganglion cells. When whole retinas or large pieces were used, the explants maintained good topographical location of neural elements despite this small-scale distortion due to rosetting. Microelectrode recordings in the retinal explants showed a discrete set of activity patterns indicating sharply defined regional functions corresponding to characteristic histological organization. Spontaneous and electrically evoked spike bursting patterns with concomitant slow waves were seen, resembling recordings of retinal ganglion cells *in situ* (Kuffler, 1953; Fig. 9B). At other sites, distinctive rhythmic negative wave complexes were seen (Fig. $9C_1$ and C_2) whose detailed properties resembled extracellular activity of amacrine cells *in situ*, i.e., the proximal negative response (Burkhardt, 1970). Strychnine and bicuculline (10^{-6}–10^{-5} M) enhanced retinal spike bursts, as occurs in freshly isolated retinas (Ames and Pollen, 1969).

Fascicles of retinal axons emerged from the explants for distances up to a millimeter or more, especially when adjacent nonneural or neural tissue was present rather than bare collagen. These fascicles showed compound action potentials to direct electrical stimulation that followed long, repetitive trains at rates well above 100/second. The conduction velocities of the thin, unmyelinated fibers in the retinal outgrowths were on the order of 0.3 m/second (cf. Stone and Fukuda, 1974, *in situ*); still finer retinal arborizations within the explants would be expected to have proportionately slower conduction velocities (George and Marks, 1974). The retinal response patterns suggest that ganglion cell axons, rather than neurites of interneurons, can sustain repetitive action potentials as utilized in our tectal mapping procedure (see below).

Evidence for functional retinotectal connections in 2- to 6-week-old cocultures was based on the presence of evoked tectal network responses to focal retinal stimulation under conditions where backfiring of tectal neurites and direct current spread could be eliminated as alternative mechanisms. The threshold was usually 0.7–5 μA (Fig. 9E, F, and J), far less than the 13–30 μA generally needed to backfire tectal neurites either in the outgrowth zone or within the retinal explant (as monitored by antidromic cell spikes in the tectum). Stimulation of retinal neurites in outgrowth zones far from the tectum also reliably evoked tectal responses. The data demonstrate the development of functional retinotectal connections *in vitro*. Further evidence for coupling was observed in recordings of spontaneous activity monitored simultaneously in the retina and tectum in several cocultures; complex slow-wave and spiking discharges occurred synchronously between the explants.

For the electrophysiological mapping experiments in cocultures

from 1.5 to 6 weeks *in vitro,* stimulating electrodes were positioned at
1–3 sites within the retina; each placement was chosen such that a
small stimulus ($< 7~\mu$A) evoked a reliable tectal response (Fig. 10).
Controls were made to ensure that alternative mechanisms of tectal
activation such as backfiring of tectal neurites, direct current spread,

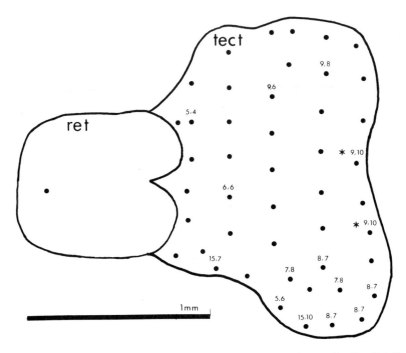

Fig. 10. Schematic diagram of microelectrode mapping sites and spike distributions in a 3.5-week-old coculture of a retinal piece (approximately 1/4) and whole tectum explanted (unoriented) from a 14-day fetal mouse. Each dot represents a recording
site within the tectum. The locations at which early direct spikes could be evoked by
retinal stimuli with thresholds $\le 10~\mu$A are plotted; e.g., 7,8 denotes a 7-msec latency
and a 8-μA threshold. Three spike patterns (7,8; 8,7; and 9,10) each of which seems to be
clustered in the same small region although higher-threshold spikes (not shown) and
some single early direct spikes are seen in other parts of the explant. The two 7,8
spikes, as well as the more centrally located nearby 8,7 spike in the target zone, were
recorded, in fact, several hours after the peripherally located 8,7 spikes were seen. Note
that the thresholds of some of the spikes located away from the target zone resemble
those of target zone spikes (6, 7, and 8) but differ in latency. If these spikes were
generated by collaterals of the fibers mapped into the target zone, they indicate that
wandering occurred (with or without arborization); on the other hand, these spikes may
be derived from different ganglion cells, since they showed different response patterns
during tests with high-frequency stimulus trains. The asterisks indicate two sites that
showed unusually large, yet remarkably similar, latency increases during long high-
frequency trains, proving their common origin from the same cell (spike shown in Fig.
9H). (From Smalheiser and Crain, 1978.)

or retinal synaptic network activation, did not complicate the mapping of directly stimulated retinal ganglion cell axons within the tectal explant (see discussion in Smalheiser and Crain, 1978). A single recording microelectrode was then systematically moved in 30–60 steps across the tectum; at each spot 4–5 depths were explored down to the collagen substrate (ca. 200 μm from the tectal surface). For each electrode placement, the retinal sites were stimulated one at a time, with pulses ranging in small steps from 0 to 30 μA, at 1-second stimulus intervals. Emphasis was placed on spikes that fulfilled the general criteria for being generated by an axon or terminal arborization from a directly stimulated retinal ganglion cell body—the "early direct spike." Such spikes had latencies appropriate to fine, unmyelinated fibers and were consistent with measurements of conduction velocities in retinal fascicles in the outgrowths of these cultures. They had sharp, low thresholds, in contrast to the complex arrays of compound action potentials evoked by large retinal stimuli distributed widely over the tectum. They followed stimulus repetition rates of at least 10/second without fatigue, latency jumps, significant changes in threshold, or intermittent failures.

When maps of the tectum were drawn and the relatively few sites containing early direct spikes were plotted, there was a clustering of spikes evoked from a single retinal stimulation site in a small region of the tectum ($\frac{1}{4}$–$\frac{1}{10}$ of the explant; Fig. 10). The distribution of these spike clusters obeys criteria suggesting that they may be "target zones" representing sites of terminal arborizations related to the retinal place of origin of their cell bodies:

1. A single early direct spike tended to occur at two to four routine adjacent mapping points ~ 120 μm apart (e.g., Fig. 10), and finer mapping carried out at loci about 50 μm apart verified that the identical spike could be observed distinctly at most spots within regions of 50–200 μm, comparable to the size of terminal arborizations *in situ* (e.g., George and Marks, 1974). The high degree of laminar localization of the spike response with depth, and other controls, showed that axonal current spread to a recording electrode at a distance was not significant.

2. Usually *several* distinct early direct spikes from the same retinal electrode placement mapped to the same region of the tectum; their individual distributions were adjacent or overlapping (Figs. 9J$_1$ and J$_2$, and 10).

3. Stimulation at a second or third retinal locus each produced its own distinct spike distribution of evoked responses in the tectal explant, including high-threshold spike arrays as well as quite different clustering patterns of early direct spikes (Smalheiser and Crain, 1978).

Our interpretation of early direct spikes as primarily generated by the terminal arborizations of locally stimulated retinal ganglion cells rests upon their distinctive electrophysiological properties, the topographies of the stimulation and recording sites that elicit them, and their correlation with tectal postsynaptic responses. The degree to which clustering of retinal early direct spikes in tectal target zones represents specificity of neurite behavior is not clear. It is possible that retinal fibers form fascicles that wander randomly within the tectum and that we merely record at their terminations; however, since different retinal stimulation sites produce different target zones, some choices (e.g., fasciculation, paths of growth, or sites of termination) are influenced by a set of cues that interact with knowledge of their retinal place of origin. Finally, at least two aspects of our culture system could tend to mask underlying specific behavior. First, the relatively low density of the retinotectal innervation when using small retinal explants with an entire tectum may reduce optic fiber–fiber interactions; and the tectum, if not fully innervated, may foster spreading and distortion of the projection upon it (Hunt and Jacobson, 1974). Second, only a subpopulation of retinal ganglion cells normally projects to the tectum. Those destined for the lateral geniculate nucleus and other targets not presented in these cocultures may form aberrant connections in inappropriate locations. We are studying the behavior of retinal explants when confronted with target and non-target brain explants to assess their possible active guiding role.

Anatomical tracing of retinal neurites in these cultures (employing HRP injections) are in progress to demonstrate directly the degree of branching, wandering, and fasciculation within the tectum. Preliminary HRP injections in retinal regions of retinocollicular cocultures have, in fact, shown elaborate neurite arborizations in the tectal explants (Smalheiser et al., 1981b). With these techniques, it should be feasible to compare the sites of terminal arborizations (and the behavior of retinal axons that do not arborize) with electrophysiological maps of the same cocultures. More systematic tests employing organized patterns of retinal stimulation sites in retinotectal cocultures are also needed to determine if they will, indeed, map topographically on the tectum in an orderly fashion (Gaze, 1970).

IV. Concluding Remarks

Our studies on the selective DRG innervation of dorsal horn target regions in spinal cord explants, and on dorsal column nuclei in medulla explants, provide a valuable *in vitro* model system for analyzing some of the cellular mechanisms regulating the formation of specific neuronal connections in the mammalian CNS. These tissue culture models set limits to speculation on the minimum factors required to produce the complex gradients and/or neuronal cell recognition codes postulated to underlie selective growth of neurites throughout the CNS, and they may help to determine the properties of these gradients and codes. Systematic alterations of the physicochemical environment and the cellular constituents of the cocultures (e.g., Coughlin and Rathbone, 1977; Ebendal and Jacobson, 1977) may lead to insights into critical factors required for the development of these specificity relationships. The possible role of functional neuronal activity during the development of specific afferent synaptic networks in DRG–cord–medulla cultures also needs to be studied. Although no significant ultrastructural or functional deficits were detected in the internuncial synaptic networks of explants of fetal mouse cerebral neocortex or spinal cord after maturation for weeks in culture media containing Xylocaine or high Mg^{2+}, at concentrations that blocked all overt neuronal impulse activity (Crain et al., 1968a; Model et al., 1971), afferent CNS networks may be more dependent upon the occurrence of sensory impulse activity (see also Gottlieb, 1973; Changeux and Danchin, 1976). Furthermore, in recent studies during the development of dissociated fetal mouse DRG and spinal cord neurons in culture media containing impulse-blocking concentrations of tetrodotoxin (10^{-7}–10^{-6} M), Bergey et al. (1978) reported that a marked reduction occurred in the number of spinal cord cells surviving after 5 weeks in vitro, whereas DRG cell counts were unaffected. On the other hand, in preliminary studies of 14-day fetal mouse cord–DRG explants chronically exposed to similar impulse-blocking levels of TTX (3×10^{-7} M), no significant deficits have been detected in DRG-evoked dorsal horn network responses, nor in ventral cord discharge patterns, after 2–3 weeks in culture (Crain, 1980). Further comparative analyses of spinal cord neurons growing in explant versus dissociated cell arrays may help to clarify the specific conditions under which chronic blockade of bioelectric activities may affect the development and survival of embryonic neurons that normally receive synaptic inputs in culture.

   Cocultures of DRGs with critically positioned arrays of target and nontarget CNS explants may provide additional clues to mechanisms of preferential growth of neurites and establishment of specific connections (Crain and Peterson, 1975b, 1981; Crain, 1976). Similar experimental paradigms are also being applied to evaluate phenotypic specificity properties of retinal ganglion cells cocultured with tectal versus nontarget tissues (e.g., Smalheiser *et al.*, 1978b). Furthermore, microelectrode mapping analyses of fetal mouse DRG–cord and retinotectal cultures may supplement and extend to the cellular level studies of locus specificity mechanisms underlying formation of the precise topographic projections that develop in the visual system and in other parts of the CNS *in situ*.

It will be of interest to apply these *in vitro* methods in analyzing the development of neuronal specificity in other types of CNS tissues. The specific histofluorescence properties of monoaminergic neurons provide a particularly valuable selective cell-labeling technique for analyses of preferential growth of central aminergic neurites *in vitro* and their connections with specific cocultured target tissues. For example, catecholaminergic neurons in fetal mouse brainstem explants containing locus ceruleus have been shown to grow into and functionally innervate cocultured hippocampal explants (Dreyfus *et al.*, 1977, 1979). Further studies are required to determine if these central noradrenergic neurons will make preferential connections *in vitro* with dentate gyrus regions in hippocampal explants, as occurs *in situ* (Blackstad *et al.*, 1967) and in transplants of fetal rat locus ceruleus into partly deafferented regions of the adult rat hippocampus (Bjorklund *et al.*, 1976; Stenevi *et al.*, 1976).

Correlative studies of the degree to which DRG, retinal, and central monoaminergic neurons develop phenotypic and locus specificity properties under controlled conditions in culture may provide significant insights into the cellular mechanisms underlying this basic problem in developmental neurobiology.

ACKNOWLEDGMENTS

This work was supported by research grants NS-06545, NS-12405, and NS-14990 from the National Institute of Neurological and Communicative Disorders and Stroke, grant BNS75-03728 from the National Science Foundation, and a grant from the Alfred P. Sloan Foundation.

I wish to express my appreciation to Edith R. Peterson for providing the DRG–CNS cultures used in these studies. The retinotectal cultures were prepared by Neil R. Smalheiser (supported by grant 5T5 GM 1674 from NIH) and by Dr. Murray B. Bornstein. All the cultures were prepared in the nerve tissue culture laboratory at Albert Einstein College of Medicine, which is under the direction of Dr. Bornstein.

Thanks are also due to Bea Crain for skillful technical assistance in carrying out many of the electrophysiological experiments.

REFERENCES

Agranoff, B. W., Field, P., and Gaze, R. M. (1976). *Brain Res.* **113**, 225–234.
Ames, A., and Pollen, D. A. (1969). *J. Neurophysiol.* **32**, 424–442.
Andersen, P., Eccles, J. C., Schmidt, R. F., and Yokota, T. (1964). *J. Neurophysiol.* **27**, 78–91.

Barker, J. L., and Nicoll, R. A. (1972). *Science* 176, 1043–1045.
Barker, J. L., and Nicoll, R. A. (1973). *J. Physiol.* 228, 259–278.
Barr-Nea, L., and Barishak, R. Y. (1970). *Invest. Ophthalmol.* 9, 447–457.
Benoist, J. M., Besson, J. M., and Boissier, J. R. (1974). *Brain Res.* 71, 172–177.
Bergey, G. K., Macdonald, R. L., and Nelson, P. G. (1978). *Soc. Neurosci. Abstr.* 4, 601.
Bjorklund, A., Stenevi, U., and Svengaard, N.-A. (1976). *Nature (London)* 262, 787–790.
Blackstad, T. W., Fuxe, K., and Hokfelt, T. (1967). *Z. Zellforsch.* 78, 463–473.
Bunge, R. P. (1976a). *In* "Neuronal Recognition" (S. H. Barondes, ed.), pp. 109–128. Plenum, New York.
Bunge, R. P. (1976b). *In* "The Nervous System" (R. O. Brady, ed.), Vol. 1, pp. 31–42. Raven, New York.
Bunge, R. P., Rees, R., Wood, P., Burton, H., and Ko, C.-P. (1974). *Brain Res.* 66, 401–412.
Burkhardt, D. A. (1970). *J. Neurophysiol.* 33, 405–420.
Chamley, J. H., Campbell, G. R., and Burnstock, G. (1973). *Dev. Biol.* 33, 344–361.
Changeux, J.-P., and Danchin, A. (1976). *Nature (London)* 264, 705–712.
Coughlin, M. D. (1975). *Dev. Biol.* 43, 140–158.
Coughlin, M. D., and Rathbone, M. P. (1977). *Dev. Biol.* 61, 131–139.
Crain, S. M. (1974a). *In* "Studies on the Development of Behavior and the Nervous System: Aspects of Neurogenesis (G. Gottlieb, ed.) Vol. 2, pp. 69–114. Academic Press, New York.
Crain, S. M. (1974b). *In* "Drugs and the Developing Brain" (A. Vernadakis and N. Weiner, eds.), pp. 29–57. Plenum, New York.
Crain, S. M. (1976). "Neurophysiologic Studies in Tissue Culture" Raven, New York.
Crain, S. M. (1978). *In* "Cell, Tissue and Organ Cultures in Neurobiology" (S. Fedoroff and L. Hertz, eds.), pp. 147–190. Academic Press, New York.
Crain, S. M. (1980). *In* "Tissue Culture in Neurobiology" (A Vernadakis and E. Giacobini, eds.), pp. 169–185. Raven, New York.
Crain, S. M., and Peterson, E. R. (1974). *Brain Res.* 79, 145–152.
Crain, S. M., and Peterson, E. R. (1975a). *Science* 188, 275–278.
Crain, S. M., and Peterson, E. R. (1975b). *Soc. Neurosci. Abstr.* 1, 751.
Crain, S. M., and Peterson, E. R. (1976). *Soc. Neurosci. Abstr.* 2, 1018.
Crain, S. M., and Peterson, E. R. (1981). *Dev. Brain Res.* Submitted.
Crain, S. M., Bornstein, M. B., and Peterson, E. R. (1968a). *Brain Res.* 8, 363–372.
Crain, S. M., Peterson, E. R., and Bornstein, M. B. (1968b). *In* Ciba Found. Symposium "Growth of the Nervous System" (G. E. W. Wolstenholme and M. O'Connor, eds.), pp. 13–40. Churchill, London.
Crain, S. M., Alfei, L., and Peterson, E. R. (1970). *J. Neurobiol.* 1, 471–489.
Crain, S. M., Peterson, E. R., Crain, B., and Simon, E. J. (1977). *Brain Res.* 133, 162–166.
Crain, S. M., Crain, B., Peterson, E. R., and Simon, E. J. (1978). *Brain Res.* 157, 196–201.
Curtis, D. R., Duggan, A. W., Felix, D., and Johnston, G. A. R. (1971). *Brain Res.* 32, 69–96.
Davidoff, R. A. (1972). *Science* 175, 331–333.
Davidson, N., and Southwick, C. A. P. (1971). *J. Physiol.* 219, 689–708.
Diamond, J., Cooper, E., Turner, C., and Macintyre, L. (1976). *Science* 193, 371–377.
Dreyfus, C. F., Gershon, M. D., and Crain, S. M. (1977). *Soc. Neurosci. Abstr.* 3, 424.
Dreyfus, C. F., Gershon, M. D., and Crain, S. M. (1979). *Brain Res.* 161, 431–445.
Ebendal, T., and Jacobson, C.-O. (1977). *Exp. Cell Res.* 105, 379–387.
Eccles, J. C. (1964). "The Physiology of Synapses." Springer-Verlag, Berlin and New York.

Fischbach, G. D. (1974). *In* "Cell Communication" (R. P. Cox, ed.), pp. 43-66. Wiley, New York.

Fischbach, G. D., Berg, D. K., Cohen, S. A., and Frank, E. (1976). *Cold Spring Harbor Symp. Quant. Biol.* 40, 347-357.

Gaze, R. M. (1970). "The Formation of Nerve Connections" Academic Press, New York.

George, S. A., and Marks, W. B. (1974). *Exp. Neurol.* 42, 467-482.

Gottlieb, G. (1973). *In* "Studies on the Development of Behavior and the Nervous System" (G. Gottlieb, ed.), Vol. 1, pp. 3-45. Academic Press, New York.

Hendelman, W. J., Marshall, K. C., Aggerwal, A. S., and Wojtowicz, J. M. (1978). *In* "Cell Tissue and Organ Cultures in Neurobiology" (S. Fedoroff and L. Hertz, eds.), pp. 539-553. Academic Press, New York.

Hild, W., and Callas, G. (1967). *Z. Zellforsch.* 80, 1-21.

Hill, C. E., Purves, R. D., Watanabe, H., and Burnstock, G. (1976). *Pflugers Arch.* 361, 127-134.

Hiller, J. M., Simon, E. J., Crain, S. M., and Peterson, E. R. (1978a). *Brain Res.* 145, 396-400.

Hiller, J. M., Simon, E. J., Crain, S. M., and Peterson, E. R. (1978b). *Fed. Proc.* 37, 238.

Hollyfield, J. G., and Witkovsky, P. (1974). *J. Exp. Zool.* 189, 357-378.

Hunt, R. K., and Jacobson, M. (1974). *Cur. Top. Dev. Biol.* 8, 203-259.

Jacobson, M. (1978). "Developmental Neurobiology," (2nd ed.). Plenum, New York.

James, D. W. (1974). *In* "Essays on the Nervous System" (R. Bellairs and E. G. Gray, eds.), pp. 31-43. Oxford Univ. Press (Clarendon), London and New York.

James, D. W., and Tresman, R. L. (1969). *Z. Zellforsch.* 101, 598-606.

Keating, M. J. (1976). *In* "Studies on the Development of Behavior and the Nervous System" (G. Gottlieb, ed.), Vol. 3, pp. 59-110. Academic Press, New York.

Kim, S. U. (1971). *Experientia* 27, 1319-1320.

Kuffler, S. W. (1953). *J. Neurophysiol.* 16, 37-68.

Landmesser, L. (1971). *J. Physiol.* 213, 707-725.

LaVail, M. M., and Hild, W. (1971). *Z. Zellforsch.* 114, 557-579.

Lucas, D. R., and Trowell, O. A. (1958). *J. Embryol. Exp. Morphol.* 6, 178-182.

Marchiafava, P. L., and Pepeu, G. C. (1966). *Arch. Ital. Biol.* 104, 406-420.

Meyer, R. L., and Sperry, R. W. (1974). *In* "Plasticity and Recovery of Function in the Central Nervous System" (D. G. Stein, J. J. Rosen and N. Butlers, eds.), pp. 45-63. Academic Press, New York.

Meyer, R. L., and Sperry, R. W. (1976). *In* "Studies on the Development of Behavior and the Nervous System" (G. Gottlieb, ed.), Vol. 3, pp. 111-149. Academic Press, New York.

Model, P. G., Bornstein, M. B., Crain, S. M., and Pappas, G. D. (1971). *J. Cell Biol.* 49, 362-371.

Nelson, P. G. (1975). *Physiol. Rev.* 55, 1-61.

Nelson, P. G., Christian, C., and Nirenberg, M. (1976). *Proc. Natl. Acad. Sci. U.S.A.* 73, 123-127.

Nurse, C. A., and O'Lague, P. H. (1975). *Proc. Natl. Acad. Sci. U.S.A.* 72, 1955-1959.

Obata, K. (1977). *Brain Res.* 119, 141-153.

Olson, M. J., and Bunge, R. P. (1973). *Brain Res.* 59, 19-33.

Peterson, E. R., and Crain, S. M. (1975). *Soc. Neurosci. Abstr.* 1, 783.

Peterson, E. R., and Crain, S. M. (1981). *Dev. Brain Res.* Submitted.

Puro, D. G., and Nirenberg, M. (1976). *Proc. Natl. Acad. Sci. U.S.A.* 73, 3544-3548.

Puro, D. G., DeMello, F. G., and Nirenberg, M. (1977). *Proc. Natl. Acad. Sci. U.S.A.* 74, 4977-4981.

Seil, F. J., and Leiman, A. L. (1977). *Exp. Neurol.* **54**, 110-127.
Sidman, R. L. (1961). *Dis. Nerv. System* **22**, 14-20.
Smalheiser, N. R., and Crain, S. M. (1978). *Brain Res.* **148**, 484-492.
Smalheiser, N. R., Crain, S. M., and Bornstein, M. B. (1977). *Soc. Neurosci. Abstr.* **3**, 432.
Smalheiser, N. R., Peterson, E. R., and Crain, S. M. (1978a). *In Vitro* **14**, 376.
Smalheiser, N. R., Peterson, E. R., and Crain, S. M. (1978b). *Soc. Neurosci. Abstr.* **4**, 479.
Smalheiser, N. R., Peterson, E. R., and Crain, S. M. (1981a). *Dev. Brain Res.* Submitted.
Smalheiser, N. R., Crain, S. M., and Bornstein, M. B. (1981b). *Brain Res.* In press.
Snyder, S. H., and Simantov, R. (1977). *J. Neurochem.* **28**, 13-20.
Sperry, R. W. (1965). *In* "Organogenesis" (R. L. DeHaan and H. Ursprung, eds.), pp. 161-186. Holt, New York.
Sperry, R. W. (1968). *In* Ciba Symposium, "Growth of the Nervous System" (G. E. W. Wolstenholme and M. O'Connor, eds.), p. 39. Churchill, London.
Stenevi, U., Bjorklund, A., and Svendgaard, N. A. (1976). *Brain Res.* **114**, 1-20.
Stevens, R. S. (1973). *Brain Res.* **49**, 309-321.
Stone, J., and Fukuda, Y. (1974). *J. Neurophysiol.* **37**, 722-772.
Strangeways, T. S. P., and Fell, H. B. (1926). *Proc. R. Soc. B.* **100**, 273-291.
Wall, P. D. (1964). *Prog. Brain Res.* **12**, 92-118.
Wojtowicz, J. M., Marshall, K. C., and Hendelman, W. J. (1978). *Neuroscience* **3**, 607-618.

CHAPTER 4

DIFFERENTIATION OF EXCITABLE MEMBRANES*

Robert L. DeHaan

DEPARTMENT OF ANATOMY
EMORY UNIVERSITY SCHOOL OF MEDICINE
ATLANTA, GEORGIA

I. Introduction

The first developing tissue to be studied with modern electrophysiological techniques was the embryonic heart (Fingl *et al.*,
1952). This was soon after microelectrodes were invented, more than a
quarter of a century ago. But early embryonic cells are often small and
fragile; the differentiating organs are microscopic in size and difficult
to handle. Thus, progress in understanding the electrophysiological
differentiation of excitable cells has been slow. It is only in the past
decade that it has been recognized that the ionic dependence of electrical activity, when it first appears at the earliest stages of development, is different in most cases from that seen in differentiated nerve,
heart, and muscle of the adult organism. It is now becoming clear that
early embryos in many species of invertebrates and vertebrates exhibit a primitive form of excitability, even in the egg membrane and
the cleavage blastomeres, that is based on a voltage-dependent Ca^{2+}
conductance mechanism. Thus, it can no longer be said that excitability is a property exclusively of differentiated cells. This was most

* Supported by NIH grant #HL 16567.

117

strikingly seen first in invertebrate embryos. Takahashi *et al.* (1971) attempted to determine how early in the embryo striated muscle cells of the tunicate became electrically excitable. To the surprise of these authors, they found electrical excitability in the gastrula, even before muscle cells had differentiated. Since then, the membrane of many vertebrate oocytes has been shown to be excitable, and to produce slow Ca^{2+}-dependent potential swings (for review, see Hagiwara and Jaffe, 1979). Most adult body tissues are not electrically active, although recent evidence has shown that some cell types such as pancreatic B cells or cultured thymocytes and L cells (traditionally thought of as nonexcitable) do produce regenerative action potential (AP)-like voltage changes and rapid shifts in ion conductance in response to appropriate stimuli (Iversen, 1976; Beigelman *et al.*, 1977; Okada *et al.*, 1979). But, adult nerve, muscle, and heart cells generate APs by means of faster Na^+ and Ca^{2+} currents and a welter of K^+ conductances. Thus, one is forced to the conclusion that the latter specific ionic conductance mechanisms may be the products of differentiation in mature excitable cells.

We begin this article with a historical introduction to the electrical events that define the AP in adult excitable tissues and explore briefly the properties of the specific ionic conductance mechanisms that underlie each part of the AP. These sections serve as a background for a more extensive review of the literature concerning developmental changes in AP characteristics and the differentiating mechanisms of excitablity. Because of my own personal bias, emphasis will be placed on cardiac tissue, but not to the exclusion of investigations of nerve or skeletal muscle.

II. Historical Review

The mechanism of impulse initiation in excitable cells has been a question of major interest for centuries, at least since William Harvey (1628) cut the heart out of a pigeon and noted that the living tissue continued its rhythmic beat when kept moist in a drop of saliva in the palm of his hand. The study of electrical properties of tissues had its origins in the observations of Galvani on "animal electricity," published in 1791, and the controversy with Volta that ensued. It was only after the development of the first galvonometers at the beginning of the nineteenth century that it became possible to measure electric currents in animal tissues. Dubois-Reymond (1848) discovered that current flowed through a primitive induction device when its two electrodes were connected to the surface of a muscle and its cut end. The surface was positive, the cut end was negative, and the "injury current" was stable for many minutes. To explain the spontaneous rhythmic generation of the impulse in heart, Englemann (1897) suggested that the tissue experienced a gradual increase in sensitivity to a weak constant internal stimulus between each beat. Early tests of this hypothesis, with extracellular electrodes, failed to demonstrate an appreciable change in either excitatory threshold or in potential during diastole (Gaskell, 1900; Eccles and Hoff, 1934). But in 1937, Arvanitaki et al. demonstrated a slow potential change during diastole of the snail heart, and shortly thereafter Bozler (1942) recorded the monophasic injury potential in the frog sinus venosus and observed a repeating phase of slow depolarization between each cardiac cycle.

The potential changes across the membrane were soon measured directly when Hodgkin and Huxley (1939) at Plymouth, and Cole and Curtis (1939) in Woods Hole, inserted glass capillaries, about 100 μm in diameter, into the cut end of the giant axon that innervates the mantle muscle of a squid. These workers (and many since) observed a resting membrane potential (V_m) on the order of -60 mV (inside negative). When the axons were stimulated electrically via a pair of external electrodes, APs were recorded with peak amplitudes of about 100 mV, that is, V_m briefly reached $+40$ mV (inside positive). The discovery of this "overshoot" was unexpected at the time, but was explained later by Hodgkin and Katz (1949) with the "sodium theory" of the action potential. The resting inside-negative potential is due to a selective permeability of the membrane mainly to K^+, as suggested by Bernstein in 1902. But the rapid depolarization ("rising") phase of the action potential represents a transient increase in conductance of the membrane to Na^+. After Ling and Gerard introduced the glass microelectrode (1949) that could be inserted into the small cells of a variety of other electrically excitable tissues, similar overshooting action potentials were recorded: from frog myelinated axons (Huxley and Stämpfli, 1951), insect axons (Narahashi, 1963), frog skeletal muscle fibers (Nastuk and Hodgkin, 1950), and mammalian heart muscle (Weidmann and Coraboeuf, 1949; Draper and Weidmann, 1951). The "sodium theory" thus seemed to be of fairly general application, at least as a starting point for further analysis.

Furthermore, Draper and Weidmann (1951) confirmed the presence of a slow diastolic depolarization phase in their tracings from a spontaneously firing mammalian Purkinje cell. Cardiac pacemaker cells were soon impaled with microelectrodes in a wide variety of species by several early investigators (Trautwein and Zink, 1952; Brady and Hecht, 1954; West, 1955) and by many in recent decades, and it is now established beyond a doubt that spontaneous APs are triggered by a slow diastolic depolarization (for reviews, see Hoffman and Cranefield, 1960; Irisawa, 1978; Bonke, 1978; Brooks and Lu, 1972; Cranefield, 1975; DeMello, 1972; DeHaan, 1980).

III. The Membrane Action Potential: Variations on a Theme

In general terms, the APs recorded from most excitable cells are similar in showing an initial rapid depolarization (the "upstroke" or "spike"), an overshooting peak, and a slower repolarization phase. However, it soon became apparent even from the early microelectrode impalements that APs recorded from different tissues, or even from different parts of a single organ such as an adult heart, had distinctively different shapes and sizes. For purposes of analysis, the different phases of a "typical" action potential can be identified (Fig. 1, trace a), in the complex shape of a cardiac impulse (Woodbury et al., 1951; Hoffman and Cranefield, 1960). The initial rapid upstroke is labeled phase 0; the peak of phase 0, where membrane potential becomes positive, is termed the overshoot. The early rapid repolarization or "notch" after phase 0 is phase 1; the prolonged phase of slow repolarization or "plateau" is phase 2, and the terminal phase of more rapid repolarization which brings the membrane potential back to its

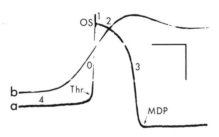

Fig. 1. Phases of the action potential. Transmembrane action potential recorded in 1.3 mM K⁺ medium from an aggregate of ventricular cells prepared from a 7-day chick embryo heart (trace a). The superimposed sigmoidal record (trace b) shows the rapid upstroke of the same action potential at a faster sweep speed. This AP resembles that of a spontaneously active adult cardiac Purkinje fiber. Vertical scale = 50 mV, horizontal scale = 100 msec (trace a), 0.5 msec (trace b), and its level indicates 0 mV (ground potential). The numbered phases of the action potential are 0, rapid upstroke; 1, early repolarization; 2, plateau; 3, rapid repolarization; 4, slow diastolic depolarization; MDP, maximal diastolic potential; OS, overshoot; Thr, threshold potential; $\dot{V}_{max}$ = 147 V/second.

maximally polarized level is phase 3. The point of greatest negativity is termed the maximal diastolic potential (MDP). Its potential level is determined by a balance of the conductances and equilibrium potentials of the ions separated by the cell membrane (for review, see Sperelakis, 1979). For most nerve cells and others that are not spontaneously active the potential during the interspike interval remains constant and is referred to as the resting potential (V_r). In spontaneously active fibers, the slow diastolic depolarization, or "pacemaker potential," termed phase 4, brings the potential from the MDP gradually up to a level from which the next AP upstroke arises. The sharp break between the gentle slope of phase 4 and the rapid upstroke defines the "threshold" potential. When phase 0 is recorded at high sweep speed on an oscilloscope, it can be seen to have a sigmoidal shape (Fig. 1, trace b) with a more or less readily definable range where the rate of voltage change is greatest. The point of maximal upstroke velocity ($\dot{V}_{max}$) is related in a complex way to the time of maximal inward current density (Strichartz and Cohen, 1978; Walton and Fozzard, 1979).

Some neural systems generate spontaneous rhythmic action potentials based on slow interspike phase 4 depolarizations (see, e.g., Connor, 1978), but most nerves exhibit a stable V_m and do not fire unless stimulated. The cardiac impulse in all vertebrate hearts arises from spontaneous action potentials in sinoatrial (SA) tissue. By the turn of the present century Gaskell (1900) had performed his classic experiments showing that the normal pacemaker of amphibian hearts is at the venous end, in the tissue of the sinus venosus. Keith and Flack (1906–1907) soon described the sinoatrial node in mammals at the atriocaval junction, and numerous other workers demonstrated that in all vertebrates the SA region is the first part of the heart to depolarize (Wybauw, 1910; Lewis et al., 1910; Eyster and Meek, 1914; for recent references see Bonke, 1978). More recent evidence indicates that the cells within the SA center are generally spindle shaped, 5–8 μm diameter by 20–30 μm long (Masson-Pevet et al., 1978) and are thus much smaller than typical working myocardial cells of the adult atrium or ventricle (McNutt and Weinstein, 1973). These dominant pacemaker cells are characterized by small slow APs which show MDP values of 50–60 mV, overshoots of about 10 mV, and very slow upstroke velocities (3–5 V/second) (Noma and Irisawa, 1974; Brooks and Lu, 1972). The phase 4 slope is steep and grades smoothly into the upstroke. Threshold potential is difficult to determine, but occurs in the 30–40 mV range. When these fibers are made quiescent by experimental conditions, they show resting potentials of -35 to -40

mV (Noma and Irisawa, 1975). Thus they rest close to threshold. More peripheral cells within the node, termed "subsidiary pacemakers" (Lipsius and Vassalle, 1978), and perinodal cells have larger APs (total amplitude 70–80 mV) in which the diastolic depolarization is less steep and the threshold transition to phase 0 is more abrupt and more negative (~ −45 mV). Moreover, the AP ustroke is faster and is composed of two parts (Paes de Carvalho *et al.*, 1966). An initial fast component ($\dot{V}_{max}$ = 20–30 V/second) is responsible for about two-thirds of the upstroke. The peak of the upstroke is completed by the second slower component (Lipsius and Vassalle, 1978).

As the impulse moves away from the sinus node and perinodal region, into the surrounding atrial tissue, the character of the AP changes markedly. Adult atrial fibers are normally quiescent and fire when excited by the SA potential or externally applied stimuli. The atrial AP has a more negative MDP (−80 to −90 mV) with little or no phase 4 depolarization, a more rapid upstroke (80–100 V/second), a characteristic "triangular" shape lacking a substantial plateau, and a high conduction velocity.

Cells of the AV node develop action potentials that resemble those of the SA node, having low resting potentials, upstroke velocities, and conduction velocity. However, like atrial cells, AV nodal fibers do not normally show automatic activity. But as the impulse emerges from the ventricular side of the node into fibers of the His bundle and Purkinje system, it again encounters fibers with large fast APs. These again are characteristically different from those in either atrial fibers or the pacemaker cells of the node. Purkinje fibers may or may not be spontaneously active; when not they normally rest near −90 mV, although in low Cl⁻ solutions a second stable voltage state is revealed near −60 mV (Gadsby and Cranefield, 1977). When a fiber exhibits automaticity, its pacemaker depolarization occurs in a potential range (−90 to −70 mV) that is much more negative than that of SA pacemaker cells. The transition at threshold is abrupt, and the upstroke velocity may be as high as 500–800 V/second (Weidmann, 1955). The AP shows a very steep slope of phase 1 repolarization, a large prolonged plateau phase (at −20 to −40 mV), and high conduction velocities.

Ventricular muscle in the adult heart is normally quiescent until stimulated by an impulse from the Purkinje fibers. The typical ventricular AP thus shows no phase 4 depolarization. Both the upstroke and phase 1 repolarization notch are fast, but not as fast as in the Purkinje fiber. There is a substantial plateau, but again it is not as prolonged or flat as in fibers of the Purkinje system.

IV. Membrane Currents that Underlie
the Action Potential

The concept that the electrical properties of excitable cells may reside primarily in the cell membrane was introduced by Bernstein (1902) and an electrical equivalent circuit for the cell membrane was developed by Fricke (1925) and Cole (1928). These workers recognized that the membrane of nerve and muscle cells behaves like a parallel resistance–capacitance (RC) circuit and follows the fundamental laws of electricity applicable to other physical systems. The concept of the membrane as an RC system was strengthened when Hodgkin and Rushton (1946) published an experimental and theoretical analysis of the subthreshold voltage responses of nerve axons to locally applied currents. But real understanding of the electrical properties of excitable cells was made possible when the voltage clamp technique, introduced by Cole and Marmont (1942), was exploited in the monumental studies of the squid giant axon by Hodgkin and Huxley (1952). The immediate and necessary conclusions from these early works were that the potential difference across the cell membrane (V_m) changes only as a result of the flow of ionic currents, that the membrane contains separate channels (conductance mechanisms) that are more or less specific for the flow of the various ions found in the cytoplasm and extracellular fluids, and that the conductance of at least some of these channels is nonlinear (that is, channel permeability changes with voltage in a manner that is not predicted by simple ohmic conductors). Hodgkin and Huxley (1952) introduced the idea of time- and voltage-dependent "gating variables" and devised a set of semiempirical equations for describing the membrane currents that underlie the AP. They also recognized that these time- and voltage-dependencies caused the membrane to exhibit inductive behavior. Earlier, Cole and Baker (1941) had applied AC impedance measurements to the membrane of the squid giant axon and discovered that the membrane behaved as if it contained an inductive element. At certain frequencies (near 100 Hz) membrane impedance increased and voltage changes preceded current changes, as occurs at the resonant frequency of a tuned resistance–inductance–capacitance (RLC) circuit. More recent workers who have investigated membrane impedance have confirmed that the inductive behavior of excitable cells arises from time- and voltage-dependent changes in specific membrane conductances (Chandler *et al.*, 1962; Mauro *et al.*, 1970; DeHaan and DeFelice, 1978a).

An important property of RLC circuits is that they tend to

oscillate when perturbed at a preferred (resonant) frequency. Early workers noted the similarities between the rhythmic voltages in heart tissue and simple electric oscillators (Van der Pol and Van der Mark, 1928). Furthermore, the first extracellular and intracellular recordings from excitable cells revealed spontaneous, subthreshold oscillatory voltage fluctuations (see DeHaan and DeFelice, 1978a, for review). The electrical activity of the normal cardiac pacemaker was variously described as a slow depolarization (Arvanitaki, 1938) or as an oscillatory potential (Bozler, 1943). It has since been shown that the Van der Pol equation for an electronic oscillator is directly related to the Hodgkin–Huxley equations (Jack *et al.*, 1975). As we have argued elsewhere (DeHaan and DeFelice, 1978a,b), the view that excitable cell properties arise from specific ionic conductances, and the view that sees those properties as manifestations of membrane impedance, are simply different ways to describe the same phenomena. We shall discuss this point further below.

A. Properties of Ionic Channels in Nerve and Muscle

In most nerves and skeletal muscle fibers the AP is brought about by the transient flow of Na^+ or Ca^{2+} into the fiber, followed by an outflow of K^+. Flow of these ions is passive, down their respective (opposite) electrochemical gradients. There is now considerable evidence that this movement of ions takes place through discrete proteinaceous pore-like structures, the ionic channels (Ehrenstein and Lecar, 1972; Armstrong, 1975; Ulbricht, 1977; Neher and Stevens, 1977).

Evidence that ionic channels are proteins penetrating the membrane lipid bilayer comes from experiments in which proteolytic enzymes are used to modify specific ion conductances (Sevcik and Narahashi, 1975). For example, alkaline endopeptidase B, a trypsin-like enzyme extracted from pronase, specifically destroys the inactivation mechanism of the sodium channel, leaving intact its activation and conducting structure (Rojas and Rudy, 1976). This enzyme is postulated to remove the voltage-dependent inactivation "gate" by cleavage at a lysine or arginine residue. Recently, the same enzyme has been shown to alter the gating properties of a K channel purified from skeletal muscle sarcoplasmic reticulum (Miller and Rosenberg, 1979).

In the membrane of the squid giant axon and many other nerve fibers, four types of ionic channels have been identified whose behavior can be deduced from the pattern of current flow that follows a sudden step depolarization from V_r to, say, 0 mV, in a voltage clamp experiment. The immediate effect is a brief surge of capacity current that is largely over when the voltage step is completed, and corresponds mainly to the charge or discharge of the membrane capacitance (about 1 $\mu F/cm^2$). Following the capacity current, a large inward current reaches its peak in less than 1 msec (at room temperature). The inward current subsides more slowly and is replaced by a steady outward current which reaches a maximum within a few milliseconds after the onset of the step. Hodgkin and Huxley (1952) showed that the pattern of current flow that follows the brief capacity transient is the sum of a rapid short-lived increase in Na^+ conductance (I_{Na}), a slower inward Ca^{2+} current (I_{Ca}), and a delayed K^+ outward current (I_K) that is more prolonged. The work of numerous investigators since, using primarily pharmacology, voltage clamp technique and noise analysis, has defined the properties of these channels in some detail (for reviews, see Ulbricht, 1977; Neher and Stevens, 1977; DeFelice, 1977; Almers, 1978; Hille and Schwarz, 1978; Cahalan and Almers, 1979; DeHaan, 1980).

The fast Na^+ channel (I_{Na}) opens promptly on depolarization but spontaneously closes again by a process termed h-inactivation even when depolarization is maintained. The opening of the channel is accompanied by a small "gating current' (Armstrong and Bezanilla, 1974), that is now believed to arise from the movement of charged components of the sodium channels as they move in the membrane (Armstrong and Bezanilla, 1974; Keynes and Rojas, 1974; Nonner et al., 1975; Cahalan and Almers, 1979). These channels are mainly selective for Na^+; the permeability ratio of the two major monovalent cations K^+ and Na^+ is about 1/12 (Chandler and Meves, 1965). The I_{Na} channel is blocked with a high degree of specificity by tetrodotoxin (TTX) or saxitoxin (STX). Experiments in frog muscle (Almers and Levinson, 1975) and in neuroblastoma cells and other excitable tissues (see Catterall and Morrow, 1978, for references) have demonstrated that TTX and STX bind to a common receptor site associated with the sodium channel. Polypeptide toxins purified from sea anemone (ATX_{II}) (Romey et al., 1976) and scorpion venom (Romey et al., 1975) slow I_{Na} inactivation and enhance the persistent activation of the channel by veratradine or batrachatoxin (Catterall and Beress, 1978). Sea anemone toxin and scorpion toxin do not share a common receptor site on the I_{Na} channel, and neither of the binding sites for these toxins is the same as the TTX–STX receptor (Catterall, 1979; 1980). The number of ATX_{II} sites in rat brain synaptosomes is twice that of TTX-receptors and 10 times greater than binding sites for scorpion toxin (Vincent et al., 1980).

The second, inward slow conductance mechanism (I_{Ca}) has been characterized in the squid axon (Baker *et al.*, 1971), but is more prominent in other molluscan neurons (see Ahmed and Connor, 1979, for references). I_{Ca} is activated 10–20 msec after the onset of a depolarizing voltage step. But upon return of V_m to rest, Ca^{2+} influx decreases slowly over a period of many seconds. Curiously, this property of I_{Ca} is shared by systems that fire very rapidly like the squid axon (Baker *et al.*, 1971), as well as slow cells such as barnacle muscle (Keynes *et al.*, 1973), neuroblastoma cells (Moolenaar and Spector, 1979a), and the unfertilized mouse oocyte (Okamoto *et al.*, 1977).

The third type of channel, common to the squid axon and most other excitable membrane, is the outward K conductance (I_K) which opens after some delay and does not inactivate. It remains open as long as the membrane is depolarized. With depolarization positive to 20 mV, Gilly and Armstrong (1980) have detected a slow-onset gating current associated with the K-channel opening. Although the K channel is permeable to some other cations (Tl^+, Rb^+, NH_4^+), the alkali cations (Na^+, Li^+) are almost totally impermeant (Hille, 1973). Both Na and K channels close promptly on repolarization. In addition to voltage-control of I_K, Ca^{2+}-dependent K^+ conductances appear to be widespread among various cell types including molluscan neurons (Meech and Standen, 1975; Heyer and Lux, 1976; Thompson, 1977), amphibian sympathetic neurons (Busis and Weight, 1976), cardiac Purkinje fibers (Isenberg, 1977b), frog skeletal muscle fibers (Meech, 1976), and mouse neuroblastoma cells (Moolenar and Spector, 1979b).

An important point recognized only in recent years is that the currents that are seen in voltage clamp experiments reflect the average behavior of populations of individual channels. The individual channels can probably exist in only two conductance states, either fully open or completely closed. The time- and voltage-dependent changes in membrane conductance represent underlying changes in the probability that individual channels will be in one state or the other (for reviews, see DeFelice, 1977; Neher and Stevens, 1977; Clay *et al.*, 1979).

After a sudden hyperpolarizing step in voltage, a steady inward current is observed, the so-called leakage current (I_l). In most nerve preparations, this current is thought to flow through a third type of channel which is permanently open, does not exhibit time- or voltage-dependences, and discriminates little among the monovalent cations (Hille, 1973). Hence, I_l presumably contributes a small ohmic current during any voltage clamp regime and during all parts of the nerve AP.

In rabbit myelinated nerve, unlike that of the frog, the ionic current at the node of Ranvier has been resolved into just two currents, I_{Na} and I_l. I_K and I_{Ca} are almost entirely absent (Chiu et al., 1979).

B. Ionic Channels in Heart Cell Membrane

The typical axonal action potential is less complicated than that in the heart. It lacks the prolonged plateau and normally shows no spontaneous repetitive activity. Thus, it should not be surprising that more than four action currents have been identified in cardiac tissue (for reviews, see Trautwein, 1973; McAllister et al., 1975; Carmeliet and Vereecke, 1979). For the nonspontaneous AP of the adult ventricle, which shows no phase 4 depolarization, five separate currents have been described (Beeler and Reuter, 1977; Trautwein and McDonald, 1978). However, the most complete description of the conductances of a "typical" cardiac membrane has been provided by McAllister et al. (1975), based on voltage clamp analyses of the cardiac Purkinje fiber. To reconstruct that AP, these authors required nine currents. In their paper, they review extensively the experimental evidence for each of them.

A major problem in the quantitative interpretation of voltage clamp data in heart tissue arises out of the multicellular nature of the preparation. Heart muscle is composed of individual cells connected by junctions of low but varying resistance, into fibers of complex geometry (DeHaan and Fozzard, 1975; Kensler et al., 1977; Sommer and Johnson, 1979). Substantial voltage gradients may exist within such fibers, and currents crossing the membrane of one cell may differ from those flowing in a distant cell. The problem of electrical inhomogeneity of cardiac tissue and the limitations it places on voltage clamp analyses have been documented and discussed at length (Johnson and Lieberman, 1971; Fozzard and Beeler, 1975; Ramon et al., 1975; Attwell and Cohen, 1977; Kass et al., 1979). It is generally agreed that, because of the multicellular nature of heart muscle, it has not been possible to control ideally the membrane potential of any preparation (Schoenberg and Fozzard, 1979). Thus, there remain substantial uncertainties regarding the absolute magnitude, kinetics, and voltage dependence of the various action currents. Nevertheless, an understanding of the ionic currents in cardiac muscle is beginning to emerge from application of voltage clamp and pharmacological techniques (see Carmeliet and Vereecke, 1979).

We have recently repeated and extended the calculations of the McAllister et al. (1975) model (the MNT model) to illustrate the dependence of the various currents on time and voltage and to explore further the relation between these currents and rhythmicity (DeHaan and DeFelice, 1978a,b).

1. Inward Currents

a. I_{Na}, the Fast Channel. There are two time-dependent inward currents that underlie the cardiac AP (for recent reviews, see DeHaan and DeFelice, 1978a; Reuter, 1979). The first resembles I_{Na} of the squid giant axon. It is abolished by removal of external Na^+ or by the drug TTX. It is activated and inactivated very rapidly, reaching its peak value in less than a millisecond. I_{Na} is activated in the range -90 to -70 mV and is completely inactivated by depolarization beyond about -50 mV. It is the main current that underlies the fast upstroke of all parts of the heart except the primary pacemaker cells of the SA node (Irisawa, 1978).

b. The Second Inward Current, I_{si}. The "slow inward current" (I_{si}) is Ca^{2+}-dependent, and is probably carried by both Ca^{2+} and Na^+. I_{si} determines the plateau phase of the cardiac AP (reviewed in Vassalle, 1979) and is also the major inward current during spontaneous activity of the SA node (Brown et al., 1977; Noma and Irisawa, 1976). At depolarized potentials, I_{si} generates repetitive activity in atrial and ventricular fibers, because it can be activated from depolarized potentials at which I_{Na} is completely inactivated (Reuter, 1979). It rises to a peak value, in the range -15 to 0 mV in about 10 msec and declines in about 50 msec. At peak its magnitude is only about 1/200 of maximal I_{Na}. I_{si} is largely, but not exclusively, carried by Ca^{2+}; it is at least 100 times more selective for Ca^{2+} than for Na^+ or K^+ (Reuter and Scholz, 1977) and is often referred to as a Ca^{2+} current. However, since both Na^+ and K^+ are much more concentrated than Ca^{2+} in the extracellular and cytoplasmic fluids, a substantial fraction of I_{si} is normally carried by these ions. Thus, I_{si} does not decrease in proportion to Ca_o^{2+}; Na^+ becomes the dominant charge carrier when Ca_o^{2+} is low (Reuter and Scholz, 1977). The slow current can be blocked by Mn^{2+} and La^+ or by drugs such as Verapamil or its methoxy derivative D600. However, none of these inhibitors is as specific for I_{si} as is TTX for the Na channels. I_{si} in ventricular muscle and Purkinje fibers is probably not identical to I_{Ca} in many neural preparations.

In the primary pacemaker cells of the rabbit SA node, the slowly rising AP upstroke is normally carried exclusively by a slow inward current which is similar to I_{si} (Noma and Irisawa, 1976). It is blocked by Mn^{2+} and D600, is insensitive to TTX, and has virtually identical kinetics and current–voltage relations (Irisawa, 1978). However, in the SA node cell, the major charge carrier is Na^+, since the current disappears upon Na^+ removal. Moreover, in voltage clamp experiments, when the membrane is held at potentials more negative than MDP for the node (i.e., more negative than -50 mV), a TTX-sensitive fast inward current is activated (Noma et al., 1978).

c. *Background Na⁺ Currents.* In addition to the two time-dependent Na currents, there is a time-independent background current (I_{Nab}) carried primarily by Na. In contrast to the fast currents, I_{Nab} is largest near rest, and its minimum value occurs at the peak of the action potential. I_{Nab} is responsible mainly for maintaining the resting potential of the quiescent Purkinje fiber near -90 mV rather than closer to the K equilibrium potential some 20 mV more negative. The large background Na-conductance in the rabbit SA-node is also the main cause of its relatively depolarized resting potential (Seyama, 1978). Since I_{Nab} is a linear function of potential, its time course mimics that of the action potential. The ionic basis of I_{Nab} is not certain. In Na-free experiments, the current may be carried by other ions such as choline (McAllister et al., 1975). This current plays an important role in pacemaking activity, since the decay of the pacemaker current can cause repetitive firing only against a steady inward depolarizing current (see discussion of I_{K_2} below).

2. Outward Currents

a. *K⁺ Currents.* Voltage clamp data for outward currents are more reliable than the inward current data because the outward currents are much slower than either I_{Na} or I_{si} (Schoenberg and Fozzard, 1979). There are two large outward currents carried by K ion. These are designated as I_{K_1} and I_{K_2} (McAllister et al., 1975). I_{K_1} is called the outward background current because it is time-independent, but shows inward rectification, i.e., it is not linear with voltage. Therefore, the time course of the current during an AP does not follow exactly the shape of the AP itself.

b. *The Pacemaker Current I_{K_2}.* During phase 4 depolarization, membrane resistance increases (Trautwein and Kassebaum, 1961). A sudden voltage clamp step activates pacemaker currents that last several seconds (Vassalle, 1966) and show reversal potentials that vary with K_o^+ according to the Nernst relation (Noble and Tsien, 1968). These observations suggest that pacemaker depolarization is due to a progressive decline in P_K. The pacemaker current in Purkinje fibers (I_{K_2}) is time- and voltage-dependent. That is, it has a gating variable (s) and is activated with a time constant τ_s. At steady-state, s_∞ is 0 at -90 mV and approaches 1 at about -60 mV. Both s and τ_s are voltage-dependent. Maximal τ_s is about 1 second at -40 mV. I_{K_2} also shows inward rectification, decreasing in magnitude at potentials more positive than -85 mV (Noble and Tsien, 1969). I_{K_2} is called the pacemaker current because it has its largest effects over the pacemaker range of voltages, i.e., during the slow depolarization from about -90 to -70 mV prior to the action potential spike (DeHaan and DeFelice, 1978a). In the steady state, the current is inactivated at -90 and fully activated at -50 mV. On depolarization from rest the current activates rapidly but shows inward-going (anomalous) rectification and is therefore small in the plateau range of potentials. However, the gating variable s attains its maximal value during the plateau. When the effects of anomalous rectification are removed upon rapid repolarizing to MDP, I_{K_2} is fully on but then begins to deactivate with a voltage-dependent time-constant of a few hundred milliseconds. Since an outward positive current hyperpolarizes the membrane, the slow decline in I_{K_2} and the anomalous rectification of both I_{K_1} and I_{K_2} set against the steady inward current of I_{Nab}, all result in the phase 4 depolarization. The slope of this depolarization determines the length of the interspike interval, and thus sets the pace of spontaneous rhythmic firing of the membranes. Cleeman and Morad (1979) have provided striking evidence that at least part of the rectification and time-dependent properties of this current in frog ventricular muscle results from the accumulation of K^+ in the intercellular clefts and the K^+-induced changes in I_K that result. The fact that phase 4 depolarization results primarily from a decrease in an outward current explains why membrane resistance increases during the diastolic interval, reaching a maximum at threshold when I_{Na} begins to activate (McAllister et al., 1975; DeHaan and DeFelice, 1978a; Clay et al., 1979).

In the SA node the mechanism of generation of the pacemaker potential is similar to that in other parts of the heart, based upon a time- and voltage-dependent outward K^+ current and a large I_1. But the range of activation of the I_K gating variable p is shifted to more positive levels than that of s for I_{K_2} (Noma et al., 1978). Moreover, $[K_o^+]$ has little effect on the activation curve. In the steady state, p_∞ is 0 at -50 mV and approaches 1 at about 10 mV (DiFrancesco et al., 1979). The longest value of τ_p (~ 0.3 sec) was at -40 mV, i.e., τ_p is shorter than τ_s. SA nodal cells also exhibit an especially large background leakage current (Seyama, 1978). Both of these factors would be expected to contribute to the fast intrinsic beat rate of SA pacemaker cells.

Ca^{2+} appears to influence the outward currents (for reviews, see Vassalle, 1979; Gelles, 1977; Clusin, 1980b). Increasing Ca_o^{2+} causes hyperpolarization of Purkinje fibers and a shortening of their AP. It is not clear whether these effects result from a direct increase in P_K (Kass and Tsien, 1976) or from more indirect mechanisms (DiFrancesco and McNaughton, 1979). In neuroblastoma cells, elevated Ca^{2+} solutions cause an inward Ca^{2+} current and the activation of a slow I_K (Moolenaar and Spector, 1978, 1979b). Increases of Ca_i^{2+} by direct injection of the ion has a similar effect (Isenberg, 1975, 1977a). Conversely, sequestration of Ca_i^{2+} by injection of EGTA depolarizes the fibers and prolongs the AP plateau (Isenberg, 1976). Voltage clamp experiments at different levels of K_o^+ show that increases in Ca_i^{2+} increase I_{K_2} but have no effect on its reversal potential. Furthermore, the effect is insensitive to potential over the voltage range in which I_{K_2} is activated. These results have prompted Isenberg to propose that K conductance is regulated by Ca_i^{2+} rather than by potential. During diastole, free Ca_i^{2+} declines, thereby reducing I_{K_2} and inactivating I_{K_1}. The apparent voltage sensitivity of the K conductance could be secondary to a voltage-sensitive I_{Ca}. In frog atrial fibers depolarization does not increase P_K until V_m reaches the level at which Ca^{2+} permeability is affected. The level of free Ca_i^{2+} may also be influenced by V_m via the electrogenic Na–Ca exchange mechanism (Mullins, 1979).

If the voltage dependence of P_K is secondary to changes in free Ca^{2+}, then the various separate voltage-sensitive K currents need not be ascribed to individual channels. Clusin (1980a,b) has argued that differences in different cell types in the ability to sequester Ca^{2+}, or variations in P_{Na} or P_{Ca} could account for the different characteristics of the pacemaker currents. Moreover, Akselrod et al. (1979) have recently suggested that intracellular Ca^{2+} and membrane I_K are involved in a feedback loop that can oscillate to produce the periodic and random voltage fluctuations that have been recorded in a number of cardiac preparations (DeHaan and DeFelice, 1978a; Lederer and Tsien, 1976; Kass et al., 1978; Akselrod et al., 1979).

c. I_{x_1} and I_{x_2}. Two small outward currents have their largest values during the plateau of the action potential and are often referred to as plateau currents. These are also carried primarily by K^+, with small contributions from other ions. They are designated I_{x_1} and I_{x_2}. I_{x_1} is entirely outward, but I_{x_2} has a small inward component during the slow depolarization phase of the action potential. The magnitude of I_{x_1}

and I_{x_2} is small compared to the currents previously discussed. Both currents are controlled by time-dependent conductances. I_{x_1} is rectified, but I_{x_2} has a linear dependence on voltage for fully activated currents.

   d. *The Early Outward Current and I_{cl}.* In Purkinje fibers the initial spike is followed by a fast repolarization notch (phase 1) prior to the plateau. The conductance change responsible for this effect has been termed the "early outward current." Dudel *et al.* (1967) reported that this current was unaffected by changes in K_o^+ but was drastically reduced when Cl_o^- was replaced by other anions. Thus, the early outward current was originally ascribed to a voltage- and time-dependent Cl^- current (Fozzard and Hiroaka, 1973; McAllister *et al.*, 1975). More recently, however, Kenyon and Gibbons (1979a) have reinvestigated the Cl^- sensitivity of the early outward current, taking care to minimize possible sources of error. They found that reduction of Cl_o^- to less than 10% of control values had minimal effects on the current, whereas agents that blocked I_K such as tetraethylammonium (TEA) or 4-aminopyridine (4-AP) caused a rapid and reversible reduction (Kenyon and Gibbons, 1979b). Thus, the role of Cl^- as a charge carrier for phase 1 is now in question. However, a time-independent background Cl^- current (I_{cl}) makes a major contribution to the leakage current in the Purkinje fiber. According to the MNT model, it represents an inward flow of negative ions, i.e., an outward current, whose magnitude is a linear function of potential (DeHaan and DeFelice, 1978a). In the rabbit SA node the inward-going rectification may result from the voltage-dependence of I_{cl} (Seyama, 1979).

V. Developmental Changes in Action Potential Mechanisms

The acquisition and development of the specialized conductance mechanisms that differentiate in excitable cells have been investigated in the frog medullary plate (Warner, 1973; Blackshaw and Warner, 1976a,b), in Rohon–Beard neurons (Baccaglini and Spitzer, 1977; Spitzer and Baccaglini, 1976) and dorsal root ganglion cells of *Xenopus* (Baccaglini, 1978) *in vivo* and in *Xenopus* neurons *in vitro* (Spitzer and Lamborghini, 1976); in cultured mouse neuroblastoma cells (Moolenaar and Spector, 1979a); and in various heart (see DeHaan, 1980) and skeletal muscle (Takahashi *et al.*, 1971; Kidokoro, 1975a,b) preparations. Spitzer (1979) and DeHaan (1980) have recently discussed some of this literature. In all cases, ionic selectivity and the voltage and time dependence of one or more of the ion channels changed as a function of developmental age or differentiated state. From a review of these studies, it may be possible to extract certain instructive generalizations about the processes of electrophysiological differentiation. We will learn, however, that little is known about the developmental mechanisms that regulate these processes.

A. NERVES AND MUSCLES

1. Amphibian Neurons

Although the oocytes of the mouse and many other diverse species are electrically excitable (see Hagiwara and Jaffe, 1979, for references), amphibian eggs have been found to be unable to generate action potentials (Warner, 1973; Slack and Warner, 1975). In the *Ambystoma* gastrula, at the time of neural induction mean V_r of the prospective medullary plate cells (overlying the archenteric roof) was -27 mV and the most highly polarized cell was -50 mV (Warner, 1973). By late neural fold stages, when specification of the neuroectoderm was complete but before overt neuronal differentiation was apparent, Warner (1973) found that mean V_r had increased to -44 mV and a small fraction of the cells impaled had much more electronegative values (-80 mV). Voltage response to depolarizing current pulses was essentially linear until late neural fold or early neural tube stages. At this time, voltage- and time-dependent responses with inward-rectifying properties began to appear. Interpretation of these results is complicated, however, by the finding that all cells of the neural plate and neural tube were electrotonically coupled by low-resistance junctions (Warner, 1973; Blackshaw and Warner, 1976a).

   a. Rohon–Beard Cells in Vivo. In *Xenopus laevis* a small group of prospective neurons destined to appear in the larval spinal cord undergo their last round of DNA synthesis before the end of gastrulation (stage 13) about 15 hours after fertilization (Spitzer and Spitzer, 1975). This group consists of Rohon–Beard cells, extramedullary neurons, and a number of large ventral neurons. These cells do not incorporate [³H]thymidine after stage 13, whereas all other cells in the medullary plate continue to do so. Rohon–Beard neuronal somata become morphologically identifiable about 4 hours later (stage 18, 19 hours), when the neural folds are closely approximated and about to form the neural tube. At 21 hours (stage 20) resting potential values more negative than -80 mV were frequently recorded, though values of -30 to -40 mV were still common. Until this stage, the Rohon–Beard cells were inexcitable. The earliest regenerative AP were seen in the cells at about stage 20. The ionic mechanism that underlies these AP, and the developmental changes they undergo, have been studied in detail by Baccaglini and Spitzer (1977). The early APs consisted of slowly rising (< 10 V/second) overshooting plateau potentials, often several hundred milliseconds in duration. They were unaffected by addition of TTX to the medium or by removal of Na^+. TEA produced a depolarization and slight prolongation of the AP, and reduced the after–hyperpolarization. But these early AP were abolished by deletion of Ca^{2+}. Moreover, the size of the AP overshoot varied linearly with the log of the external Ca^{2+} concentration, as predicted by the Nernst equation. The results argued strongly that the inward charge carrier was Ca^{2+}.

At intermediate stages of development (from about 28 to 60 hours), Rohon–Beard cells produced cardiac-like AP in which the long plateau response was replaced by a fast spike followed by a plateau, usually 20–80 msec in duration. The inward current responsible for the spike was demonstrated to be I_{Na}, since it could be blocked by TTX or by exposure to Na^+-free medium. The plateau could be selectively blocked by La^{3+}, Co^{2+}, or Mn^{2+}, leaving the spike shape unchanged.

The AP characteristic of Rohon–Beard somas in the larval spinal cord (from 2.5 to 20 days of development) consisted only of a rapid (0.5–2.0 msec) spike with no plateau. This was a purely I_{Na}-dependent AP, as demonstrated by the fact that it was blocked by TTX or Na^+ deletion, was unaffected by Co^{2+}, La^{3+}, or Mn^{2+}, and that its OS size varied with Na^+ concentration according to the Nernst equation.

While the ionic basis of the inward current underwent the developments described, no qualitative changes were observed in the outward current. As soon as APs could be elicited, TEA was able to reduce phase 3 repolarization and the after-hyperpolarization; this effect was maintained into the late larval stages. These results are consistent with Warner's (1973) finding that delayed rectification (due to voltage-dependent outward current channels) appeared in the medullary plate of *Ambystoma* embryos in mid- to late-neurula stages, before regenerative APs could be produced.

The inward current that carries the APs of dorsal root ganglion cells in *Xenopus* larvae (4.5 to 50 days) was studied by Baccaglini (1978), using techniques similar to those employed for the investigations of Rohon–Beard cells. She impaled a range of cell sizes, 7–70 μm in diameter, and found long duration, slow Ca^{2+}-AP only in small cells.

Ca^{2+}/Na^+ potentials similar to the intermediate stage Rohon–Beard cells were obtained from intermediate-sized cells. Relatively pure Na^{2+}-dependent APs were recorded from cells of all sizes. Assuming that cell size is in some way proportional to age, Baccaglini (1978) proposed that the different kinds of APs represent stages in a developmental sequence similar to that of Rohon–Beard cells.

b. Neurons in Vitro. Since it is not usually possible to visualize the axonal processes of neurons in living tissue, the studies described above were limited to the cell bodies. By dissociating the medullary plate of stage 15 *Xenopus* embryos in Ca^{2+}–Mg^{2+}-free saline and placing the isolated cells in monolayer tissue culture, Spitzer and Lamborghini (1976) were able to compare the electrical developments of the cell bodies *in vitro* with those *in vivo,* and Spitzer (1979) has recorded simultaneously from the soma and neurite. The cultured neurons produced APs in their somata when stimulated with a depolarizing pulse and exhibited the same sequence of stages in excitability along a similar time course as the Rohon–Beard neurons *in vivo.* Since each embryo yielded about 30 neurons, and it was not possible to distinguish which among these were Rohon–Beard, extramedullary, or ventral neurons, it may be assumed that all three cell types were progressing through the same differentiative sequence approximately in unison.

It was not possible to record AP with intracellular electrodes inserted into the neurites. However, APs that could be elicited by extracellular stimulation of the axonal fiber were manifested as small responses in the cell body after a delay reflecting the conduction time. Spitzer (1979) has reported that APs in the neurites were dependent on Ca^{2+}, blocked by Co^{2+}, and unaffected by Na^+ deletion for the first 6–11 hours in culture. From 11 hours on, APs in the neurites depended on Na^+, they were blocked by TTX or Na deletion, and were unaffected by Co^{2+}. Of particular interest is the period of about 3 hours between 11 and 14 hours in culture, when APs in the neurites were Na^+-dependent while those in the cell body resulted largely from I_{Ca}, suggesting that the ion channels of the neurites and those of the cell body may be separately regulated.

2. Mouse Neuroblastoma Cells

Cells of the adrenergic neuroblastoma clone N1E-115 can be caused to undergo synchronous electrophysiological differentiation when grown in the presence of dimethylsulfoxide. After 2–3 weeks of such exposure neurite-like processes appear, the cells achieve a soma diameter up to 150 μm, and they generate overshooting fast AP when stimulated (Kimki *et al.*, 1976). Voltage clamp analysis of differentiated cells in normal medium containing 1.8 mM Ca^{2+} revealed three voltage-dependent currents: I_{Na}, I_K, and I_{Ca} (Moolenaar and Spector, 1978, 1979a). In medium containing elevated Ca^{2+} and reduced Na$^+$, the magnitude of I_{Ca} was considerably enhanced. It could be separated clearly from I_{Na} and studied in detail (Moolenaar and Spector, 1979a). I_{Ca} was activated at -55 mV while I_{Na} became prominent at potentials positive to -25 mV. At -15 mV a delayed outward current (I_K) developed. I_{Ca} in neuroblastoma cells was found to resemble that in the presynaptic terminal of the squid giant synapse (Katz and Miledi, 1969) and barnacle muscle fibers (Hagiwara *et al.*, 1969). Its pharmacological and kinetic behavior were similar to those of many other preparations: current was carried readily by Ba^{2+} and Sr^{2+}, whereas La^{3+}, Mn^{2+}, and Mg^{2+} all blocked the channel, presumably by competition for the Ca^{2+}-binding site. Of special interest was the resemblance in I-V relation, kinetic behavior and pharmacological properties of I_{Ca} in the neuroblastoma cells and the mouse egg membrane (Okamoto *et al.*, 1977). This suggests a remarkable conservation of the molecular structure of the I_{Ca} channel from the unfertilized oocyte to mature mammalian sympathetic nerve cell (the source of the original neuroblastoma tumor). Since similar Ca^{2+} channels have been identified in rapidly growing undifferentiated neuroblastoma cells (Moolenaar and Spector, 1978; Miyake, 1978) as well, it is clear that I_{Ca} is not limited to cells in the differentiated state. In contrast, there was no evidence for a functional fast inward Na$^+$ current, blockable by TTX, either in the mouse egg (Okamoto *et al.*, 1977) or in undifferentiated neuroblastoma cells.

In 20 mM Ca^{2+} medium differentiated neuroblastoma cells exhibited a slowly rising outward current that contrasted sharply with voltage-dependent delayed I_K characteristic of many other excitable preparations and of neuroblastoma in normal medium. The Ca^{2+}-dependent K$^+$ current ($I_{K(Ca)}$) has slower kinetics, is not blocked by tetraethylammonium ion (TEA), and is abolished by any manipulation that blocks or inactivates I_{Ca} (Moolenaar and Spector, 1979b). The status of $I_{K(Ca)}$) in undifferentiated cells is not known, but will be discussed again, below, in relation to the cardiac pacemaker current.

3. Physiological Differentiation of Skeletal Muscle

a. *Chick Muscle in Vivo.* At about 13 days of incubation, most of the myoblasts in chicken leg muscle have stopped dividing and many have become incorporated into myotubes (Hermann *et al.*, 1970). By inserting electrodes into the intact muscle, Kano (1975) has shown that these early myotubes are still inexcitable, yielding only a passive response to a depolarizing stimulus. By 16 days, when virtually all cells have fused into myotubes, the first regenerative APs can be evoked. These consist of slowly rising plateau-like potentials, often several seconds in duration, which are blocked by Mn^{2+} or Co^{2+} and are usually completely resistant to TTX. Between 19 and 21 days (hatching) the AP is composed of a TTX-sensitive fast spike followed by a plateau, which can be blocked by Co^{2+} or Mn^{2+}. Within the first few days after hatching, AP duration decreases from several hundred milliseconds to 2–5 msec, as the plateau component disappears. At the end of the first week of posthatch life, the AP consists of a rapidly rising (> 200 V/second) Na^+-dependent spike a few milliseconds in duration, which is TTX-sensitive. Thus, the AP in chick skeletal muscle appears to progress through a sequence from Ca^{2+}-dependent to Ca^{2+}/Na^+-dependent to Na^+-dependent, which resembles that in amphibian neurons. A similar progression has been recorded in chick skeletal muscle cultured *in vitro* (Kano and Shimada, 1973; Kano and Yamamoto, 1977).

  b. Tunicate Muscle. Because tunicates undergo mosaic development, the blastomeres destined to give rise to skeletal muscle are readily identifiable in the early embryo and they can be traced through all stages of development. Moreover, these cells remain large at all stages; in the fully differentiated larva the tail muscle consists of six chains of seven to eight mononucleate muscle cells, each cell 15–20 μm in diameter. The ionic basis of excitability has been studied in these forms in great detail (Takahashi *et al.*, 1971; Miyazaki *et al.*, 1972). From cleavage through the early gastrula (128 cell stage) prospective muscle cells were inexcitable and had low resting potentials similar to nonmuscle cells of about -19 mV. In the early gastrula, the prospective muscle cells rapidly increased their electronegativity ($V_r = -71$ mV). In this condition, they respond to depolarizing stimuli with 5–10 second plateau potentials, which were both Na^+- and Ca^{2+}-dependent and were not blocked by TTX. During the period from mid-gastrula to mid-tadpole stage, the TTX-insensitive Na^+ permeability disappeared, and inward rectification increased. This was manifested as an increase in membrane resistance as the cell was depolarized. At later stages, from young tadpole to hatchling, the muscle AP consisted of an overshooting spike and plateau that diminished from several seconds to about 100 msec during this period. However, according to Miyazaki *et al.* (1974) neither the spike nor the plateau components were TTX-sensitive or affected much by Na^+ deletion. Both were apparently Ca^{2+}-dependent. Thus in this form, contrary to those discussed above, the ionic dependence of the AP inward current changed from Na^+/Ca^{2+} to Ca^{2+}. The Na^+ component disappeared some time before hatching, and the time- and voltage-dependent I_K became prominent relatively late in development.

c. *Cloned Muscle Cell Line.* An unusual progression of ionic dependencies has been reported in a clonal rat skeletal muscle line (Kidokoro, 1975a,b), which contrasts with the events described for primary cultures of chick muscle. In this preparation, the rapidly dividing mononucleate myoblasts were excitable. They responded to anode-break stimulation with small, slowly rising (3–13 V/second) APs which showed no overshoot. Multinucleate myotubes responded to similar stimuli with large overshooting APs with faster rise-times (93 ± 28 V/second) and a distinct plateau component 100–200 msec in duration. Both the myoblast AP and the myotube spike were Na^+-dependent; both were blocked in Na-depleted medium, but the myotube AP was TTX-resistant. The plateau was shown to result from a Ca^{2+} current in the myotubes. Under similar maintenance conditions, fibers of adult rat soleus muscle showed no sign of Ca^{2+} current. Thus, in the cloned cell line the ionic dependence of the inward current changes from Na^+ to Na^+/Ca^{2+}.

B. INITIATION OF FUNCTION IN THE EMBRYONIC HEART

1. Action Potential Shape and Rhythmicity

The vertebrate heart begins beating at a very primitive stage in its development, when only the conoventricular portion of the cardiac tube has differentiated (Johnstone, 1925; Patten and Kramer, 1933; Davis, 1927). APs at this stage—despite their ventricular origin—resemble those of the primary pacemaker cells of the adult SA node showing small amplitudes, slow upstrokes, and a rounded shape (Bernard, 1976). The posterior portions of the ventricle, the atria, and the sinoatrial tissue are each added in turn, later (see DeHaan, 1968, for early references). As each new region of the heart forms, it brings to the organ tissues with different physiological properties. The shape and parameters of the APs recorded from cells in ventricular, atrial, and SA tissues are already different within hours after they form in the heart tube in ways that anticipate the definitive differences of the adult organ (Fingl *et al.*, 1952; Meda and Ferroni, 1959; Lieberman and Paes de Carvalho, 1965; Coraboeuf *et al.*, 1965).

Each new region is also different in its intrinsic rate of contraction.

In the chick the beat starts at about 36 hours of incubation in the right margin of the newly formed ventricular myocardium, near its caudal end. The rat embryo heart exhibits its first contractions on the left side, at about 8.5 days of development before fusion of the paired left and right rudiments is complete. But within a few hours, the site of origin of the beat shifts to the right posterior end of the heart tube (for references, see DeHaan, 1959). Initially irregular and spasmodic in its beats, the heart soon develops a rhythmic slow rate of 30–40 beats per minute. Gradually, as more posterior regions of the heart tube differentiate, the rate increases. By the time dextral looping of the ventricle has occurred and a distinct atrioventricular sulcus is observable, the heart rate in the chick has increased to 80–90 beats per minute. By about 60 hours of incubation, after the sinoatrial tissue has differentiated, the heart normally beats 110–120 times per minute.

This gradient of rhythmicity is built into each portion of the heart tube. If the tube is cut transversely into three fragments at a stage when it is beating 120 beats per minute, the sinoatrial piece continues to beat at that rate, whereas each of the more rostral pieces reverts to its earlier rhythm. That is, the ventricular piece slows to about 70 beats per minute and the conoventricular portion takes on a rate of only 30–40 beats per minute (Barry, 1942; DeHaan, 1965). In fact, these rate differences must reside ultimately in the cells that comprise the heart. Cells isolated in tissue culture from the embryonic ventricle beat more slowly, on the average, than those from the atria (Cavanaugh, 1955). Under these conditions, with each cell isolated from contact with any neighbors, every beating cell determines its own pulsation rate (DeHaan and Gottlieb, 1968).

The gradual increase in rate of the entire heart tube suggests that as each new segment of the heart differentiates, with a higher intrinsic rate, it acts as pacemaker for the rest of the organ, driving the heart at its own rate. This has been confirmed by Van Mierop (1967), who determined the location of the pacemaker region in the tubular chick heart before and during the time of initiation of the beat, using both intracellular and surface exploring electrodes. He reported that when the first region of tissue began to contract in the right posterior portion of the ventricle, the electrical stimulus for that contraction arose about 100 msec prior to each beat from a more caudal point. Furthermore, even at earlier stages, 3–6 hours before actual contractile activity begins, Van Mierop could record rhythmic action potentials from the caudalmost tissue. These findings indicate that pacemaker activity is localized at the posterior end of the tubular heart and suggest that cells in the myocardial troughs caudal to the formed heart tube at each stage begin to function as primary pacemakers before they fuse in the midline and are themselves incorporated into the beating myocardium. Their action potentials are presumably conducted to the more rostral contractile muscle via electrotonic junctions that connect the cells of the splanchnic mesoderm. We have recently reviewed the evidence for electrotonic coupling and early formation of nexal junctions between embryonic heart cells (DeHaan et al., 1980; Ypey et al., 1980).

There is good evidence that this gradient of rhythmicity—high rate caudally, low rostrally—is coded into the cells of the embryo well before the heart itself forms. The presomite embryo can be cut into fragments containing either the anterior, middle, or posterior portion of the cardiogenic crescent, that is, the regions destined to form, respectively, the conoventricular, ventricular, and sinoatrial parts of the heart. Such fragments isolated in culture medium form vesicles of heart tissue that begin beating spontaneously at an average rate of 35, 65, and 115 beats per minute, respectively (DeHaan, 1963). When they are impaled with intracellular microelectrodes, the slowly beating tissue shows action potentials typical of conus, the vesicles with intermediate rates have ventricle-like action potentials, and the fastest fragments exhibit pacemaker potentials characteristic of sinoatrial tissue (Le Douarin et al., 1966). Thus, not only prospective rate, but the eventual physiological character of the beat, appear to be determined in the cells of the premyocardial mesoderm at early stages. However, these experiments also show that before atrium or SA tissue have differentiated, when the ventricular tube is acting as its own pacemaker and even for some time after, it generates slowly rising APs similar to those of the primary pacemaker cells in the adult SA node. Bernard (1976) has shown, for example, that the AP of the embryonic rat heart at 10 days of gestation has an MDP of 50–60 mV with little or no overshoot and an upstroke $\dot{V}_{max}$ of 5–8 V/second. At that stage, the phase 0 rise-time was unaffected by a 90% substitution of Na^+ by Tris, and the overshoot declined by only 3–6 mV. However, the AP peak was reduced by 20–25 mV in medium containing 1/10 the control Ca^{2+} (0.2 mM). The 10 day AP was not altered by 10 μg/ml TTX but was completely abolished by 2 mM Mn^{2+}. Moreover, V_{max} could not be increased above 10 V/second by holding V_m at -85 to -90 mV with injected current to remove inactivation before firing an AP. These results contrasted with those from 13- or 20-day embryonic hearts in which rise-time or phase 0 was progressively less affected by reduction of Ca_o^{2+} or exposure to Mn^{2+}, but was dramatically reduced and AP ceased altogether in TTX.

In the chick heart, also, the mechanisms that underlie the generation of AP during the second day of incubation and for the next few days are markedly different from those that characterize the adult (for review, see DeHaan et al., 1976; Galper and Catterall, 1978; Iijima and Pappano, 1979; DeHaan, 1980). Hearts of embryos aged 2–3 days fire action potentials with slow rise-times (10–30 V/second) and small amplitudes and continue to beat in the presence of concentrations of TTX up to 10 μg/ml. In contrast, electrical activity in hearts from 7-day embryos or older is completely blocked at a dose of TTX a thousand times lower (McDonald et al., 1972; Shigenobu and Sperelakis, 1971), and $\dot{V}_{max}$ reaches values of 100–200 V/second (Iijima and Pappano, 1979). Prior to about 4 days of development, the hearts are blocked by 0.1 μM D600, a concentration which affects AP duration only at later stages (Galper and Catterall, 1978). Furthermore, the early chick heart is relatively impermeable to K^+ (Sperelakis and Shigenobu, 1972; McDonald and DeHaan, 1973; Carmeliet et al., 1976) and therefore continues to generate APs at elevated K^+ concentrations that suppress activity by hyperpolarization at later stages, when P_K is greater (DeHaan, 1970; Pappano, 1976). This insensitivity to $[K_o^+]$ resembles that of the rabbit SA-node (DiFrancesco et al., 1979).

At the end of the first week of cardiac function, the rate of rise of the action potential and its TTX sensitivity are still readily modified by environmental conditions. Slowly rising action currents which are unresponsive to TTX can be induced in the intact heart by catecholamines or methylxanthines (Shigenobu et al., 1974) or in many cells by dissociation from the tissue with trypsin (McDonald et al., 1973; Sachs et al., 1973; McLean and Sperelakis, 1974). If reassociated into spheroidal aggregates, or if fibroblasts are excluded from the cultures (Lompre et al., 1979), the myocytes regain their TTX sensitivity. Furthermore, the process of resensitization can be enhanced by exposure to insulin (Le Douarin et al., 1974) and can be blocked by inhibition of protein synthesis (McDonald et al., 1973; DeHaan et al., 1976).

It has been suggested that for the first 2 to 3 days after the heart starts to beat, the sodium-specific, TTX-sensitive, fast conductance mechanism is absent or nonfunctional. At that time, action currents might depend exclusively on the TTX-insensitive slow pathway (I_{si}). Between days 4 and 7 in the chick embryo, or 13 to 20 days in the rat, functional fast channels appear to differentiate in the myocardial cell membrane, and the action potential becomes progressively more dependent upon that mechanism (McDonald et al., 1972; Shigenobu and Sperelakis, 1971; Sperelakis and Shigenobu, 1972; Shigenobu et al., 1974; DeHaan et al., 1976; Bernard, 1976). Thus it appears that the ionic basis of the cardiac AP progresses through a sequence from Ca^{2+} to Ca^{2+}/Na^+ dependence. But unlike skeletal muscle and amphibian neurons, the Ca^{2+} current does not disappear at fully differentiated stages. Instead it continues to carry most of the plateau component of the AP and to play an important role in the rate-setting mechanism of the heart. This sequence of membrane differentiative events appears to be retarded or progress more slowly in the parts of the heart that are embryologically more posterior. APs in the atria of chick hearts are still insensitive to TTX at 6 days and changes in $[Ca^{2+}]_o$ have large effects on the overshoot while alteration in $[Na^+]_o$ has little effect (Pappano, 1972, 1976; Ishima, 1968).

2. *Membrane Currents in Heart Cell Aggregates under Voltage Clamp*

All of the above evidence concerning the identity and kinetics of ionic currents in the embryonic heart has been indirect, based upon AP parameters and the sensitivity of electrical events to agents whose inhibitory effects have been identified accurately only in adult tissues. The fact that TTX, which specifically blocks I_{Na} in adult nerve and heart (review by Narahashi, 1974; Catterall, 1980), has no effect on the AP of the 2-day chick heart is only suggestive; it is not a direct test of the idea that I_{Na} is absent from the early embryo. Such direct evidence can come only from the specific identification of conductance mechanisms with the use of the voltage clamp technique, or with the isolation of the channel molecule itself. We cited earlier the problems in applying the voltage clamp to heart tissue that arise out of its multicellular nature. One way to minimize the geometrical problems of heart muscle would be to study single isolated cells. Some success has recently been achieved in this approach with single adult heart cells (Lee *et al.*, 1979), but the technique has not yet been applied to developmental studies. We have recently reported on a voltage clamp analysis of a heart tissue model system consisting of a spheroidal aggregate of embryonic ventricular cells in culture (Nathan and DeHaan, 1978; 1979). We have shown previously that the cells within such an aggregate are tightly coupled electrically. The entire aggregate membrane appears to be virtually isopotential during the voltage changes produced by injecting small current pulses through an intracellular micropipet (DeHaan and Fozzard, 1975; Clay *et al.*, 1979), although it deviates somewhat from uniformity during the fast rise-time of an action potential (DeHaan and Fozzard, 1975). In these preparations, total junctional impedance between cells is $< 10\%$ of transmembrane impedance and is independent of intercellular voltage gradients in the range of 10^{-5}–10^{-3} V, from dc to 160 Hz (DeFelice and DeHaan, 1977). With the aid of an exploring voltage electrode, we demonstrated that deviation from voltage homogeneity during a clamp step in an aggregate was comparable to that seen in adult cardiac preparations (Nathan and DeHaan, 1979). Aggregates prepared from 7-day chick ventricle exhibited two kinetically and pharmacologically distinct components of inward current and a delayed outward current with properties similar to those of I_{Na}, I_{si}, and I_{K_2} in adult heart tissue (Nathan and DeHaan, 1979). In contrast, spheroidal aggregates from 3-day embryos showed only a single I_{si}-like conductance; the TTX-sensitive I_{Na}-like current was absent at that stage (Nathan and DeHaan, 1978).

Three criteria were used to test for *in vitro* differentiation of TTX-sensitive fast Na$^+$ channels: (a) intracellular recording of action potentials with upstroke velocities > 90 V/second, which could be reduced by TTX to < 20 V/second before spontaneous activity had stopped completely; (b) microscopic observation of cessation of spontaneous contraction in the presence of TTX; and (c) recording of a fast, TTX-sensitive component of inward current in voltage clamp, with properties similar to those observed in aggregates prepared from 7-day hearts that had developed *in ovo*. These electrophysiological properties were compared for 3 + 2 (3-day heart *in ovo* and 2 days in gyration culture), 3 + 5, and 7 + 3 aggregates. Under control conditions in this study, all 3 + 2 aggregates beat spontaneously, even in high concentrations of TTX (10 μg/ml). Action potentials had maximum upstroke velocities ranging from 10 to 20 V/second (Fig. 2A). Allowing these aggregates to gyrate an additional 3 days (for a total of 5 days *in vitro*) resulted in records similar to those obtained from ag-

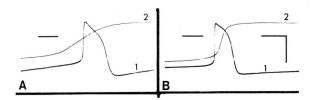

FIG. 2. *In vitro* differentiation of a fast inward current. Spontaneous action potentials recorded in medium containing 1.3 mM K$^+$ from ventricular aggregates prepared from 3-day heart, after 2 days (A) or 5 days (B) of gyration culture. Horizontal scale = 100 msec (trace 1) or 2 msec (trace 2), and its level represents 0 mV; vertical scale = 50 mV. V_{max} = 16 V/second (A); 96 V/second (B). (Reproduced by permission of *Proc. Natl. Acad. Sci. U.S.A.;* from Nathan and DeHaan, 1978.)

gregates derived from 7-day hearts that had differentiated *in ovo* (DeHaan and Fozzard, 1975; Nathan and DeHaan, 1979). The 3 + 5 aggregates exhibited characteristic fast rise times averaging 110 V/second (Fig. 2B) which were slowed dramatically by exposure to TTX (0.02–10 μg/ml) before cessation of spontaneous activity. The more rapid rate of depolarization of 3 + 5 aggregates was not due to greater sodium activation secondary to an increase in the maximum diastolic potential (MDP); on the contrary, 3 + 2 and 3 + 5 aggregates differed only slightly in the magnitudes of their maximum diastolic potentials.

Under voltage clamp, inward current in 3 + 2 aggregates differed markedly from that seen in 7 + 2 or 7 + 3 controls (Nathan and DeHaan, 1978, 1979). The 3 + 2 aggregates had only a single component of inward current (Fig. 3A and B) while those from 7-day heart had two distinguishable currents (Fig. 4A and B). The kinetics of the single component in 3 + 2 preparations were slower (time constant for current decay was 12–44 msec) than those of the I_{Na}-like component in 7 + 3 controls (decay time constant, 0.7–1.7 msec). Moreover, this slow current in 3 + 2 aggregates was unaltered by TTX (Fig. 3C and D).

After 3 additional days *in vitro,* the inward current recorded in 3 + 5 aggregates resembled that obtained in 7 + 3 controls that had

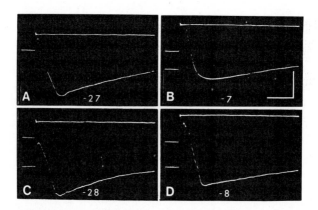

FIG. 3. Insensitivity of the slow inward current to TTX. Voltage clamp experiment showing voltage step (top trace) and current (lower trace) recorded from 3 + 2 aggregates in control medium (A,B) and in medium containing 10 μg/ml TTX (C,D). The number in each panel represents the voltage to which the membrane potential was stepped from the holding potential (−56 mV). Horizontal scale = 10 msec. Vertical scale = 50 mV (upper trace); 1.2 μA (lower trace). The magnitude, kinetics, and voltage dependence of the current are essentially unaffected by TTX. (Reproduced by permission of *Proc. Natl. Acad. Sci. U.S.A.;* from Nathan and DeHaan, 1978.)

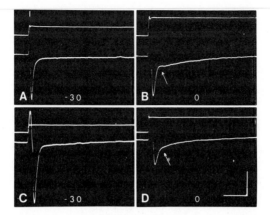

FIG. 4. Inward currents in differentiated aggregates. Voltage steps (top trace) and current (lower trace) recorded from a 7 + 3 aggregate (A,B) and 3 + 5 aggregate (C,D). Vertical scale: 80 mV (A,B), 100 mV (C,D,) 0.5 μA (A–D). Note the similarity of the inward current magnitudes, kinetics, and voltage dependence in the two preparations. A small delayed inward current is activated at 0 mV (arrows). (Reproduced by permission of *Proc. Natl. Acad. Sci. U.S.A.*; from Nathan and DeHaan, 1978.)

developed *in ovo*, exhibiting two kinetically distinguishable components of inward current, each with a different voltage dependence (Fig. 4C and D). The fast component reached a peak near -30 mV, and the slow component at 0 mV.

In order to test whether protein synthesis was required for the differentiation *in vitro* of I_{Na}-like channels, cycloheximide (CHX) was added to the medium bathing the 3 + 2 aggregates at a concentration (1 μg/ml) sufficient to reduce amino acid incorporation into protein to < 15% of controls (DeHaan *et al.*, 1976). After 2 additional days gyrating in CHX, the 3 + 4 aggregates were washed with fresh medium and intracellular recordings were obtained. All continued to beat rhythmically and were not visibly altered by the prolonged exposure to the inhibitor. However, the treated 3 + 4 aggregates had failed to differentiate fast Na$^+$ channels. Their electrophysiological properties were much like 3 + 2 controls; they had upstroke velocities < 20 V/second, which were not influenced significantly by TTX. On the other hand, replicate cultures of untreated 3 + 4 controls had V_{max} > 90 V/second which were reduced to < 20 V/second by TTX. From these studies we could conclude that in aggregates of chicken heart cells from 3-day embryos after 2 days of gyration culture, TTX-sensitive, fast Na$^+$ conductance channels were not functional. However, after 2–3 days of additional culture *in vitro*, these aggregates satisfied three different criteria for the existence of such a conductance mechanism: rapidly rising action potentials, cessation of spontaneous activity in the presence of TTX, and a fast component of inward current recorded during voltage clamp. The membrane properties of these 3 + 5 aggregates were similar to those of aggregates derived from 7-day hearts that had developed *in ovo*. The difference in potential at which the slow component of inward current was maximally activated (-35 mV at 3 + 2 days and 0 mV at 3 + 5 days) may explain why the slow component was unable to generate spontaneous action potentials in TTX-treated 3 + 5 aggregates but was able to support spontaneous activity at 3 + 2 days. If the pacemaker mechanism is assumed to function within the same voltage range (around -60 mV) in both preparations, then the more positive threshold for initiating action potentials in TTX-treated 3 + 5 aggregates would be the deciding factor. The observed changes in conductance characteristics during development could take place in two ways: (a) the TTX-insensitive slow conductance at 3 + 2 days might increase its TTX sensitivity, as well as its rate of activation and inactivation, to become the "fast" component at 3 + 5 days; or (b) the conductance at 3 + 2 days might retain its TTX insensitivity but shift its voltage dependence during development to become the "slow" component at 3 + 5 days, while a new I_{Na} channel becomes operative.

We also demonstrated that the development of the fast Na^+ conductance mechanism in embryonic cardiac muscle is dependent upon the products of protein synthesis, since fast-rising, TTX-sensitive action potentials failed to develop in 3 + 4 aggregates that had been incubated 2 days in cycloheximide. These results support and extend previous observations of organ-cultured 4-day hearts (DeHaan *et al.*, 1976) in showing that such preparations can continue to beat rhythmically for several days with protein synthetic pathways blocked, and that protein synthesis is required for *in vitro* development of TTX sensitivity and rapid upstroke velocity. In this regard, they are consistent with results of UV-irradiation studies (Nathan *et al.*, 1976) and with other data (Shrager, 1975; Rojas and Rudy, 1976) that suggest that the fast Na^+ conductance channels may be associated with protein structures. These data do not necessarily indicate, however, that fast Na^+ channels are normally synthesized *de novo* after the third day of incubation. An alternative is that such channels exist in cardiac myocytes even at the earliest stages in a masked or nonfunctional form. By this hypothesis, the protein synthetic events that take place after 3 days would be those required to unmask the fast conductance mechanism (see below).

W. T. Clusin (personal communication) has been able to demonstrate electrophysiological differentiation of embryonic cardiac muscle *in vitro* with media and techniques similar to our own. However, Sperelakis and his colleagues (Sperelakis and Shigenobu, 1974; Shigenobu and Sperelakis, 1974; McLean *et al.*, 1976) have reported that development ceases when the heart is removed from an embryo and maintained in culture. Hearts from 2- or 3-day embryos survived for 1–2 weeks when cultured in test tubes with medium containing 15% horse serum and 5.4 mM KCl (Sperelakis and Shigenobu, 1974; Shigenobu and Sperelakis, 1974), when cultured on the chorioallantoic membrane of 6- to 7-day-old eggs, or when grafted onto the chorioallantoic membrane of a host chicken (6-day), where they were revascularized by the host circulation (Renaud and Sperelakis, 1976). In none of these conditions did the cardiac cells develop a TTX-sensitive, fast Na$^+$ conductance; potentials retained their slow upstroke velocities (10–20 V/second) throughout the culture period. Similar results were obtained with ventricles from 48- to 60-hour hearts and even with aggregates prepared from such hearts by methods similar to our own and cultured for 6 to 18 days (McLean *et al.*, 1976). Only when these preparations were incubated with adult heart mRNA for 6 or more days did the average upstroke velocity increase to 79 ± 4 V/second, and the action potential become sensitive to TTX blockade (McLean *et al.*, 1976). These mRNA-induced changes could be prevented by addition of cycloheximide (1 μg/ml) at the time of mRNA addition.

Our data indicate that aggregates formed by gyration of 3-day heart cells do not require an enriched nutrient medium in order to differentiate functional fast Na$^+$ channels. Since our 3 + 2 aggregates incubated for 2–3 additional days in simple balanced salt solution exhibited fast, TTX-sensitive upstroke velocities and an I_{Na}-like current, the only components that appear to be necessary are the basic ionic constituents of the extracellular milieu. We attempted to duplicate some of the culture conditions used by Sperelakis and Shigenobu (1974), i.e., elevation of potassium levels to 5.4 mM and horse serum concentration to 12%; nonetheless, electrical development proceeded in every case.

3. Pacemaker Currents

There is relatively little reliable information concerning outward current channels in the embryonic heart. Indirect measurements of membrane permeability to K$^+$ have been made in terms of the ratio of permeabilities P_{Na}/P_K, calculated from the Goldman equation. Membrane potential in hearts from 3 to 18 days of development have been measured, and K_i^+ and Na_i^+ have been determined along with water spaces (McDonald and DeHaan, 1973) or have been assumed (Sperelakis et al., 1976) to provide the required values to make those calculations. These measurements indicate that K_i^+ falls from 170 mM/kg cell H_2O (at 3 days) to about 120 mM/kg cell H_2O (in the 18-day heart), while the driving force for K^+ current $(I_K\text{-}V_M)$ declined from 32 mV at 3 days to about 6 mV at 18 days, and Na_i^+ remained constant during that period. These data were interpreted to mean that P_{Na} was unchanged while P_K increased by about 10-fold during embryonic development (McDonald and DeHaan, 1973). Direct measurement of ion fluxes in strips of tissue from 7-day and 20-day chick embryo hearts confirmed that P_{Na} remained constant but P_K increased 2-fold, between those two times (Carmeliet et al., 1976). However, serious questions have been raised about the validity of the latter determinations (Horres et al., 1979).

We have recently begun a voltage clamp analysis of developmental changes in the currents in the pacemaker range of voltages in spheroidal aggregates of embryonic ventricle cells (Shrier et al., 1979). The spontaneous activity of these preparations varies greatly, depending on the age of heart from which they are prepared. In low-K^+ (1.3 mM) medium, for example, aggregates from 7-day hearts beat spontaneously and rhythmically, 12-day preparations usually beat spasmodically or are quiescent, and 17-day aggregates normally are not active at all. In the presence of TTX (~ 3 μM), 7-day aggregates rest near -50 mV but can be shifted to a second stable V_r (~ 70 mV) by increasing K_o^+ (DeHaan and DeFelice, 1978b). The 12-day preparations exhibit two resting states at constant K_o^+ (near -90 and -60 mV) and undergo large nonlinear oscillations between these two levels upon application of a small steady current. (Shrier et al., 1979). Under voltage clamp, between V_r and -80 mV, the 7-day aggregates exhibit a substantial inward current opposing two distinctive outward components, one time-dependent and one a background (time-independent) current. The time-dependent conductance resembles the pacemaker current I_{K_2} in the MNT Purkinje fiber model (Shrier and Clay, 1980). It becomes smaller in 12-day preparations and disappears from the 17-day cells. This presumably underlies the declining rhythmicity of the older preparations. The steady state IV curve for 7-day aggregates in the low-K_o^+ medium is essentially linear between -80 mV and V_r with a slope resistance of 40–80 kΩ cm^2 (Fig. 5). By 12 days, the IV relation has developed a pronounced N shape with zero current points at about -90, -75, and -60 mV. The middle crossing at -75 mV is

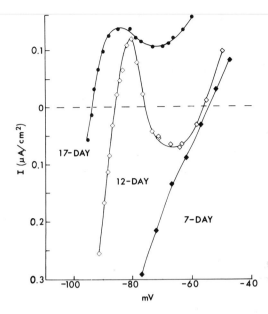

F<small>IG</small>. 5. Developmental changes in the voltage dependence of the delayed outward current. Steady-state current–voltage curves for aggregates prepared from ventricles of 7-, 12-, and 17-day embryo hearts. Measurements were made from the current required to maintain holding potentials or from the current at the end of 4-second or 10-second voltage clamp steps to the voltage indicated on the abscissa. (Courtesy of A. Shrier and J. R. Clay.)

on the negative slope limb of the curve and thus explains the tendency of the 12-day preparation to oscillate between the two stable states represented by the positive-slope crossings at -60 and -90 mV. This condition is analogous to that found in the adult Purkinje fiber in low Cl_o^- solutions (Gadsby and Cranefield, 1977). Aggregates of 17-day heart cells also exhibited an N-shaped IV curve, but this was shifted in an outward direction, with only one zero-current crossing point at about -95 mV. Nothing is known about the molecular mechanisms that underlie any of these developmental changes in I_K.

Aggregates of embryonic chick ventricular cells have also provided evidence that K currents may be regulated by Ca_i^{2+} (Clusin, 1979, 1980a,b). Blockage of Ca influx by application of Co^{2+} or EGTA produces a sustained depolarization to about -30 mV and causes the membrane to become electrically linear with a high resistance (Clusin, 1979). When free Ca_i^{2+} was assessed by a measure of myofibrillar contraction (Clusin, 1980), it was found to correlate well with V_m.

Therefore, Ca_i^{2+} could control diastolic K currents and would result in their apparent voltage-dependence.

4. Voltage Fluctuations and Interbeat Interval

The phase 4 pacemaker potential was described by its earliest observers either as a slow depolarization (Arvanitaki, 1938) or as an oscillatory potential (Bozler, 1943). As described above, the first view sees the phase 4 depolarization as a time-dependent decline in an outward current set against an overriding inward current, which brings the cell from MDP gradually to threshold. The alternative model which is based on a small-signal analysis of the heart cell membrane (Clay et al., 1979) shows that it—like all excitable membranes—behaves like a parallel circuit containing resistive, capacitive, and inductive elements. According to this impedance model, the RLC characteristics of the excitable membrane are so tuned that any small perturbation causes the voltage to oscillate at a natural resonant frequency. In this view, the pacemaker depolarization represents the depolarizing phase of one oscillatory half-cycle. If threshold is not achieved at the crest of the oscillation, it does not continue to depolarize, but merely completes the cycle. Recordings from a wide variety of excitable cell preparations have revealed spontaneous, subthreshold oscillatory voltage fluctuations (see DeHaan and DeFelice, 1978a, for review). The squid axon has a natural oscillation frequency at about 100 Hz, close to the rate at which it fires spontaneously when depolarized or exposed to low-Ca_o^{2+} solutions (Guttman and Barnhill, 1970). In contrast, V_m in cardiac Purkinje fibers oscillates at about 0.5 Hz (Cranefield, 1975) which is near their normal beat frequency. When embryonic heart cell aggregates are made quiescent by exposure to TTX they exhibit two classes of subthreshold behavior, pronounced voltage oscillations at about 1 Hz of 1–10 mV amplitude, and random voltage fluctuations ranging from 50 to 500 μV. There is a large body of evidence (reviewed in DeFelice, 1977; DeHaan and DeFelice, 1978a; Clay et al., 1979) that these fluctuations of potential ("noise") result from the random opening and closing of individual current channels in the membrane—presumably the same channels that carry the macroscopic currents of the action potential, I_{Na}, I_{si}, I_K, etc. The "phenomenological inductance" that represents the L of the RLC membrane circuit is, in fact, a manifestation of the voltage- and time-dependences of these conductances. When random microcurrents flow across the impedance of the membrane, they produce voltage perturbations that set the membrane oscillating at the resonant frequency of its RLC circuit. When such oscillations achieve threshold they may activate a regenerative current (I_{Na} or I_{si}) and initiate an AP.

We have recently exploited this view of the pacemaker mechanism —as random voltage fluctuations superimposed on an oscillating pacemaker potential—to explain variations in rhythmicity of cultured heart cell preparations (Clay and DeHaan, 1979). The rhythm of the intact heart appears, at first glance, to be strikingly regular from one beat to the next. But when measured carefully consecutive interbeat intervals (IBIs) exhibit small, apparently random variations about some mean value. Gustafson *et al.* (1978) measured a standard deviation of 0.018 second in trains of heartbeats in a population of young adults with a mean interval of 0.728 second, i.e., a variation of about 2%. The variability of IBI in clusters of heart cells in culture appears to depend on the number of cells in the cluster (Fig. 6). Single cells and

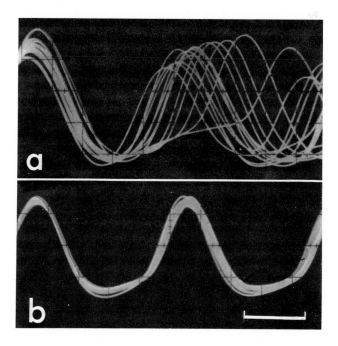

Fig. 6. Spontaneous fluctuation in interbeat interval. Fifteen consecutive interbeat intervals recorded optically from two heart-cell clusters in an undisturbed culture dish under identical conditions. (a) Two-cell group, mean interbeat interval, 0.465 second, coefficient of variation, 23.5%; (b) 75-cell group, mean interbeat interval 0.490 second, coefficient of variation 3.9%. Scale = 100 msec. (Reproduced by permission of *Biophysical Journal;* from Clay and DeHaan, 1979.)

groups of a few cells had coefficients of variation of 10–20%, whereas variation in IBI in clusters of 100 cells was only 3–4%. We have shown that the amplitude of voltage noise recorded from heart cell aggregates should vary as $N^{-\frac{1}{2}}$ (where N = cell number), i.e., a small cluster of cells (with a large R_i) should exhibit a large amplitude of voltage noise. If random fluctuations in membrane voltage, superimposed on the phase 4 depolarization, cause variations in the time it takes for V to reach threshold, then the variation in IBI should be directly related to voltage noise amplitude, or inversely related to $N^{-\frac{1}{2}}$. To test this hypothesis, we measured directly the variance in IBI from trains of beats recorded from clusters of 1–125 cells in culture, and confirmed the $N^{-\frac{1}{2}}$ relationship (Clay and DeHaan, 1979). These results support the model and confirm that electrical noise plays an important role in the rate-setting mechanism in cultured heart cell preparations. It is not known whether random voltage fluctuations differ in any characteristics in cells taken from different aged hearts.

C. Do New Channels Differentiate in Excitable Membranes?

The evidence cited above indicates that the ionic basis of the AP changes with developmental state in a wide variety of excitable cells. The processes of differentiation of the inward current mechanisms fall into four categories.

Type I: Xenopus Rohon–Beard neurons (Baccaglini and Spitzer, 1977), dorsal root ganglion cells (Baccaglini, 1978), and chick skeletal muscle (Kano, 1975) exemplify the first category. At the earliest stages of differentiation these cells are inexcitable, when regenerative AP can first be elicited, the inward current is carried entirely by I_{Ca}. Later in differentiation, the rising phase of the AP is carried by a mixed Na/Ca current. In the definitive differentiated state, I_{Ca} is much reduced or disappears altogether, and the upstroke depends entirely on I_{Na}. It is blocked by TTX or Na^+ depletion and is unaffected by Co^{2+} or La^{2+}.

Type II: In the mouse unfertilized oocyte (Okamoto *et al.*, 1977), in rapidly growing neuroblastoma cells, and in neuroblastoma in a highly differentiated state (Moolenaar and Spector, 1979a) a common prominent La^{2+}-sensitive I_{Ca} carries excitable activity. A TTX-sensitive I_{Na} appears in the most developed stages, and Na^+–Ca^{2+} AP result. Excitability can be blocked only by inhibiting both kinds of channels at the same time.

Type III: Tunicate muscle exemplifies the third class of developmental sequence (Miyazaki *et al.*, 1972). Here the ionic dependence of the AP shifts from Na^+/Ca^{2+} in the egg and early myoblasts to an almost purely Ca^{2+}-dependent AP in mature muscle.

Type IV: A clonal line of rat skeletal muscle cells exhibits the fourth class of behavior (Kidokoro, 1975a,b). At the earliest stages of differentiation, myoblasts were already excitable and the small APs were almost completely Na^+-dependent. In differentiated myotubes, I_{Na} remained the major carrier of the AP upstroke, but a plateau-like shoulder on the AP was carried by I_{Ca}.

The vertebrate heart differentiates mainly according to Type I behavior. At the earliest excitable stages the AP is almost exclusively Ca^{2+}-dependent. But the sequence from Ca^{2+} to Na^+/Ca^{2+} to Na^+ seems to arrest at different times in different parts of the organ, and thus results in the variety of AP shapes described above. The primary pacemaker cells of the sinoatrial node appear to progress through the sequence only slightly beyond the Ca^{2+}-dependent condition. Even in the adult I_{Na} is hardly represented. I_{Na} becomes very prominent in most of the working muscle and Purkinje fiber cells. However, I_{Ca} is retained as a major plateau current. In atrial fibers, where the plateau is much reduced, I_{Ca} also decreases to small but not negligible proportions.

1. Ligands for I_{Na} in Nerve and Muscle

Several neurotoxins are now known that act specifically on voltage-sensitive sodium channels in excitable membranes (for review, see Catterall, 1980). To test whether cells at TTX-insensitive stages nonetheless retain fast I_{Na} channels, several groups have applied the alkaloid veratradine. In nerve (Ulbricht, 1969) and neuroblastoma cells (Catterall and Niremberg, 1973) veratradine causes a fraction of the I_{Na}-like channels to open irreversibly. There is substantial evidence that the channels opened by veratradine (and the structurally related alkaloid, batrachotoxin) are the same as those normally activated by depolarization. Both drug-activated and depolarization-activated Na^+ channels are inhibited by TTX with a K_d of about 3 nM (Catterall, 1975). Both channels exhibit a conductance of approximately 3 pS (Catterall, 1977a; Conti *et al.*, 1976). Variant neuroblastoma cells lacking a phase 0 upstroke do not respond to veratradine or batrachotoxin (Catterall, 1977b). And finally, in denervated rat muscle and cultured muscle cells the increase in TTX-resistance of depolarization-induced AP is the same as that for veratradine-activated increase in Na flux (Albuquerque and Warnick, 1972; Catterall, 1976).

154   ROBERT L. DEHAAN

That is, the effects of denervation on drug-activated and voltage-sensitive channels are identical.

Similar results were obtained with the sea anemone toxin ATX_{II}. However, this drug stimulates influx of both ^{22}Na and ^{45}Ca. The increased Na flow is inhibited by TTX, while the Ca^{2+} entry system is blocked by verapamil or D600. With a combination of ATX_{II} and veratradine, which act synergistically when applied to cultured neuroblastoma cells, Jacques et al. (1978) have obtained results indicating that there may be two populations of Na^+ channels normally present in this cell type, one group functional and the other normally "silent." The first population of channels (about 27% of the total) would open transiently upon depolarization and have a high affinity (in the open state) for ATX_{II}. The silent channels would not be activatable by depolarization and would have a much lower affinity for ATX_{II}. Veratradine would activate (and prevent the inactivation of) both types of Na^+ channels, thereby explaining the synergistic action of the two drugs. We noted above that several inexcitable (i.e., voltage-insensitive) cell types can respond to chemical stimuli with rapid changes in ion conductance: thymocytes (Iversen, 1976), L cells (Okada et al., 1979), and pancreatic B cells (Beigelman et al., 1977). In B cells (Donatsch et al., 1977) and in the C_9 cell line, which is derived from a bladder metastasis of a rat brain tumor (Romey et al., 1979), the conductance change appears to be mediated by silent I_{Na} channels. Both cell types respond to veratradine with increased Na-influx, and in the C_9 cell line this effect was augmented by ATX_{II} and blocked by TTX (Romey et al., 1979).

2. Evidence for "Silent" I_{Na}-Channels in Heart

Galper and Catterall (1978) have compared the effect of D600 and of TTX on APs and ion fluxes in intact chick heart or cells cultured as monolayers or spheroidal aggregates. They confirmed earlier observations (Sperelakis, 1972; McDonald et al., 1972) that monolayer cultures are relatively insensitive to TTX regardless of the age of the hearts from which they are derived, whereas aggregates regain the TTX sensitivity of the tissue of origin (McDonald et al., 1972; DeHaan et al., 1976). They also showed—as have others—that the time course of the increase in TTX sensitivity from 2 to 7 days of development in intact hearts and aggregates was paralleled by a concomitant decrease in sensitivity to D600. They noted, however, that the latter change in response to the slow Ca^{2+} current was also observable in monolayer cells. That is, the developmental events responsible for the progressive decrease in sensitivity to D600 persisted in monolayer cultures, while those responsible for the increase in TTX sensitivity were lost.

We noted above that when fully differentiated, TTX-sensitive heart tissue is dissociated, the resultant cells in monolayer culture are found to generate Ca^{2+}-dependent AP. In order to test whether the TTX-insensitive monolayered cardiac myocytes had nonfunctional Na^+ channels, several groups have applied veratradine (Sperelakis and Pappano, 1969; Fosset et al., 1977; Galper and Catterall, 1978). In 10- or 12-day heart monolayer cultures, for example, veratradine markedly stimulated the rates of $^{22}Na^+$ and $^{45}Ca^{2+}$ flux. This increase was completely inhibited by TTX (Fosset et al., 1977; Galper and Catterall, 1978). The concentration of TTX that gave half-maximal inhibition (1–30 nM, depending on veratradine concentration) was similar to the K_d for APs in nerve and muscle (Evans, 1972) and to the concentration of TTX required to reduce $\dot{V}_{max}$ by 50% in chick ventricle (Iijima and Pappano, 1979) or ventricular aggregates (Shrier and De-Haan, unpublished). The drug-induced increases in Na^+ and Ca^{2+} influx were unaffected by verapamil or D600. The veratradine-stimulated Ca^{2+} uptake was ascribed to a separate electroneutral carrier, linked to Na accumulation (Fosset et al., 1977). Of greatest significance was that cultures prepared from 3-day and 12-day embryonic hearts exhibited no difference in the magnitude of veratradine-stimulated ^{22}Na flux, the activation curve for veratradine, or the dose–response curve for TTX. Since veratradine-stimulated Na flux is ostensibly proportional to the density of I_{Na}-like channels, the authors concluded that 3-day embryonic heart cells contain nonfunctional fast Na^+ channels. Since these are not normally involved in beating, they must be physiologically inactive but retain the response to veratradine and TTX.

Romey et al. (1980) have recently suggested that 3-day chick embryo heart cells contain only the silent channels in their membranes, and that they are not normally functional because they are permanently (or too rapidly) h-inactivated. When h-inactivation was removed in 3-day heart-cell aggregates by exposure to ATX_{II}, a verapamil-insensitive inward Na^+-flux was produced. According to this logic, the differentiative process that leads to the appearance of a fast Na current in the chick heart between 3 and 7 days of development would represent a modification of the inactivation properties of the channels to convert an increasing fraction from the silent to the functional state.

Since the ionic flux mediated by slow Na^+/Ca^{2+} channels in cultured cells becomes progressively less sensitive throughout development to inhibition by D600, these observations suggest that the primary developmental event determining the changing pharmacological sensitivity of the embryonic heart is a progressive decrease in density of the I_{si}-like channels which prevents them from maintaining beating without the simultaneous activity of I_{Na} channels. This is reflected in a reduced affinity for D600 (Galper and Catterall, 1978) and the declining ability of divalent cations to carry the AP (Pappano, 1976). The increase in TTX sensitivity of beating would then result from the increasing dependence on functional I_{Na} channels to maintain beating and from the increase in the affinity of those channels for TTX, as they are converted from silent to functional. The fact that veratradine-activated channels in early heart exhibit half-maximal inhibition at TTX concentrations that are similar to those reported at later stages argues against the hypothesis that TTX insensitivity of early heart cells results from inaccessibility of the binding site to TTX due to the surface glycocalyx, as suggested by Lieberman et al. (1976).

Evidence that I_{Na}-like channels are present at early developmental stages in the embryonic chick heart has also been obtained from measurements of the effects of TTX on $\dot{V}_{max}$. Iijima and Pappano (1979) recorded $\dot{V}_{max}$ from intact ventricle preparations, and reported an increase from 112 to 217 V/second between the fourth and twelfth incubation day. Neither the steady-state h-inactivation of $\dot{V}_{max}$ (determined by depolarization in high K_o^+ solutions), the temperature-dependent shift of h, nor the time constant for recovery of $\dot{V}_{max}$ changed significantly during that developmental period. Moreover, the K_d for 50% inhibition of $\dot{V}_{max}$ by TTX was the same in hearts from 4-, 5-, and 18-day embryos. On the assumption that $\dot{V}_{max}$ is a valid measure of I_{Na}, that I_{Na} is, in turn, proportional to $\overline{G}_{Na}$ (the maximal sodium conductance), and that $\overline{G}_{Na}$ reflects the total number of activatable fast sodium channels, the authors conclude that I_{Na} channels with the same physicochemical properties and TTX affinity are present in the heart from 4 to 18 days of development. The developmental increase in beat sensitivity to TTX is ascribed by them to an increase in $\overline{G}_{Na}$, i.e., I_{Na} channel density.

3. Is $\dot{V}_{max}$ a Valid Measure of $\overline{G}_{Na}$?

A note of caution is required here, however, since the linear dependence of $\dot{V}_{max}$ on $\overline{G}_{Na}$ has been the subject of recent controversy (Cohen and Strichartz, 1977; Strichartz and Cohen, 1978; Hondeghem, 1978; Walton and Fozzard, 1979). We have further evaluated this matter by using the MNT Purkinje fiber model to compute the theoretical relationship between $\overline{G}_{Na}$ and $\dot{V}_{max}$ and by using 7-day embryo heart cell aggregates to provide experimental evidence for that prediction (Shrier and DeHaan, 1978). According to the MNT model, the dependence of $\dot{V}_{max}$ on $\overline{G}_{Na}$ is influenced by at least four factors: (a) the rate and voltage dependence of h-inactivation during depolarization; (b) the electrochemical driving force; (c) the relationship between G_{Na} at $\dot{V}_{max}$ and $\overline{G}_{Na}$; and finally, (d) the contribution of nonsodium currents to total membrane current. A marked nonlinearity between $\dot{V}_{max}$ and $\overline{G}_{Na}$ is introduced by h-inactivation that takes place during pacemaker depolarization in computed MNT action potentials. However, h-inactivation did not appreciably influence the relationship between $\dot{V}_{max}$ and $\overline{G}_{Na}$ in the aggregates. Nonetheless, V_m at $\dot{V}_{max}$ clearly hyperpolarized in a nonlinear way as $\dot{V}_{max}$ was reduced. There is also evidence that substantial inward currents other than I_{Na} flow during the AP, and that the fraction is greater at early stages of development, and under pharmacological regimes that reduce I_{Na}. Moreover, even the best cardiac preparations fail to achieve true isopotentiality. Although aggregates deviate only slightly from voltage homogeneity during small slow voltage changes near rest (Nathan and DeHaan, 1979; Clay et al., 1979), they depart appreciably from isopotentiality with rapid voltage changes. A 50-μsec latency can be detected between potentials recorded from widely separated cells in an aggregate during the most rapid part of action potential upstroke phase (DeHaan and Fozzard, 1975; Clapham, 1979). This delay reflects a voltage gradient across the aggregate at the time of $\dot{V}_{max}$ that can be illustrated with a third microelectrode during large voltage clamp steps (Nathan and DeHaan, 1979). The gradient is proportional to current flow across the aggregate and represents the fraction of I_{Na} that does not charge the local membrane capacitance.

We concluded (Shrier and DeHaan, unpublished) that changes in $\dot{V}_{max}$ that result from development of the heart, or those produced in a cardiac preparation with TTX or other pharmacological agents, do not necessarily reflect accurately changes in $\overline{G}_{Na}$.

D. MECHANISMS OF CHANNEL DIFFERENTIATION

There is little information about the cellular or molecular mechanisms that regulate the ionic conductances in the plasma membrane. Virtually nothing is known about how the number of specific channels or their selectivity is determined. A few observations seem pertinent.

1. Blackshaw and Warner (1976a) demonstrated that the normal rapid increase in V_r in neural plate cells could be blocked by brief exposure of axolotl embryos at the mid-neural fold stage to ouabain or strophanthidin. This treatment resulted in a 65% reduction in number of neurons differentiating in culture but had no effect on muscle differentiation (Messenger and Warner, 1979). Moreover, the hearts of chick embryos grown in 140 mM K^+ developed normally, and began to beat within minutes after the medium was replaced by standard Tyrodes (Manasek and Monroe, 1972). Thus, either undisturbed Na-pump activity or an appropriate level of electronegativity appears to be required for proper electrophysiological development of nerves, but heart and muscle cells may be less sensitive to this requirement.

2. Active protein synthesis is required for differentiation of a functional fast Na^+ conductance in the heart (DeHaan et al., 1976); Nathan and DeHaan, 1978). The protein required for differentiation is apparently not the I_{Na} channel itself but a modulating factor that causes the gate to become voltage-sensitive.

3. Agents that disrupt the submembranous microtubules disturb the AP mechanism in squid axons (Matsumoto and Sakai, 1979a), and reagents that activate tyrosine-tubulin ligase restore excitability (Matsumoto and Sakai, 1979b). Thus, differentiation of the AP mechanism may require an intact cytoarchitecture, or be regulated by cytoskeletal elements. The TTX-desensitization of heart cells by trypsin (Sachs et al., 1973) might be explained by this observation, since trypsin penetrates cells and disrupts the cytoarchitecture.

Clearly, an understanding of how ionic channels develop awaits further experimental analysis.

REFERENCES

Ahmed, X., and Connor, J. A. (1979). J. Physiol. (London) 286, 61–82.
Akselrod, S., Landau, E. M., and Lass, Y. (1979). J. Physiol. (London) 290, 387–397.
Albuquerque, E. X., and Warnick, J. E. (1972). J. Pharm. Exp. Ther. 180, 683–697.
Almers, W. (1978). Rev. Physiol. Biochem. Pharmacol. 82, 96–190.
Almers, W., and Levinson, S. R. (1975) J. Physiol. (London) 247, 483–509.
Armstrong, C. M. (1975). Biophys. J. 15, 932–933.
Armstrong, C. M., and Bezanilla, F. (1974). J. Gen. Physiol. 63, 533–552.
Arvanitaki, A. (1938). "Proprietes rhythmiques de la matiere vivante. II. Etude experimentale sur le myocarde d'helix." Herman and Cie, Paris.
Arvanitaki, A., Fersard, A., Kruta, A., and Kruta, V. (1937). C. R. Soc. Biol. 124, 165–167.
Atwell, D., and Cohen, I. (1977). Biophys. Mol. Biol. 31, 201–245.
Baccaglini, P. I. (1978). J. Physiol. (London) 283, 585–604.
Baccaglini, P. I., and Spitzer, N. C. (1977). J. Physiol. (London) 271, 93–117.

Baker, P. F., Hodgkin, A. L., and Ridgway, E. B. (1971). *J. Physiol. (London)* **218**, 705-755.

Barry, A. (1942). *J. Exp. Zool.* **91**, 119-130.

Beeler, G. W., and Reuter, H. (1977). *J. Physiol. (London)* **268**, 177-210.

Beigelman, P. M., Ribalet, B., and Atwater, I. (1977). *J. Physiol. (Paris)* **73**, 201-217.

Bernard, C. (1976). In "Developmental and Physiological Correlates of Cardiac Muscle" (M. Lieberman and T. Sano, eds.), pp. 169-184. Raven, New York.

Bernstein, J. (1902). *Pflugers Arch. Gesamte Physiol.* **92**, 521-562.

Blackshaw, S. E., and Warner, A. E. (1976a). *J. Physiol. (London)* **255**, 209-230.

Blackshaw, S. E., and Warner, A. E. (1976b). *J. Physiol. (London)* **255**, 231-247.

Bonke, F. I. M. (1978). "The Sinus Node. Structure, Function and Clinical Relevance." Martinus Nijhoff, The Hague.

Bozler, E. (1942). *Am. J. Physiol.* **136**, 543-552.

Bozler, E. (1943). *Am. J. Physiol.* **139**, 477-480.

Brady, A. J., and Hecht, H. H. (1954). *Am. J. Med.* **17**, 110.

Brooks, C. M., and Lu, H. H. (1972). "The Sinoatrial Pacemaker of the Heart." Thomas, Springfield, Illinois.

Brown, H. F., Giles, W., and Noble, S. J. (1977). *J. Physiol. (London)* **271**, 783-816.

Busis, N. A., and Weight, F. F. (1976). *Nature (London)* **263**, 434-436.

Cahalan, M. D., and Almers, W. (1979). *Biophys. J.* **27**, 39-56.

Carmeliet, E., and Vereecke, J. (1979). *In* "Handbook of Physiology" (R. M. Berne, ed.), Sect. 2, Vol. 1, pp. 269-356. American Physiological Society, New York.

Carmeliet, E., Horres, C. R., Lieberman, M., and Vereecke, J. S. (1976). *J. Physiol. (London)* **254**, 673-692.

Catterall, W. A. (1975). *J. Biol. Chem.* **250**, 4053-4059.

Catterall, W. A. (1976). *Biochem. Biophys. Res. Commun.* **68**, 136-142.

Catterall, W. A. (1977a). *J. Biol. Chem.* **252**, 8660-8668.

Catterall, W. A. (1977b). *J. Biol. Chem.* **252**, 8668-8678.

Catterall, W. A. (1979). *J. Gen. Physiol.* **74**, 375-391.

Catterall, W. A. (1980). *Ann. Rev. Pharmacol. Toxicol.* **20**, 15-43.

Catterall, W. A., and Beress, L. (1978). *J. Biol. Chem.* **253**, 7387-7396.

Catterall, W. A., and Morrow, C. S. (1978). *Proc. Natl. Acad. Sci. U.S.A.* **75**, 218-222.

Catterall, W. A., and Nirimberg, M. (1973). *Proc. Natl. Acad. Sci. U.S.A.* **70**, 3759-3763.

Cavanaugh, M. W. (1955). *J. Exp. Zool.* **128**, 573-589.

Chandler, W. K., and Meves, H. (1965). *J. Physiol. (London)* **180**, 788-820.

Chandler, W. K., Fitzhugh, R., and Cole, K. S. (1962). *Biophys. J.* **2**, 105-217.

Chiu, S. Y., Ritchie, J. M., Ragart, R. B., and Stagg, D. (1979). *J. Physiol. (London)* **292**, 149-166.

Clapham, D. E. (1979). Ph.D. Thesis, Emory University, Altanta, Georgia.

Clay, J. R., and DeHaan, R. L. (1979). *Biophys. J.* **28**, 169-184.

Clay, J. R., DeFelice, L. J., and DeHaan, R. L. (1979) *Biophys. J.* **28**, 377-390.

Cleeman, L., and Morad, M. (1979). *J. Physiol. (London)* **286**, 113-143.

Clusin, W. T. (1979). *Biophys. J.* **25**, 80a.

Clusin, W. T. (1980a). *Proc. Natl. Acad, Sci. U.S.A.* **77**, 679-683.

Clusin, W. T. (1980b). *In* "Drug-Induced Heart Disease" (M. Bristow, ed.). Elsevier, Amsterdam (in press).

Cohen, I., and Strichartz, G. R. (1977). *Biophys. J.* **17**, 275-279.

Cole, K. S. (1928). *J. Gen. Physiol.* **12**, 29-36.

Cole, K. S. and Baker, R. F. (1941). *J. Gen. Physiol.* **24**, 771-778.

Cole, K. S., and Curtis, H. J. (1939). *J. Gen. Physiol.* **22**, 649.

Cole, K. S., and Marmont, G. (1942). *Fed. Proc.* 11, 15-16.

Connor, J. A. (1978). *Fed. Proc.* 37, 2139-2145.

Conti, F., Hille, B., Neumcke, B., Nonner, W., and Stämpfli, R. (1976). *J. Physiol. (London)* 262, 699-727.

Coraboeuf, E., Le Douarin, G., and Obrecht, G. G. (1965). *C. R. Soc. Biol.* 159, 110-114.

Cranefield, P. F. (1975). "The Conduction of the Cardiac Impulse." Futura Press, Mt. Kisco, New York.

Davis, C. L. (1927). *Carnegie Inst. Wash. Contrib. Embryol.* 19, 245-284.

DeBarry, J., Fosset, M., and Lazdunski, M. (1977). *Biochemistry* 16, 3850-3855.

DeFelice, L. J. (1977). *Int. Rev. Neurobiol.* 29, 169-208.

DeFelice, L. J., and DeHaan, R. L. (1977). *Proc. IEEE Spec. Issue Biol. Signals* 65, 796-799.

DeHaan, R. L. (1959). *Dev. Biol.* 1, 586-602.

DeHaan, R. L. (1963). *J. Embryol. Exp. Morphol.* 11, 65-76.

DeHaan, R. L. (1965). *Ann. N. Y. Acad. Sci.* 126, 7-18.

DeHaan, R. L. (1968). *Dev. Biol. (Suppl.)* 2, 208-250.

DeHaan, R. L. (1970). *Dev. Biol.* 23, 226-240.

DeHaan, R. L. (1980). *In* "Physiology of Atrial Pacemakers and Conductive Tissues" (R. C. Little, ed.), pp. 21-53. Futura Press, Mt. Kisco, New York.

DeHaan, R. L., and DeFelice, L. J. (1978a). *In* "Theoretical Chemistry: Advances and Perspectives" (H. Eyring, ed.), pp. 181-233. Academic Press, New York.

DeHaan, R. L., and DeFelice, L. J. (1978b). *Fed. Proc.* 37, 2132-2138.

DeHaan, R. L., and Fozzard, H. A. (1975). *J. Gen. Physiol.* 65, 207-222.

DeHaan, R. L., and Gottlieb, S. H. (1968). *J. Gen. Physiol.* 52, 643-665.

DeHaan, R. L., McDonald, T. F., and Sachs, H. G. (1976). *In* "Developmental and Physiological Correlates of Cardiac Muscle" (M. Lieberman and T. Sano, eds.), pp. 155-168. Raven, New York.

DeHaan, R. L., Williams, E. H., Ypey, D. L., and Clapham, D. E. (1980). *In* "Mechanisms of Cardiac Morphogenesis and Teratogenesis" (T. Pexieder, ed.), pp. 299-316. Raven, New York.

DeMello, W. C., ed. (1972). "Electrical Phenomena in the Heart." Academic Press, New York.

DiFrancesco, D., and McNaughton, P. A. (1979). *J. Physiol.* 289, 347-373.

DiFrancesco, D., Noma, A., and Trautwein, W. (1979). *Pflugers Arch. Gesamte Physiol.* 381, 271-279.

Donatsch, P., Lowe, D. A., Richardson, B. P., and Taylor, P. (1977). *J. Physiol. (London)* 267, 357-376.

Draper, M. H., and Weidmann, S. (1951). *J. Physiol. (London)* 115, 74-94.

Dubois-Reymond, E. (1848). "Untersuchungen uber thierische Elektricität," Vol. 1. Verlag von G. Reimer, Berlin.

Dudel, J., Peper, K., Rubel, R., and Trautwin, W. (1967) *Pflugers Arch. Gesamte Physiol.* 295, 197-212.

Eccles, J. C., and Hoff, H. E. (1934). *Proc. R. Soc. London, Ser. B,* 115, 307-327.

Ehrenstein, G., and Lecar, H. (1972). *Annu. Rev. Biophys. Bioeng.* 1, 347-368.

Englemann, T. W. (1897). *Pflugers Arch. Gesamte Physiol.* 65, 109-114.

Evans, M. H. (1972). *Int. Rev. Neurobiol.* 15, 83-166.

Eyster, J. A. E., and Meek, W. J. (1914). *Heart* 5, 119-134.

Fingl, E., Woodbury, L. A., and Hecht, H. H. (1952). *J. Pharmacol. Exp. Ther.* 104, 103-114.

Fosset, M., de Barry, J., Leonoir, M.-C., and Lazdunski, M. (1977). *J. Biol. Chem.* 252, 6112-6117.

Fozzard, H. A., and Beeler, G. W. (1975). *Circ. Res.* **37**, 409-413.
Fozzard, H. A., and Hiraoka, M. (1973). *J. Physiol. (London)* **234**, 569-586.
Fricke, H. (1925). *Phys. Rev.* **26**, 678-781.
Gadsby, D. C., and Cranefield, P. F. (1977). *J. Gen. Physiol.* **70**, 725-746.
Galper, J. B., and Catterall, W. A. (1978). *Dev. Biol.* **65**, 216-227.
Galvani, L. (1791). "Commentary on the Effects of Electricity on Muscular Motion" (M. G. Foley, trans., 1953). Burndy Library, Norwalk, Connecticut.
Gaskell, W. H. (1900). *In* "Textbook of Physiology" (E. A. Schaefer, ed.), Vol. 2, pp. 169-277. Pentland, Edinburgh.
Gelles, J. M. (1977). *Circ. Res.* **41**, 94-98.
Gilly, W. F., and Armstrong, C. M. (1980). *Biophys. J.* **29**, 485-492.
Gustafson, D. E., Willsky, A. S., Wang, J. Y., Lancaster, M. C., and Trilbwasser, J. (1978). *Inst. Electron. Electr. Eng. Biomed.* **25**, 344-353.
Guttman, R., and Barnhill, R. (1970). *J. Gen. Physiol.* **55**, 104-118.
Hagiwara, S., Hayashi, H., and Takahashi, K. (1969). *J. Physiol. (London)* **205**, 115-129.
Hagiwara, S., and Jaffe, L. A. (1979). *Annu. Rev. Biophys. Bioeng.* **8**, 385-416.
Harvey, W. (1628). *In* "Classics of Cardiology" (1941) (F. A. Willius and T. E. Keys, eds.), pp. 18-79. Dover, New York.
Hermann, H., Heywood, S. M., and Marchok, A. C. (1970). *Curr. Top. Dev. Biol.* **5**, 181-234.
Heyer, C. B., and Lux, H. D. (1976). *J. Physiol.* **262**, 319-348.
Hille, B. (1973). *J. Gen. Physiol.* **61**, 669-686.
Hille, B., and Schwarz, W. (1978). *J. Gen. Physiol.* **72**, 404-442.
Hodgkin, A. L., and Huxley, A. F. (1939). *Nature (London)* **147**, 710.
Hodgkin, A. L., and Huxley, A. F. (1952). *J. Physiol. (London)* **117**, 500-544.
Hodgkin, A. L., and Katz, B. (1949). *J. Physiol. (London)* **108**, 37-77.
Hodgkin, A. L., and Rushton, W. A. H. (1946). *Proc. R. Soc. London, Ser. B* **133**, 444-479.
Hoffman, B. F., and Cranefield, P. F. (1960). "Electrophysiology of the Heart." McGraw-Hill, New York.
Hondeghem, L. M. (1978). *Biophys. J.* **23**, 147-152.
Horres, C. R., Aiton, J. F., and Lieberman, M. (1979). *Am. J. Physiol. Cell Physiol.* **5**, C163-C170.
Huxley, A. F., and Stämpfli, R. (1951). *J. Physiol. (London)* **112**, 496-608.
Iijima, T., and Pappano, A. J. (1979). *Circ. Res.* **44**, 358-367.
Irisawa, H. (1978). *Physiol. Rev.* **58**, 461-498.
Isenberg, G. (1975). *Nature (London)* **253**, 273-274.
Isenberg, G. (1976). *Pflugers Arch. Gesamte Physiol.* **365**, 99-106.
Isenberg, G. (1977a). *Eur. J. Physiol.* **371**, 51-59.
Isenberg, G. (1977b). *Eur. J. Physiol.* **371**, 77-85.
Ishima, Y. (1968). *Proc. Jpn. Acad.* **44**, 170-175.
Iversen, J. G. (1976). *J. Cell. Physiol.* **89**, 367-376.
Jack, J. J. B., Noble, D., and Tsien, R. W. (1975). "Electric Current Flow in Excitable Cells." Oxford University Press, Oxford.
Jacques, Y., Fosset, M., and Lazdunski, M. (1978). *J. Biol. Chem.* **253**, 7383-7392.
Johnson, E. A., and Lieberman, M. (1971). *Annu. Rev. Physiol.* **33**, 479-532.
Johnstone, P. N. (1925). *Bull. Johns Hopkins Hosp.* **36**, 299-311.
Kano, M. (1975). *J. Cell Physiol.* **86**, 503-510.
Kano, M., and Shimada, Y. (1973). *J. Cell. Physiol.* **81**, 85-90.
Kano, M., and Yamamoto, M. (1977). *J. Cell. Physiol.* **90**, 439-444.

Kass, R. S., Lederer, W. J., Tsien, R. W., and Weingart, R. (1978). *J. Physiol. (London)* **281,** 187–208.

Kass, R. S., Siegelbaum, S. A., and Tsien, R. W. (1979). *J. Physiol. (London)* **290,** 201–226.

Katz, B., and Miledi, R. (1969). *J. Physiol. (London)* **203,** 459–487.

Keith, A., and Flack, M. (1906/1907). *J. Anat. Physiol.* **41,** 172–189.

Kensler, R. W., Brink, P., and Dewey, M. M. (1977). *J. Cell Biol.* **73,** 768–782.

Kenyon, J. L., and Gibbons, W. R. (1979a). *J. Gen. Physiol.* **73,** 118–138.

Kenyon, J. L., and Gibbons, W. R. (1979b). *J. Gen. Physiol.* **73,** 139–157.

Keynes, R. D., and Rojas, E. (1974). *J. Physiol. (London)* **239,** 393–434.

Keynes, R. D., Rojas, E., Taylor, R. E., and Vergara, J. (1973). *J. Physiol. (London)* **229,** 409–455.

Kidokoro, Y. (1975a). *J. Physiol. (London)* **244,** 129–143.

Kidokoro, Y. (1975b). *J. Physiol. (London)* **244,** 144–159.

Kimhi, Y., Palfrey, C., Spector, I., Barak, Y., and Littauer, U. Z. (1976). *Proc. Natl. Acad. Sci. U.S.A.* **73,** 462–466.

Lederer, W. J., and Tsien, R. W. (1976). *J. Physiol. (London)* **263,** 73–100.

Le Douarin, G., Obrecht, G. G., and Coraboeuf, E. (1966). *J. Exp. Embryol. Morphol.* **15,** 153–167.

Le Douarin, G., Renaud, J.-F., Renaud, D., and Coraboeuf, E. (1974). *J. Mol. Cell. Cardiol.* **6,** 523–529.

Lee, K. S., Weeks, T. A., Kao, R. L., Akaike, N., and Brown, A. M. (1979). *Nature (London)* **278,** 269–271.

Lewis, T., Oppenheimer, B. S., and Oppenheimer, A. (1910). *Heart* **2,** 147–169.

Lieberman, M., and Paes de Carvalho, A. (1965). *J. Gen. Physiol.* **49,** 351–363.

Lieberman, M., Sawanabori, T., Shigeto, N., and Johnson, E. A. (1976). *In* "Developmental and Physiological Correlates of Cardiac Muscle" (M. Lieberman and T. Sano, eds.), pp. 139–154. Raven, New York.

Ling, G., and Gerard, R. W. (1949) *J. Cell. Comp. Physiol.* **34,** 383–396.

Lipsius, S. L., and Vassale, M. (1978). *In* "The Sinus Node. Structure, Function and Clinical Relevance" (F. I. M. Bonke, ed.), pp. 233–244. Martinus Nijhoff, The Hague.

Lompre, A. M., Poggioli, J., and Vassort, G. (1979). *J. Mol. Cell. Cardiol.* **11,** 813–825.

McAllister, R. E., Noble, D., and Tsien, R. W. (1975). *J. Physiol. (London)* **251,** 1–59.

McDonald, T. F., and DeHaan, R. L. (1973). *J. Gen. Physiol.* **61,** 89–109.

McDonald, T. F., Sachs, H. G., and DeHaan, R. L. (1972). *Science* **176,** 1248–1250.

McDonald, T. F., Sachs, H. G., and DeHaan, R. L. (1973). *J. Gen. Physiol.* **62,** 286–302.

McLean, M. J., Renaud, J. F., Sperelakis, N., and Niu, M. C. (1976) *Science* **191,** 297–299.

McLean, M. J., and Sperelakis, N. (1974). *Exp. Cell Res.* **86,** 351–364.

McNutt, N. S., and Weinstein, R. S. (1973). *Prog. Biophys. Mol. Biol.* **26,** 45–101.

Manasek, F. J., and Monroe, R. G. (1972). *Dev. Biol.* **27,** 584–588.

Masson-Pevet, M., Bleeker, W. K., Mackaay, A. J. C., Gros, D., and Bouman, L. N. (1978). *In* "The Sinus Node. Structure, Function and Clinical Relevance" (F. I. M. Bonke, ed.), pp. 195–211. Martinus Nijhoff, The Hague.

Matsumoto, G., and Sakai, H. (1979a). *J. Membr. Biol.* **50,** 1–14.

Matsumoto, G., and Sakai, H. (1979b). *J. Membr. Biol.* **50,** 15–22.

Mauro, A., Conti, F., Dodge, F., and Schor, R. (1970). *J. Gen. Physiol.* **55,** 497–523.

Meda, E., and Ferroni, A. (1959). *Experientia* **15,** 427–428.

Meech, R. W. (1976). *Symp. Soc. Exp. Biol.* **30,** 161–191.

Meech, R. W., and Standen, N. B. (1975). *J. Physiol. (London)* **249,** 211–239.

Messenger, E. A., and Warner, A. E. (1979). *J. Physiol. (London)* **292**, 85-105.
Miller, C., and Rosenberg, R. L. (1979). *J. Gen. Physiol.* **74**, 457-478.
Miyake, M. (1978). *Brain Res.* **143**, 349-354.
Miyazaki, S., Takahashi, K., and Tsuda, K. (1972). *Science* **176**, 1441-1443.
Miyazaki, S., Takahashi, K., and Tsuda, K. (1974). *J. Physiol. (London)* **238**, 37-54.
Moolenaar, W. H., and Spector, I. (1977). *Science* **196**, 331-333.
Moolenaar, W. H., and Spector, I. (1978). *J. Physiol. (London)* **278**, 265-286.
Moolenaar, W. H., and Spector, I. (1979a). *J. Physiol. (London)* **292**, 297-302.
Moolenaar, W. H., and Spector, I. (1979b). *J. Physiol. (London)* **292**, 307-323.
Mullins, L. J. (1979). *Am. J. Physiol. Cell Physiol.* **5**, C103-C110.
Narahashi, T. (1963). *Adv. Insect. Physiol.* **1**, 176-256.
Narahashi, T. (1974). *Physiol. Rev.* **54**, 813-889.
Nastuk, W. L., and Hodgkin, A. L. (1950). *J. Cell. Comp. Physiol.* **35**, 39.
Nathan, R. D., and DeHaan, R. L. (1978). *Proc. Nat. Acad. Sci. U.S.A.* **75**, 2776-2780.
Nathan, R. D., and DeHaan, R. L. (1979). *J. Gen. Physiol.* **73**, 175-198.
Nathan, R. D., Pooler, J. D., and DeHaan, R. L. (1976). *J. Gen. Physiol.* **67**, 27-44.
Neher, E., and Stevens, C. F. (1977). *Annu. Rev. Biophys. Bioeng.* **6**, 345-382.
Noble, D., and Tsien, R. W. (1968). *J. Physiol. (London)* **195**, 185-214.
Noble, D., and Tsien, R. W. (1969). *J. Physiol. (London)* **200**, 205-231.
Noma, A., and Irisawa, H. (1974). *Jpn. J. Physiol.* **24**, 617-632.
Noma, A., and Irisawa, H. (1975). *Jpn. J. Physiol.* **25**, 287-302.
Noma, A., and Irisawa, H. (1976). *Pflugers Arch. Gesamte Physiol.* **374**, 45-52.
Noma, A., Yanagihara, K., and Irisawa, H. (1978). *In* "The Sinus Node. Structure, Function and Clinical Relevance" (F. I. M. Bonke, ed.), pp. 301-310. Martinus Nijhoff, The Hague.
Nonner, W., Rojas, E., and Stämpfli, R. (1975). *Phil. Trans. R. Soc. London Biol. Sci.* **270**, 483-492.
Okada, Y., Wakoh, T., and Inouye, A. (1979). *J. Membr. Biol.* **47**, 357-376.
Okamota, H., Takahashi, K., and Yamashita, N. (1977). *J. Physiol. (London)* **267**, 465-495.
Paes de Carvalho, A., Hoffman, B. F., and Langan, W. B. (1966). *Nature (London)* **211**, 938-940.
Pappano, A. J. (1972). *Circ. Res.* **31**, 379-388.
Pappano, A. J. (1976). *Circ. Res.* **39**, 99-105.
Patten, B. M., and Kramer, T. C. (1933). *Am. J. Anat.* **53**, 349-375.
Ramon, F., Anderson, N., Joyner, R. W., and Moore, J. W. (1975). *Biophys. J.* **15**, 55-69.
Renaud, J. F., and Sperelakis, N. (1976). *J. Mol. Cell. Cardiol.* **8**, 889-900.
Reuter, H. (1977). *Ann. Rev. Physiol.* **41**, 413-424.
Reuter, H., and Scholz, H. (1977). *J. Physiol. (London)* **264**, 49-62.
Rojas, E., and Rudy, B. (1976). *J. Physiol. (London)* **263**, 501-531.
Romey, G., Abita, J. P., Schweitz, H., Wunderer, G., and Lazdunski, M. (1976) *Proc. Natl. Acad. Sci. U.S.A.* **73**, 4055-4059.
Romey, G., Chicheportiche, R., Lazdunski, M., Rochat, M., Miranda, F., and Lissitsky, S. (1975). *Biochem. Biophys. Res. Commun.* **64**, 115-121.
Romey, G., Jacques, Y., Schweitz, H., Fosset, M., and Lazdunski, M. (1979). *Biochim. Biophys. Acta* **556**, 344-353.
Romey, G., Renaud, J. F., Fosset, M., and Lazdunski, M. (1980). *J. Pharmacol. Exp. Ther.* **213**, 607-615.
Sachs, H. G., McDonald, T. F., and DeHaan, R. L. (1973). *J. Cell Biol.* **56**, 255-258.
Schoenberg, M., and Fozzard, H. A. (1979). *Biophys. J.* **25**, 217-234.

Sevcik, C. T., and Narahashi, T. (1975). *J. Membr. Biol.* **24**, 329–339.

Seyama, I. (1978). *In* "The Sinus Node. Structure, Function and Clinical Relevance" (F. I. M. Bonke, ed.), pp. 339–347. Martinus Nijhoff, The Hague.

Seyama, I. (1979). *J. Physiol. (London)* **294**, 447–460.

Shigenobu, K., Schneider, J. A., and Sperelakis, N. (1974). *J. Pharmacol. Exp. Ther.* **190**, 280–288.

Shigenobu, K., and Sperelakis, N. (1971). *J. Mol. Cell. Cardiol.* **3**, 271–286.

Shigenobu, K., and Sperelakis, N. (1974). *Dev. Biol.* **39**, 326–330.

Shrager, P. (1975). *Ann. N. Y. Acad. Sci.* **264**, 293–303.

Shrier, A., and Clay, J. R. (1980). *Nature (London)* **283**, 670–671.

Shrier, A., and DeHaan, R. L. (1978). *Biophys. J.* **21**, 63a.

Shrier, A., Clay, J. R., and DeHaan, R. L. (1979). *Biophys. J.* **25**, 299a.

Slack, C., and Warner, A. E. (1975). *J. Physiol. (London)* **248**, 97–120.

Sommer, J. R., and Johnson, E. A. (1979). *In* "Handbook of Physiology" (R. M. Berne, ed.), Sec. 2, Vol. 1, pp. 113–186. American Physiological Society, New York.

Sperelakis, N. (1972). *In* "Electrical Phenomena in the Heart" (W. C. DeMello, ed.), pp. 1–62. Academic Press, New York.

Sperelakis, N. (1979). *In* "Handbook of Physiology" (R. M. Berne, ed.), Sec. 2, Vol. 1, pp. 187–267. American Physiological Society, New York.

Sperelakis, N., and Shigenobu, K. (1972). *J. Gen. Physiol.* **60**, 430–453.

Sperelakis, N., and Shigenobu, K. (1974). *J. Mol. Cell. Cardiol.* **6**, 449–471.

Sperelakis, N., Shigenobu, K., and McLean, M. J. (1976). *In* "Developmental and Physiological Correlates of Cardiac Muscle" (M. Lieberman and T. Sano, eds.), pp. 209–234. Raven, New York.

Sperelakis, N., and Pappano, A. J. (1969). *J. Gen. Physiol.* **53**, 97–114.

Spitzer, N. (1979). *Annu. Rev. Neurosci.* **2**, 1–83.

Spitzer, N., and Baccaglini, P. I. (1976). *Brain Res.* **107**, 610–616.

Spitzer, N., and Lamborghini, J. E. (1976). *Proc. Natl. Acad. Sci. U.S.A.* **73**, 1641–1645.

Spitzer, N., and Spitzer, J. L. (1975). *Am. Zool.* **15**, 781.

Strichartz, G., and Cohen, I. (1978). *Biophys. J.* **23**, 153–156.

Takahashi, K., Miyazaki, S., and Kodokoro, Y. (1971). *Science* **171**, 415–418.

Thompson, S. H. (1977). *J. Physiol. (London)*. **265**, 465–488.

Trautwein, W. (1973). *Physiol. Rev.* **53**, 793–835.

Trautwein, W., and McDonald, T. F. (1978). *J. Mol. Cell. Cardiol.* **10**, 387–394.

Trautwein, W., and Zink, K. (1952). *Pflugers Arch. Gesamte Physiol.* **265**, 68–84.

Ulbricht, W. (1969). *Ergeb. Physiol. Biol. Chem. Exp. Pharm.* **61**, 18–71.

Ulbricht, W. (1977). *Annu. Rev. Biophys. Bioeng.* **6**, 7–32.

Van der Pol, B., and Van der Mark, J. (1928). *Phil. Mag.* **6**, 763–775.

Van Mierop, L. H. S. (1967). *Am. J. Physiol.* **212**, 407–415.

Vassalle, M. (1966). *Am. J. Physiol.* **210**, 1335–1341.

Vassalle, M. (1979). *Annu. Rev. Physiol.* **41**, 425–440.

Vincent, J. P., Balerna, B., Barhanin, J., Fosset, M., and Lazdunski, M. (1980). *Proc. Natl. Acad. Sci. U.S.A.* **77**, 1646–1650.

Walton, M., and Fozzard, H. A. (1979). *Biophys. J.* **25**, 407–420.

Warner, A. E. (1973). *J. Physiol. (London)* **235**, 267–286.

Weidmann, S. (1955). *J. Physiol. (London)* **127**, 213–224.

Weidmann, S., and Coraboeuf, E. (1949). *C. R. Soc. Biol. (Paris)* **143**, 1329–1331.

West, T. C. (1955). *J. Pharmacol. Exp. Ther.* **115**, 283–290.

Woodbury, L. A., Hecht, H. H., and Christopherson, A. R. (1951). *Am. J. Physiol.* **164**, 307–318.

Wybauw, R. (1910). *Arch. Int. Physiol.* **10**, 78–89.

Ypey, D. L., Clapham, D. E., and DeHaan, R. L. (1980). *J. Membr. Biol.* **51**, 75–79.

CHAPTER 5

REGULATION OF THE ELONGATING NERVE FIBER

Randal N. Johnston and Norman K. Wessells

DEPARTMENT OF BIOLOGICAL SCIENCES
STANFORD UNIVERSITY
STANFORD, CALIFORNIA

I. Introduction

In 1907, Harrison first described the extension of nerve fibers by fragments of frog nervous tissue *in vitro*. This observation confirmed the proposals of His and Cajal that nerve fibers are formed as outgrowths of single cells and rendered untenable the theories of Hensen and Held of the syncytial nature of these fibers. In addition, it represented the first study *in vitro* of the motile tip of the nerve fiber, the ameboid "growth cone," that Cajal had earlier described from histological sections of chicken embyros (1890). The extraordinary behavior of this organelle led Harrison to suggest the existence of two main types of factors, acting primarily on the growth cone, that are of fundamental importance to the development of precise pathways of

165

CURRENT TOPICS IN
DEVELOPMENTAL BIOLOGY, Vol. 16

connection in the nervous system. These were, first, "forces immanent in the neuroblast itself" and, second, the ability of the growth cone to sense and respond to such features of its environment as "paths of low mechanical resistance" or "factors of a chemotactic nature" (Harrison, 1910). Since then, the properties of nerve fibers and growth cones have been extensively studied both *in vivo* and *in vitro*. In this chapter, we will address the general question of how growing nerve fibers may be governed by both intracellular properties and extracellular cues and how the activities of these fibers may be important to the development of the nervous system.

II. Morphology of Nerve Fibers and Growth Cones

A. MAJOR FEATURES

When fragments of nervous tissue are dissociated into suspensions of single cells and then cultured under appropriate conditions (Fischbach and Nelson, 1977), one can initially observe cells of different types. Nonneuronal cells, which may include several varieties of supportive cells, generally assume a fibroblastic or epithelial morphology. Freshly dissociated neurons, however, are distinctly different in morphology, appearing smooth and round or ovoid in contour, with a large, often eccentrically placed nucleus and perhaps one or more stubs of nerve fibers broken during the process of dissociation. [Although neuronal cells *in vivo* generally have morphologically distinguishable dendrites and axons, it is difficult to identify them unambiguously when cells are cultured *in vitro* (although, see Landis, 1977). Thus, except when identification is certain, we refer to these cellular extensions collectively as nerve fibers or neurites.]

After periods of one to several days in culture, neurons can extend neurites long distances over the culture substratum, as much as a millimeter or more, and these neurites may branch extensively. At the tip of each actively extending neurite may be observed a growth cone. Typically, a growth cone appears as a fan-shaped or leaf-shaped expansion of the neurite ending (Fig. 1, inset), from which extend filopodia (long, thin cellular projections, also called microspikes, approximately 0.1–0.2 μm in diameter) and lamellipodia (broad, thin expanses of cell surface, also called veils, which may extend independently or in association with existing filopodia). Time-lapse films have demonstrated that single growth cones may extend both types of projections, simultaneously or at different times (Ludueña and Wessells, 1973), and that these projections may play an active role in the elongation of neurites. Although filopodia and lamellipodia are usually transient structures that may form, wave about, and retract within a few minutes, they occasionally persist after appearing to attach to the substratum, and some filopodia may then enlarge by spreading laterally. During elongation of the neurite, such long-lived sites of attachment may then become new zones of motile growth cone activity, while previously active parts of the growth cone apparently become quiescent, round up, and contribute to the lengthening of the neurite.

B. ULTRASTRUCTURE

The ultrastructure of filamentous organelles in the neurite and growth cone has been extensively studied (Fig. 1; for a review, see Palay and Chan-Palay, 1977). Major structural features of the neurite include microtubules (24–28 nm in diameter) and neurofilaments (9–11 nm in diameter). These are longitudinally arranged and exhibit some tendency to cluster, with neurofilaments often lying centrally (Tennyson, 1970; Yamada et al., 1971). Occasionally, periodic fibrous projections from microtubules may be identified (Fernandez et al., 1971). These projections may well correspond to the microtubule-associated proteins (MAPs; Binder and Rosenbaum, 1978) that have recently been isolated. These proteins are hypothesized to promote lateral interactions of microtubules with each other and with other organelles and may permit transport of vesicles (see below) or mitochondria, which often appear in close association with the microtubules (see Haimo et al., 1979; Kim et al., 1979). The distribution of microfilaments (4–6 nm in diameter) within the neurite has also been described (Tennyson, 1970; Yamada et al., 1971; Ludueña and Wessells, 1973). After standard fixation (aldehydes and osmium) and sectioning for transmission electron microscopy, microfilaments interpreted to be arrayed as a meshwork can be seen. This meshwork is most prominent just beneath the plasma membrane and may serve as a site for fixation-induced agglutination of the "microtrabecular lattice" (see Heuser and Kirschner, 1980) seen by electron microscopy of whole mounts of neurons (Nuttall and Wessells, 1979) and of many other types of cells (Buckley and Raju, 1976; Wolosewick and Porter, 1979; Porter et al., 1979).

In the growth cone, the distribution of these filamentous organelles differs from that in the neurite. Microtubules and neurofilaments splay out within the cone and may curve and intertwine in a complex fashion (Fig. 1). Though these will occasionally extend a short distance into a filopodium or lamellipodium, in general they are

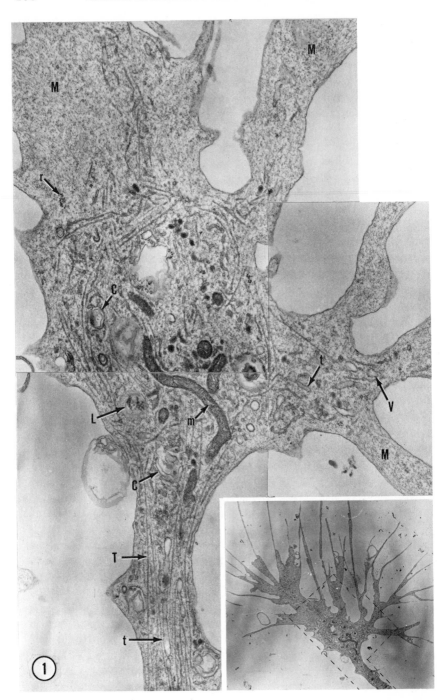

restricted to the central portion of the growth cone (Bunge, 1973; Ludueña and Wessells, 1973). Microfilaments, however, are major components within filopodia and lamellipodia and are also found beneath the plasma membrane of the leading active edge of the growth cone (Tennyson, 1970; Yamada et al., 1971; Hinds and Hinds, 1972; Bunge, 1973; Gray, 1973; Kuczmarski and Rosenbaum, 1979). Although in a given section the microfilaments usually appear as a meshwork or "cytonet" (Gray, 1973), in favorable sections parallel arrays of filaments are often preserved which may extend the length of filopodia and project well into the growth cone. These observations, and the similar organization of filaments seen in whole mounts of critical point-dried growth cones (Fig. 2; also, see Kuczmarski and Rosenbaum, 1979), suggest that microfilaments are a major component of the submembranous cytoplasm, especially in regions of great surface motility. At least with neuroblastoma cells, decoration with heavy meromyosin indicates a uniform orientation of the parallel filopodial microfilaments, as the meromyosin "arrowheads" all point toward the growth cone (Isenberg and Small, 1978; this implies that the cell surface may be equivalent to the Z band in skeletal muscle cells; the cell surface may represent a site from which actin takes origin and upon which tension may be exerted, or toward which "flow" may be directed). Although the parallel arrays of microfilaments in growth cones are reminiscent of actin cables seen in many other cell types (Kuczmarski and Rosenbaum, 1979), the cables of such cells usually exhibit a much greater density of microfilament packing, and the microfilaments within a single cable can even run in opposite directions (Begg et al., 1978). Whether the presence of parallel microfilamentous arrays in neurons reflects filopodial extension, retraction, attachment, or some other behavior is unknown.

Also unknown is the degree to which the filamentous systems of neurites and growth cones may be altered by fixation and dehydration procedures. Paula-Barbosa and Gray (1974), Gilbert et al. (1975) Kuczmarski and Rosenbaum (1976), Buckley and Raju (1976), and

---

FIG. 1. A thin section through the growth cone of a sensory ganglion cell. The microtubules (T) and neurofilaments of the neurite (below) splay out through the central portion of the cone and only occasionally extend into the base of lamellipodia or filopodia. The large size of the latter projections may be appreciated by comparison with the inset (lower right), where the main figure is outlined. The microfilamentous material (M) occupies most of the volume of the motile projections from the cone surface. Agranular vesicles (V), tubular membranes (t), cup-shaped bodies (C), mitochondria (m), occasional ribosomal clusters (r), and presumed lysosomes (L), are seen. Courtesy of Elsevier and Academic Press. [From "Locomotion of Tissue Cells." Ciba Symp. 14 (new series) (1973) and Dev. Biol. 30, 427–440, 1973.]

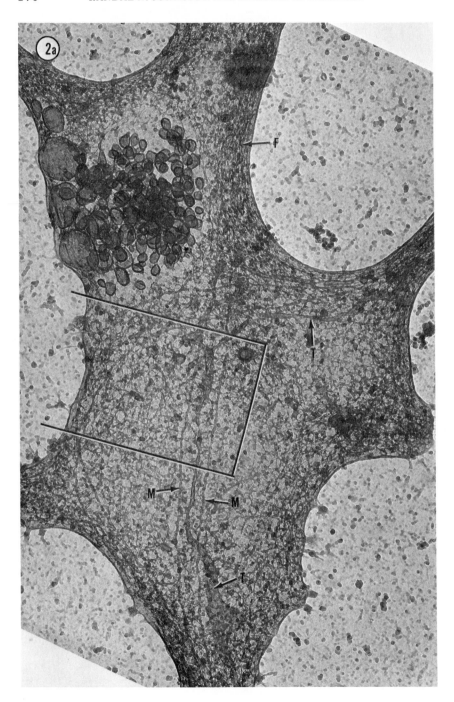

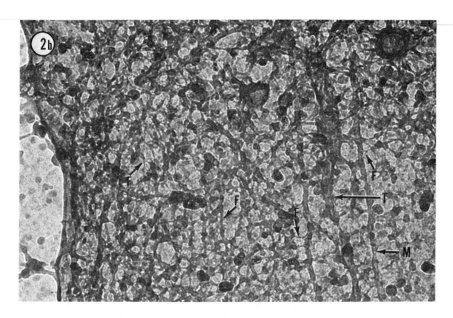

Fɪɢ. 2. A growth cone of a ciliary ganglionic neuron after fixation and critical-point drying, viewed as a whole mount with transmission electron microscopy [the distal part of the neurite is shown at the bottom of (a)]. The "ground cytoplasm" appears as a complex intertwining of various-sized filaments (F) and is shown at higher magnification in (b) [the area of which is outlined in (a)]. Microtubules (M) are present, and in this growth cone are associated with an elongate membranous tubular system (T) which is seen to branch just below the top of (b) and to extend toward the base of the filopodium to the right. A parallel alignment of filaments is seen in filopodia and just beneath the plasma membrane near filopodia [as at the top of (a)]. In (b) note the wide variation in diameter of the filamentous material and apparent connections of the space-filling lattice with the microtubules. Note also the aggregation of agranular vesicles at the upper left of (a); the lattice materials tend to be absent from the regions around such vesicles (see Nuttall and Wessells, 1979). When viewed with scanning microscopy, the cell surface over such aggregations may bulge upward as a "mound." (a) × 25,000. (b) × 55,000. (Courtesy of R. P. Nuttall.)

Maupin-Szamier and Pollard (1978) have all provided evidence that some alterations in filaments may occur during processing. A complete explanation of growth cone and neurite structure and function will not be achieved until we better understand how their various filamentous organelles may interact with each other and with membranous systems, how the assembly and disassembly of filaments are controlled, and how the diverse associated proteins may contribute to these events.

A variety of membranous components occurs in neurites and growth cones. In many types of preparations, one can identify as a major component elements of the smooth endoplasmic reticulum (Tennyson, 1970; Hinds and Hinds, 1972), often arrayed as a branched system which may in places appear continuous with the plasma membrane (Bunge, 1973). In addition, agranular electron-lucent vesicles of varying size may be present singly, in linear arrays associated with microtubules or with the long axis of filopodia, or as large accumulations, some of which protrude outward from the surface of fixed cells as "mounds" (Bunge, 1973, 1977; Spooner et al., 1974). Other vesicles have electron-dense cores and, in cells from sympathetic ganglia, adrenergic staining properties (Landis, 1978). Some membranous tubules may also have electron-dense contents. Shortly after the addition of a marker such as horseradish peroxidase or Thorotrast (Bunge, 1977; Wessells et al., 1974) to the extracellular medium, the marker can be seen within some of the membranous tubular systems and in many of the so-called cup-shaped bodies. Later, such markers are frequently seen within multivesiculate bodies or structures resembling lipofucsin granules (Wessells et al., 1974; Tsukita and Ishikawa, 1980). Such vesicles of the lysosomal series are common along newly elongated neurites of cultured cells. From their size and distribution near microtubules and neurofilament bundles, they may be among the particles that appear opaque with phase-contrast microscopy and that can be seen in time-lapse films to move distally or proximally in neurites (Nakai, 1956; Leestma and Freeman, 1977). Finally, phase-lucent vesicles, presumably pinocytotic vesicles (Hughes, 1953), may be observed in some living motile growth cones.

Intramembranous particles observed in growth cone and neurite plasmalemmas (Pfenninger and Bunge, 1974) constitute another feature of the neuronal surface. Noteworthy is the finding that the membranes of "young" nerve fibers and growth cones have a remarkably low density of particles ($75–100/\mu m^2$) when compared with "older" nerve fibers or glial cells ($400–700/\mu m^2$).

Though one might expect to find differences in the overall morphologies of neurites and growth cones in vivo and in vitro (for instance, because of the special chemical environment in vitro or the restriction of neurite elongation to two dimensions in vitro as opposed to three dimensions in vivo), there is nevertheless a good correspondence between morphologies under the two conditions (Harrison, 1910; Tennyson, 1970; Pfenninger and Bunge, 1974). In particular, growth cones reconstructed from sections of retinal ganglion cell fibers (Fig. 3; Hinds and Hinds, 1974) and of olfactory bulb dendrites (Hinds and Hinds, 1972) bear a strong resemblance to growth cones of nerve fibers that have extended in plasma clots (Hughes, 1953) or in collagen matrices (Fig. 7c; Ebendal, 1976). Similarly, the growth cones and axons of Rohon-Beard cells, growing just beneath

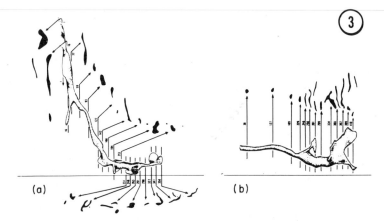

FIG. 3. Reconstructions of optic nerve growth cones from intact mouse embryos. The profiles of the individual sections through the neurites and growth cones are indicated in black for each of the numbered positions. The authors describe the various extensions as "sheetlike or threadlike" filopodia. Growth cones *in vitro* tend to resemble these when grown on nonadhesive substrata. Courtesy of J. W. Hinds and P. L. Hinds, and Academic Press. (From *Dev. Biol.* **37**, 381–416, 1974.)

larval frog epidermis (Fig. 4; Roberts, 1976), resemble growth cones and neurites of similar amphibian neurons (Spitzer and Lamborghini, 1976) or of chicken embryonic neurons (Fig. 8) *in vitro*. On the basis of these comparisons, and on other comparisons that will be presented below, it appears that the properties of growth cones and neurites *in vitro* resemble, in large part, the properties of equivalent cellular structures *in vivo*.

III. Behavior of Neurites and Growth Cones

A. PRIMARY INITIATION VERSUS REGENERATION OF NEURITES

We now address the questions of how neurites and growth cones are generated and how they function in time. In most experimental studies, it has proved simplest to examine the reextension of neurites *in vitro* by cells that had already developed neurites *in vivo,* prior to removal from the embryo. It is not yet known, however, whether this "regeneration" of neurites accurately resembles the *primary* initiation by recently postmitotic neurons *in vivo*. Typically a large, round neuron *in vitro* may give rise to several neurites originating from seemingly random sites on the cell circumference (Collins, 1978a;

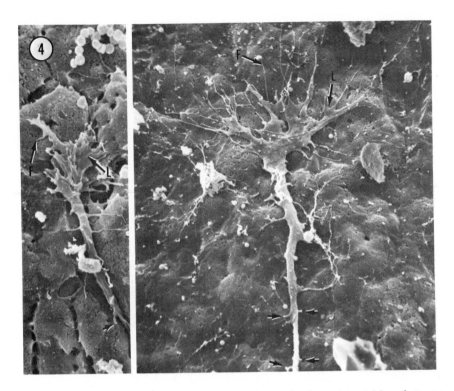

FIG. 4. Growth cones in *Xenopus laevis* embryos that have emerged from between somite cells (left) and moved along the inner surface of the epidermal basal lamina (right). All such growth cones extend filopodia (F) and about half have lamellipodia (L). The structure of these growth cones is remarkably like that of avian or mammalian cones *in vitro*. Note the occasional "branchlets" (arrows; Roberts, 1976) along the neurites, which are interpreted to serve as points of stabilization. Time-lapse films have shown that similar structures can form as retraction fibers *in vitro* (Letourneau, unpublished); whether they can also arise by extension of the cell surface, as do filopodia, is not yet certain. Courtesy of A. Roberts and Elsevier. (From *Brain Res.* 118, 526–530, 1976.)

Bray, 1973a). In at least some cases, however, neurons *in vivo* that are interpreted to be "young" typically appear spindle-shaped, with two extensions that may run in predictable directions (Tennyson, 1965; Hinds and Hinds, 1974). Under these two conditions, it is not clear to what extent the number of initiation sites or their placement on the cell surface reflects properties of the cell or of the extracellular environment. One approach to this question would be to study the behavior of very young nerve cells *in vitro*. Indeed, conditions of culture have recently been described that permit neural crest cells cultured in aggregates to differentiate certain neuronal properties (Greenberg and Schrier, 1977; Cohen, 1977). However, an analysis of the primary differentiation of single cells under such conditions is not yet available. At present, therefore, we do not know whether there is any fundamental difference in mechanism between the true primary initiation and the regeneration of neurites.

B. GROWTH OF NEURITES

In the absence of a system for studying primary initiation, the analysis of neurite regeneration has offered a more manageable (though perhaps imperfect) model for the extension of nerve fibers *in vivo*. The early stages of neurite regeneration have recently been carefully described (Collins, 1978a). When neurons from the ciliary ganglion are cultured on a substratum to which they are highly adhesive (see below), they will attach readily and extend numerous active filopodia from the entire cell circumference (Fig. 5). Some of these filopodia appear to attach to the substratum and fail to retract as swiftly as others. These adherent filopodia may then serve as sites where broad, flattened lamellae, the incipient growth cones, develop. As these localized motile regions grow out from the cell body, their proximal portions form the neurites by rounding up and becoming relatively quiescent. In regions between the newly forming growth cones, cell surface motility of the neuronal soma also becomes reduced. Interestingly, this paralysis of motility occurs at different times for different parts of the cell circumference, indicating that the event leading to the localization of cell surface activity to the growth cones is itself localized and not a global cellular process (Wessells *et al.*, 1978). It is also of interest that at no time during this process of neurite regeneration do the morphology and motile behavior of a dissociated and cultured neuron revert to those typical of neuronal precursor cells [for example, to the quasi-fibroblastic shape and locomotory pattern typical of neural crest cells *in vivo* or *in vitro* (Tosney, 1978; Cohen and Konigsberg, 1975)]. Apparently, neuronal morphology and pattern of motility are stable characteristics, of the type usually associated with the determined condition. How the behaviors and morphologies of neuronal precursors are changed during differentiation, perhaps irreversibly, is unknown.

The rate of elongation of neurites *in vitro* varies considerably and can typically be 50 μm/hour (Hughes, 1953; Bray, 1973a) to 100 μm/hour (Ludueña, 1973a), comparable to the average rate of about 40 μm/hour measured by Speidel (1941) for axons in the tail fin of amphib-

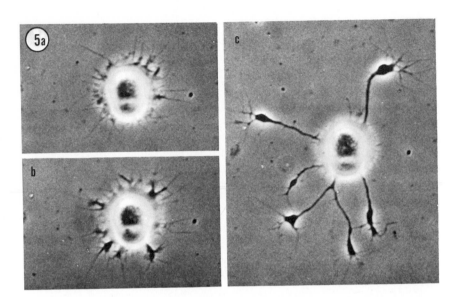

FIG. 5. A neuron from the ciliary ganglion of a chicken embryo showing the process of growth cone initiation. Note the many filopodia and the flattened regions around the periphery in (a), which was photographed 25 minutes after the conditioned medium that triggers initiation (Collins, 1978a) was added to the dish. Seven minutes later (b), localized foci of motility are apparent, and intervening regions of the periphery are becoming quiescent. After 24 minutes more (c), discrete growth cones have moved outward, and the surface of the soma is no longer actively moving. Courtesy of F. Collins and Academic Press. (From *Dev. Biol.* 65, 50–57, 1978.)

ian tadpoles. Neurite extension, however, is not necessarily a steady process, but instead can include periods of relative quiescence in which the growth cones fail to advance, even though filopodial and lamellipodial activity may be retained, or in which the neurites may even partly retract; later, extension can be resumed in the same or in another direction (Hughes, 1953; Ludueña, 1973a).

The overall pattern of neurite extension can, to a certain extent, vary with the adhesiveness of the substratum (Letourneau, 1975a). For example, when neurons are cultured on glass coverslips in medium containing serum, the few neurites that are extended adhere poorly, and long segments of neurite can even detach from the glass. Under these conditions, neurites tend to stretch linearly between the more adhesive cell somata and growth cones and are swept in arcs as the growth cone wanders over the substratum. If, however, neurons are cultured under conditions of greater adhesion (for example, in medium without serum; Ludueña, 1973b), the often sinuous contours of the resulting tightly adherent neurites then describe the path taken by the growth cones during their migration. In addition, each neuron will extend relatively many neurites, these appearing flattened and having many fine lateral projections and branching sites.

Branching of neurites *in vitro* can occur through the bifurcation of an active growth cone, or through the development of collateral fibers. In the latter case, regions of neurites without apparent motility, but which may exhibit local flattening or other evidence of adhesion, can on rare occasions regain motility and form a new growth cone that can extend independently of the original (Nakai, 1956; Bray, 1973a). Collateral production can also be observed in single nerve fibers in the tadpole tail (Speidel, 1941). More frequently, branching of a neurite follows the bifurcation of its growth cone (Bray, 1973a; Wessells and Nuttall, 1978). A growth cone that is about to split will first broaden. Then, a leading edge will become quiescent, while motility is retained on either side of it. Finally, the quiescent region will retract slightly as the sides round up, forming a Y-shaped or T-shaped branch point.

Not yet well established is the extent to which different types of neurons, cultured under identical conditions in isolation from other cells, yield different branching patterns. Such an analysis may be one way to evaluate the degree of genetic versus environmental regulation of neuronal morphology. A different approach to this issue would be to examine the patterns of axonal branching by sister neurons. Such an analysis has been attempted by Solomon (1979), who followed the behaviors of recently divided sister neuroblastoma cells when these were stimulated to differentiate by deprivation of serum. The processes extended by these cells, in up to 60% of the cell pairs, were related in morphology either directly or as mirror images. One interpretation of this observation is that the cellular morphologies are not necessarily determined in detail by genetic information, but rather by some aspect of the cytoskeletal organization that persists, at least partially, through mitosis. If this model is correct, it is then not clear why a large fraction of cell pairs should fail to correspond in their morphologies. It is also not clear to what extent the results with neuroblastoma cells might apply to the differentiation of non-transformed neuronal precursor cells. In cases where it has been possible to follow the differentiation of progeny of identified neuroblasts *in vivo* (for example, see Goodman *et al.*, 1979, 1980, and references therein), it has become evident that closely related cells can exhibit strikingly different morphologies and electrical properties, though at least some of these differences are undoubtedly due to variations in neuronal environments.

C. Interactions of Nerve Fibers with Other Fibers or Cells

The mutual interactions of neurites and growth cones constitute another important aspect of their behavior. When fragments of nervous tissue are cultured as intact explants, or when suspensions of dissociated cells are cultured at high density, such interactions may be quite frequent. Under such conditions of culture, neurites will then frequently form fascicles, close lateral associations with other neurites (Nakai, 1960; Bunge and Bunge, 1978; Rutishauser et al., 1978). Indeed, it is often impossible to establish with the light microscope whether even fine nerve fibers in vitro represent single or multiple neurites (Bray, 1973a; Bunge, 1973), so close is this association. Under other culture conditions (for example, those that employ firm plasma clots), fasciculation of neurites occurs less frequently (Dunn, 1971). Perhaps related to this is the apparent inhibition of growth cone extension within the clot after contact with another neurite; this inhibition is then followed by reextension of the cone in a new direction (Dunn, 1971). However, it appears that contact inhibition of extension of growth cones need not always occur, as one can sometimes observe growth cones that collide with other neurites yet are unimpeded in their motility and extension (see Wessells et al., 1980, and below).

Interactions between neurons and nonneuronal cells are also important elements of the behavior of the elongating nerve fiber. For example, it occasionally happens that a neurite will become tightly associated, often at its tip, with a nonneuronal cell in vitro. As the nonneuronal cell moves about, it may then pull the neurite long distances over the culture substratum, causing a lengthening of the central part of the neurite without the direct participation of motility of a growth cone. It would be interesting to know whether this "intercalary" growth is mechanistically similar to that which must occur in vivo during the growth of an organism while neuronal somata and synaptic endings remain fixed. Contact between the neuronal soma and a nonneuronal cell can also enhance the survival and frequency of subsequent extension of neurites by a neuron (Ludueña, 1973a). The nature of the support provided by the nonneuronal cell is unknown, but it may in part be mediated by local "conditioning" of the extracellular medium, perhaps through the release of growth factors required by the neuron (Varon et al., 1974; Helfand et al., 1978). These simple observations are relevant to experiments that attempt to analyze neuronal behavior (see below). When interactions between neurons and nonneuronal cells are frequent, it is often impossible to determine whether a given experimental treatment directly influences neurite behavior, or whether a neuronal response is indirect and mediated by a nonneuronal cell.

The remainder of this chapter is devoted to a discussion of mechanisms or factors that may regulate the structures and behaviors of neurites and growth cones that we have just described.

IV. "Intrinsic" Regulation of Neuronal Behavior

A. POSITION IN THE CELL CYCLE

As Harrison (1910) suggested, there are indeed limits or guides to neuronal behavior that are determined by the neuron's own properties and state. A fundamental question is how a neuron's position in the cell cycle is related to its state of differentiation. For example, studies using tritiated thymidine autoradiography have repeatedly confirmed that young neurons undergoing growth cone and neurite extension are postmitotic (for a review, see Eccles, 1970). In addition, in a detailed study of the development of retinal ganglion cells, Hinds and Hinds (1974) found that axons with distinct growth cones were never observed on cells undergoing mitosis but were consistently found on cells judged to be recently postmitotic on the basis of nuclear position, cytoplasmic properties, and proximity to other cells. It has been suggested that neuronal DNA synthesis may sometimes continue in the absence of mitosis and result in polyploidy (for example, in Purkinje cells; Lapham, 1968). More recent analyses (Cohen et al., 1973), however, have demonstrated that the cytophotometric techniques used earlier are subject to artifact and that the Purkinje cells in question are only diploid. Nevertheless, at least some invertebrate neurons may be polyploid. For example, nuclei from giant neurons of the marine mollusk Aplysia can have up to 200,000 times the haploid DNA content (Lasek and Dower, 1971). Pertinent to this issue are the reports of Cone and Cone (1976, 1978) that DNA synthesis and mitosis can occur in mature neurons in vitro. In their experiments, neuronal and nonneuronal cells are cultured at high cell densities and then chronically depolarized by such treatments as exposure to the drug ouabain. Cone and Cone report that a variable fraction of neurons is thereby stimulated to incorporate thymidine, and some may become binucleate or undergo mitosis. These results are controversial, however, as Chalazonitis and Fischbach (1980) report that prolonged depolarization induces morphological differentiation, but not DNA synthesis, by dorsal root ganglion neurons.

Nevertheless, neuronal properties other than morphological differentiation, such as catecholamine synthesis, specific uptake of norepinephrine, and the development of subcellular storage granules, can indeed be present in actively dividing neuronal precursors in sympathetic ganglia (Rothman et al., 1978). Synthesis of a neurotransmitter can even be demonstrated in migrating mesencephalic neural crest cells in vivo (Smith et al., 1979). Thus the transition from a dividing neuronal precursor cell to a mature neuron does not appear to be abrupt but may take place in discrete stages. An interesting question, not yet addressed experimentally, is whether the change in motile activity of neuronal precursors (from the fibroblastic to the neuronal type) correlates with the cessation of mitosis.

B. Synthesis of RNA and Protein

The degree to which the synthesis of RNA or specific proteins (such as tubulin) is required for the extension of neurites and growth cones has been the subject of several investigations, with conflicting results (Yamada and Wessells, 1971; Hier et al., 1972; Mizel and Bamburg, 1975, 1976; Burnham and Varon, 1974; Burstein and Greene, 1978). These reports have variously demonstrated that the stimulation of neurite growth by nerve growth factor (NGF) (Levi-Montalcini, 1966; Varon and Bunge, 1978; Harper and Thoenen, 1980) does (Mizel and Bamburg, 1976) or does not (Burnham and Varon, 1974; Yamada and Wessells, 1971) require RNA synthesis and is (Hier et al., 1972) or is not (Yamada and Wessells, 1971; Mizel and Bamburg, 1975) mediated by specific increases in neurotubule protein. It is unclear why such different conclusions from similar experiments should be obtained, although small differences in culture conditions, such as the exclusion of chick embryo extract (Hier et al., 1972), may be relevant. The importance of such factors is emphasized by the work of Mizel and Bamburg (1976) who found that the dependence of neurite outgrowth on RNA synthesis, but not on protein synthesis, could be bypassed by culturing ganglia in plasma clots or on collagen or polylysine-treated dishes, as opposed to untreated dishes. Working with PC12 rat pheochromocytoma cells, Burstein and Greene (1978) have recently obtained results that may help clarify this matter. These cells, when stimulated by NGF, will cease division and acquire many characteristics of differentiated neurons (Greene and Tischler, 1976), including the presence of structures that closely resemble neurites. Whereas the initial extension of these neurites is dependent upon continued RNA synthesis, the regeneration of new neurites after mechanical removal of the original ones can proceed in the absence of normal rates of RNA synthesis (Burnstein and Greene, 1978).

A simple model that is at least consistent with these diverse findings is as follows. During the initial differentiation of neurons from their precursor cells, it seems likely that increases in the amount (or changes in types; Burnstein and Green, 1978) of RNA and protein are required for the extension of neurites. Once an intracellular pool of such components has been established, then further growth (or regrowth) of neurites may proceed, at least for a short while, even when the synthesis of new components is inhibited. Ultimately, of course, in the absence of such synthesis, these intracellular pools would be depleted and elongation would stop. Nevertheless, the period of elongation might be prolonged by the presence of certain substances in the extracellular environment, substances that might replace the function of the cellular products depleted soonest.

C. ROLES OF INTRACELLULAR FILAMENTS

Other intracellular "limits" to neuronal behavior may be determined by the properties of the filamentous organelles, the microtubules, neurofilaments, and microfilaments. Most enigmatic in this sense are the neurofilaments (see Lazarides, 1980, for a review). At present, there exists no experimental approach for modifying directly the distribution or behavior of these filaments. Even the biochemistry of neurofilaments is not yet well understood. In different species, for example, one, two, or three neurofilament components, of varying molecular weight, have been isolated (Lazarides, 1980). Some of this uncertainty may arise from the fact that neurofilaments, and the integrity of the axoplasm to which they contribute, are extremely sensitive to calcium-stimulated proteolysis (Schlaepfer, 1974, 1978; Gilbert et al., 1975; Day, 1980). Such observations, and the distribution of neurofilaments within the neurite and growth cone seem to suggest an important structural or functional role (for example, in axonal transport, discussed below). Although the apparent absence of neurofilaments from crayfish and some other invertebrate axons (Fernandez et al., 1971) might seem inconsistent with this possibility, these axons have unusual properties (such as their prolonged survival after axotomy; Krasne and Lee, 1977) that may yet provide indirect clues to the true role of neurofilaments.

The precise function of microfilaments in growing nerve fibers is also somewhat controversial. In early experiments with the drug cytochalasin B (CB) (Yamada et al., 1970; Wessells et al., 1971), its striking inhibition of growth cone motility and neurite extension was attributed to disruption of the function of microfilaments. This interpretation was questioned (Estensen et al., 1971), however, especially when it became evident that one action of CB was to inhibit glucose transport (Cohn et al., 1972; Kletzien et al., 1972). Nevertheless, it was later shown that the inhibition by CB of growth cone motility was not mediated by an altered transport of glucose (Yamada and Wessells, 1973), that cells may have several distinct classes of binding sites for CB (Lin and Snyder, 1977), and that CB can alter viscosity and rates of polymerization of actin solutions in vitro (Flanagan and Lin, 1979; Brenner and Korn, 1980).

That microfilaments may play an important role in growth cone motility remains a likely possibility. The presence of microfilaments in filopodia, lamellipodia, and the leading edges of growth cones (Tennyson, 1970; Hinds and Hinds, 1972; Bunge, 1973; Yamada et al., 1971; Isenberg et al., 1977), the organization of microfilaments into parallel arrays projecting from filopodia back into the growth cone (Fig. 1; Lueduña and Wessells, 1973; Letourneau, 1979), and the ubiquity and apparent involvement of actin in the motility of nonmuscle cells (Hitchcock, 1977; Mooseker, 1976) all argue in favor of this hypothesis. One model for the extension of cell surface protrusions that may provide insights to the mechanism of growth cone motility arises from experiments with sperm and eggs. Here, the extension of surface microvilli of eggs (Begg and Rebhun, 1979) or of acrosomal filaments of sperm (Tilney et al., 1978), both of which are membrane-bound structures containing prominent arrays of microfilaments, can be induced by elevating cellular pH. The extension is associated with the polymerization of actin from preexisting reservoirs, and is apparently not dependent on the function of myosin-like proteins. Might such a mechanism explain filopodial extension? The available evidence is inconclusive, but does not strongly support such a model. First, electron micrographs show that not all filopodia contain parallel arrays of microfilaments (Letourneau, 1979). Second, studies using polarization, fluorescence, and electron microscopy (Chen, 1977; also see Heath and Dunn, 1978) suggest that parallel arrays of microfilaments in motile cells are more closely associated with retraction or tension from a site of adhesion than with extension outward of cell surface. At present, therefore, the most useful class of models in explaining movements of growth cones and of their surfaces are those that invoke interactions of actin with myosin-like proteins (for example, Edds, 1977 and Condeelis and Taylor, 1977), regulated perhaps by calcium concentration.

Another potential role for microfilaments may be in rapid axonal transport, in a mechanism based on that of the sliding filaments of muscle (Ochs and Worth, 1978). Pertinent to this is the finding that both actin and myosin can be detected in neurites and growth cones of sensory neurons, using fluorescently labeled myosin fragments or fluorescent antibody to brain myosin (Kuczmarski and Rosenbaum, 1979). In addition, extracts of whole sciatic nerve (including both nerve fibers and nonneuronal cells) contain a relatively high concentration of a Mg^{2+}-Ca^{2+}-activated ATPase with actomyosin-like properties (Khan and Ochs, 1974). However, experimental support has also been obtained for a role (perhaps complementary to that of actin and myosin) for microtubules in axonal transport. For example, disruption of microtubules with a drug such as colchicine or Colcemid or by exposure to cold results in a reversible inhibition of transport (for reviews, see Ochs and Worth, 1978, and the contribution by Kelly to this series). Nevertheless, the relationship between microtubules and axonal transport does not seem simple since, under some conditions, transport may be blocked by treatment with drugs without apparent alteration in microtubular structure (Fernandez et al., 1970). The possibility that these drugs may also act on membranes (Ochs and Worth, 1978), on other filaments (Fernandez et al., 1971; Lazarides, 1980), or on proteins associated with tubules or filaments (Kim et al., 1979; Nunez et al., 1979) cannot yet be excluded.

While it is not clear whether microtubules are directly involved in axonal transport, they do seem to be required for the integrity of extending neurites (for a review, see Daniels, 1975). When colchicine is added at a low concentration to cultures of nerve cells, the extension of neurites is reversibly inhibited; at higher concentrations of colchicine, neurites will retract toward the cell soma (Yamada et al., 1970; Daniels, 1973; Bray et al., 1978). Electron microscopy of neurites undergoing retraction has demonstrated a simultaneous reduction in the density of microtubules (Daniels, 1973). Interestingly, even during the retraction due to alkaloids, motility of the growth cone can persist (Yamada et al., 1970; Bray et al., 1978), suggesting that this motility is not directly dependent on the function of microtubules. Bray et al. (1978) have further noted that the restriction of motility to the growth cone is lost during this retraction and that long portions of the neurite may then become actively motile. When the retraction of neurites into the cell soma is complete, active filopodia extend from the surface and move about (Wessells, unpublished), so that the cell's behavior is reminiscent of that of a freshly dissociated neuron (see above).

These kinds of observations, coupled with results of experiments in which neurites are cut or treated with cold, led Bray to suggest that some property of microtubules is normally responsible for the restriction of motility to the tip of the neurite, in which this property may be absent. An intriguing possibility is that the difference in organization of microtubules in the neurite and in the growth cone (Fig. 1; Ludueña and Wessells, 1973; also, see Vasiliev et al., 1970) may by causally related to this regulation. Recently, Spiegelman et al. (1979), working with mouse neuroblastoma cells, have shown that sites of subsequent neurite extension by these cells are first indicated by prominent aggregates of previously dispersed microtubule initiation sites. The existence and aggregation of these microtubule initiation sites can be demonstrated by brief exposure of cells to Colcemid or vinblastine, followed by fluorescent staining for tubulin after varying periods of recovery. It is not clear, however, to what extent the process of neurite initiation by neuroblastoma cells resembles in mechanism the analogous events occurring in nontransformed nerve cells.

It is also not yet clear whether microtubules are strictly necessary for the maintenance of axons in vivo. Microtubules are apparently absent from giant axons of the marine worm Myxicola (Gilbert et al., 1975). One possibility is that microtubules may be required during the extension of axons, but that in some special cases the functions of microtubules may later be assumed by supportive nonneuronal cells, or other features of the axonal environment.

Two other ways in which microtubules may limit or direct the extension of nerve fibers deserve mention. First is the question of whether the number of microtubules in a neurite may set a limit to later branching by that neurite. Although it was hypothesized that all microtubules might initiate in the neuronal soma and that the number of microtubules per neurite would be reduced distally through division at each branch point (when only one microtubule remained in a given branch, further branching would be impossible), in at least some nerve cells the total number of microtubules in all axonal branches can exceed by 11-fold that in the proximal axon (Zenker and Hohberg, 1973). This observation implies that initiation sites for microtubules must exist in the peripheral parts of neurites, or even in the growth cone proper, and that the number of microtubules in a neurite need not limit the neurite's later branching. Direct evidence for the existence of peripheral initiation sites for microtubules comes from serial sections of axons of Caenorhabditis (Chalfie and Thomson, 1979). Here, the number of microtubules seen in cross sections of identified axons can vary along the length of the axons, and ends of individual microtubules can easily be identified. In axons of different neurons, both the average number of microtubules per section and the average length of the microtubules can vary considerably (from 5 to 50 per section and from 6 to 27 μm in length). Interestingly, the two ends of each microtubule differ in their ultrastructure, one staining more densely than the other, thus establishing a morphological polarity. Approximately 75% of the microtubules are oriented with the densely staining termini directed toward the cell body; the remaining microtubules are of opposite polarity. If microtubules are indeed related to axonal transport, their extension in opposite directions within a single neurite may well correspond to the existence of orthograde and retrograde axonal flow (see below). However, distinct morphological indicators of microtubule polarity have not yet been observed in axons of higher organisms. It may nevertheless be possible to explore such questions using techniques recently developed by Rosenbaum and his associates (Haimo *et al.*, 1979), in which polarity is determined by measuring the angle of binding to microtubules by added flagellar dynein.

The second main question of interest is whether some aspect of the large-scale organization of microtubules or other axoplasmic filaments may influence the characteristics of the growing neurite. Longitudinal sections of neurites in the electron microscope (Fig. 6; Ludueña and Wessells, 1973; Bunge, 1973) sometimes convey the distinct impression that, in three dimensions, the microtubules may follow an approximately helical path within the neurite. For instance, rare sections show a single microtubule passing from just internal to the plasma membrane on one side of a neurite, all the way across the axoplasm so as to lie directly beneath the plasma membrane on the opposite side. Consistent with this, a helical organization of fibrillar components is visible by differential interference microscopy within squid axoplasm (Metuzals and Izzard, 1969). In *Myxicola* axoplasm, neurofilaments (in the absence of microtubules; see above) also display a striking helical organization (Gilbert, 1975). [Such a helical organization is not, however, observed in axons of *Caenorhabditis* (Chalfie and Thomson, 1979), where microtubules are instead linearly arrayed.] If a helical organization of axoplasmic components is typical, and if one assumes that it may exert a torque or some other asymmetric influence upon the neurite (Gilbert, 1975), this organization may in part explain the finding that, when neurites are cut or growth cones are released from distal sites of attachment, the neurites sometimes adopt a markedly sinuous or helical shape during their subsequent retraction (Shaw and Bray, 1977). Further, if the helicity in microtubules or other filaments

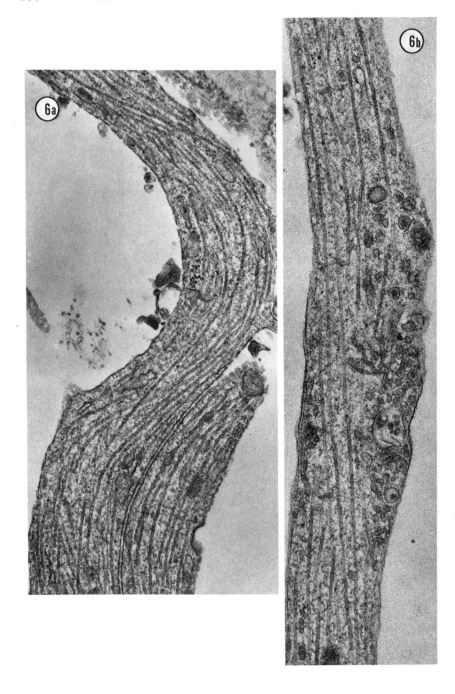

is of a consistent orientation (that is, clockwise or anticlockwise from the soma), this might be related to the extraordinary observation that neurites extending from retinal explants tend to curve in a clockwise fashion (Heacock and Agranoff, 1977).

It is apparent that the various filamentous organelles may play important roles in the extension of neurites. Little information is available, however, on how the organization or function of these organelles is regulated. Further study of the mechanisms of their assembly and disassembly and of the associations of other proteins (such as MAPs; Binder and Rosenbaum, 1978) is required.

D. AXONAL TRANSPORT

Simple calculations show that the volume of axoplasm in a mature neuron *in vivo* may easily be a thousandfold greater than the volume of the cell body. How is such a disparity in volume generated and maintained? One possibility is that glial cells may directly contribute to the substance of the axoplasm. For example, a transfer of protein from glia to squid axons has recently been demonstrated (Lasek *et al.,* 1977). However, the extension and survival of neurites does not necessarily require the participation of glia, since neurons cultured in the absence of glia can nevertheless produce apparently normal neurites of substantial length (McCarthy and Partlow, 1976; Pettman *et al.,* 1979; Tuttle *et al.,* 1980). A second possible explanation for the supply of axonal components during growth is that these components are synthesized within the axon itself. It is the case, however, that most protein, RNA, and glycolipid synthesis occurs in the neuronal soma (for example, see Sherbany *et al.,* 1979); the small fraction of total synthesis occurring in the axoplasm is largely mitochondrial (Hamburger *et al.,* 1970). Thus, transport of components synthesized in the cell body remains the most likely candidate for support of the growth of neurites (Grafstein, 1977).

The role that transport of materials within axons might play during the growth or regeneration of nerves was first explored by Weiss and Hiscoe (1948). They observed that, when a nerve trunk of a rat was crushed and a constricting cuff placed around the trunk distal to

---

FIG. 6. Longitudinal sections of neurites [sensory ganglion neurons (a) and ciliary ganglion neuron (b)] in which microtubules are seen to extend from one side of the neuritic cylinders toward the other. This arrangement is commonly observed at points where neurites bend [as in (a)]. How this arrangement is related to the bending of neurites or to the helical configuration neurites sometimes assume during retraction (see text) is an intriguing question.

the site of crushing, the regenerating trunk underwent a swelling just proximal to the cuff. If the constriction were released, then the swelling would move distally, at a rate of about 1 mm/day. Weiss and Hiscoe concluded that the constriction caused a damming of the natural bulk axoplasmic flow along the trunk and that such flow was necessary for the normal growth and maintenance of nerve fibers. These and similar experiments led to a model for axonal elongation, still widely presented in introductory texts, that depends upon the "movement of the whole axon itself as a semisolid column."

However, the transport of materials within the axon is a more complex phenomenon than that originally proposed by Weiss and Hiscoe (1948). Different axonal components may be transported at different velocities, and transport is not unidirectional but proceeds both toward (retrograde flow) and away from (orthograde flow) the cell body (Nakai, 1956; Leestma and Freeman, 1977; Ochs and Worth, 1978; Kelly, this series). Although much of the material transported in the orthograde direction continues to the axonal terminal, Droz et al. (1973) have shown that about 95% of the slowly transported component is deposited along the length of the axon. Such a mechanism may account for much of the increase in diameter and length of axons as these stretch during the normal growth of animals. Increasingly sophisticated methods have recently permitted surprising demonstrations of asymmetries in axonal transport. By following movements of labeled polypeptides from adult rat dorsal root ganglia, Mori et al. (1979) were able to show that different structural proteins, including tubulin and actin, were transported at different rates, and that transport was more rapid in peripheral than in central branches of the sensory axons. It would be of great interest to know whether such asymmetries are present even in elongating or embryonic neurites, or whether asymmetric transport is characteristic of established neurites only, and thus perhaps related to requirements for maintence of the differing volumes of axoplasm in the central versus peripheral neurites.

A related question is whether the rate of elongation of neurites is limited by the supply of components through axonal transport, or whether other factors (such as rates of protein synthesis or of assembly of axonal components) are limiting. Perhaps relevant to this is the observation that alkaloids inhibiting axonal transport can also block neurite elongation (see Daniels, 1975, and above). However, the primary effect of these drugs might instead be to alter structural elements of neurites, and the inhibition of transport could be a secondary response. Rates of axonal transport have not yet been measured for neurites *in vitro* under conditions that permit measurements of the rate of neurite elongation. Interestingly, although the rate of neuronal protein synthesis can increase when neurites elongate rapidly (Ludueña, 1973a), the longest neurites are often the thinnest (e.g., less than 1 μm in diameter). It may well be, therefore, that under conditions of unusually rapid growth the elongation of a neurite and the motility of a growth cone are not strictly dependent upon "adequate" supplies from the neuron and may proceed through the redistribution of previously assembled axonal components. This interpretation is indirectly supported by experiments in which neurites are severed, separating the growth cone and a portion of the distal neurite from the cell body (Hughes, 1953; Bray *et al.*, 1978; Rieske and Kreutzberg, 1978; Wessells *et al.*, 1978). In spite of the absence of transport from the cell body, motility of the growth cone and elongation of the fragment of neurite may persist for several hours. This observation eliminates the possibility that neurite elongation is a consequence of bulk transport from the cell body, as toothpaste is squeezed from a tube (Hughes, 1953). Although these fragments of neurite eventually degenerate, at least their short-term function apparently does not depend on normal axonal transport (other than that related to the redistribution of components within the distal stump).

Retrograde transport may also play an important role in the developing nerve fiber (Heslop, 1975). This is suggested by experiments that demonstrate both the uptake and retrograde transport of NGF (Stoeckel *et al.*, 1975). The physiological importance of this transport is indicated by the work of Campenot (1977) who grew neurons in dishes with multiple chambers, such that the cell bodies and neuronal tips were in different chambers. Neurons survived when their tips, but not when only their somata or proximal neurites, were exposed to NGF. This observation implies that retrograde transport may provide neurons growing *in vivo* with growth factors or other different types of macromolecules (Harper and Thoenen, 1980).

E. THE CELL SURFACE

In addition to increases in axoplasmic volume, it is readily apparent that the growing neurite must also increase its surface area. For example, a neuron can easily exhibit a fivefold increase in surface area after extending neurites for only 1 day *in vitro*. Transport of membrane components (for example, glycolipids; Sherbany *et al.*, 1979), by endoplasmic reticulum (Droz *et al.*, 1975) or vesicles, is probably an important factor in this increase. One possible mechanism for the increase in cell surface area invokes an imbalance of membrane cycling, whereby there is an excess of fusion of intracellular vesicles or other membranous structures with the plasma membrane, relative to the deletion of surface membrane by invagination (Bray, 1973b; Bunge, 1973, 1977; Spooner et al., 1974; see also Abercrombie et al., 1977).

Pertinent to the question of cycling are instances in which the branched tubular membrane system is seen in continuity with the plasma membranes of growth cones (Bunge, 1973) and of other types of cells (Wessells et al., 1974). Also, strings of flattened membranous sacs are sometimes seen in association with filopodia at sites where changes in surface area might be expected to occur (Bunge, 1973; Bray, 1973b; Spooner et al., 1974). It is unknown whether these structures are related to the addition or the deletion of cell surface.

Attempts to trace membrane during the hypothesized cycling have rested on the use of agents that bind to the cell surface, such as thorium dioxide, horseradish peroxidase, and ferritin (Birks et al., 1972; Wessells et al., 1974, 1976; Bunge, 1977; Pfenninger, 1978). In these experiments, label is added to the extracellular fluid, and the distribution of bound label later examined in the electron microscope. These studies have revealed that, after brief exposure, cells are labeled fairly uniformly on their surface (Wessells et al., 1976; Bunge, 1977). After prolonged exposure to label, various cytoplasmic organelles (particularly uncoated vesicles and agranular reticulum; Birks et al., 1972; Wessells et al., 1974; Bunge, 1977) become increasingly heavily labeled, suggesting that these organelles may represent elements of the mechanism for recycling cell surface. Cell surface label is not found on regions of membrane that immediately overlie mounds (Wessells et al., 1976), sites of densely packed agranular cytoplasmic vesicles, suggesting together with the observation that the density of intramembranous particles is especially low at these sites (Pfenninger and Bunge, 1974) that such regions may be zones of active membrane addition. However, recent experiments have indicated that the swelling associated with such mounds on neurites and growth cones (Nuttall and Wessells, 1979) and on corneal fibroblasts (Hasty and Hay, 1978) may be artifactual. At present, one simple interpretation of these structures is that they represent local reservoirs of membrane in an unstable state.

If the hypothesis of recycling of cell surface in an elongating neurite is correct, and if the addition and removal of surface occur at different sites (Bray, 1973b), then one would predict that net "flow" might occur in the plane of this surface. Indeed, Bray (1970) found that particulate material touched by the leading edge of the growth cone of a sympathetic neuron was rapidly moved rearward over the upper surface of the cone and then remained stationary or moved slightly in the direction of elongation. In contrast, particles stuck to the sides of elongating neurites failed to move appreciably relative to the soma. [Koda and Partlow (1976) found, however, that adherent polystyrene beads could move in the retrograde direction along the surface of neurites *in vitro*.] The movements of particles adhering to the surfaces of many types of locomotory cells (Abercrombie *et al.*, 1970; Heaysman, 1978) are similar to those observed for the growth cone. Thus, the growth cone, considered as a relatively isolated motile organelle, may function much like a fibroblast (Bray and Bunge, 1973), but one with a long, adhering "tail," the neurite. In each case, addition of membrane components might occur at the leading motile edge, and deletion of membrane might occur centrally. [The similarity in function of growth cones and fibroblasts is not absolute. Strassman *et al.* (1973) have shown that, under at least some conditions of culture, growth cone motility and neurite extension can continue while fibroblast locomotion is inhibited.]

Another cell surface movement is observed if neurons are subjected to labeling with an agent such as cationic ferritin (Wessells *et al.*, 1976). Label initially distributed over the neurite or growth cone surface aggregates into large, thick patches in a few minutes, followed by "clearing" of the aggregated patches, just as occurs from the surfaces of some normal tissue cells (Skutelsky and Danon, 1976). Also, when a growth cone of a retinal neuron is exposed to fluorescently labeled concanavalin A (Letourneau, 1978), fluorescence is initially distributed over the entire cone and its filopodia. Within minutes, however, the fluorescence becomes restricted to the central part of the cone, leaving the filopodia bare of label (related studies have employed a variety of lectins; for a review, see Pfenninger and Maylié-Pfenninger, 1978).

It is not yet known to what extent the movements of these molecular labels, or of the particles or beads, are induced by binding of the foreign agent or particle (see the literature on dispersed, clustered, patched, and capped lectins or antibodies; Nicolson, 1976), or whether they reflect redistributions of surface-associated molecules (e.g., fibronectins and CAM; Rutishauser *et al.*, 1978), which might occur as a normal part of growth cone motility or neurite adhesion to objects in the environment. That movements of components within the plane of the neurite membrane may normally occur is strongly suggested by the measurements of de Laat *et al.* (1978), from which the mobility of a lipid analog can be shown by fluorescence polarization techniques to increase upon stimulation of labeled neuroblastoma cells to extend neurites. The precise interpretation of this and similar experiments is subject to some uncertainty, however, as it is not yet clear whether such exogenously applied labels are truly incorporated into membranes (Conrad and Singer, 1979).

The many uncertainties about these phenomena emphasize that this is an important area for future study. In particular, probes must be developed that allow unequivocal identification of sites of incorporation of new lipid molecules, new integral proteins, and new surface-associated macromolecules into the surface of the elongating neurite and its growth cone. For the moment, Bray's (1970, 1973b) suggestion that the growth cone likely represents the site of net incorporation of new cell surface remains the primary hypothesis worthy of test. In making such tests, it may be important to distinguish between regions of movement of cell surface and the sites where surface macromolecules are inserted or accumulate, since there is no reason a priori to assume that the two necessarily correspond.

V. "Extrinsic" Regulation of Neuronal Behavior

A. Embryonic Gradients

While it is clear that much of the behavior of growing nerve fibers may be limited or influenced by the cellular properties of neurons, it is equally apparent that the neuronal environment also regulates this behavior. Isolated neurons cultured in featureless plastic culture dishes will readily extend neurites, but the pattern of extension exhibits little order (Bray, 1973a). However, where it has been investigated, extension of neurites *in vivo* can sometimes exhibit accurate and regular orientation (Hinds and Hinds, 1974; Tennyson, 1965). What cues in its environment are important to the developing nerve fiber, and what neuronal responses can these cues elicit? In part, the problem of how fiber outgrowth is oriented is made simpler by introducing the concept of "pioneering nerve fibers" (Harrison, 1910), where the neurites that initially establish pathways act as guides, through fasciculation (Nakai, 1960; Rutishauser *et al.*, 1978), for the growth of later fibers (for example, in the development of optic nerve axons of *Daphnia*, Lopresti *et al.*, 1973; see also Meinertzhagen, 1975). This mechanism would not, however, resolve the issue of how the initial fibers choose the correct pathways. For example, in the developing limb, even the earliest fibers that can be detected follow anatomically appropriate courses (Landmesser, 1978).

Indirect evidence in support of the existence of directional gradients in the embryo, to which nerve fibers can respond, it provided by experiments in which components of the nervous system are implanted in unnatural positions or orientations in host embryos (Constantine-Paton and Capranica, 1976). For example, when segments of amphibian medulla containing young Mauthner cells are grafted into hosts in a reversed anteroposterior orientation, axons from these cells first extend anteriorly, the correct direction relative to the implant but wrong relative to the host (Hibbard, 1965). Then, either directly or after traversing a convoluted path, the axon usually turns and extends posteriorly (for a discussion of how this result may vary with factors such as the size of the implant or time of implantation, see Kimmel and Model, 1978). It is apparent that, in some sense, the Mauthner cell can respond to the major axis of orientation of the embryo.

Several different kinds of gradients that may exist in embryos have been proposed, including electrical, chemical (diffusible and nondiffusible substances), and physical gradients (for a review, see Jacobson, 1978). In testing such proposals, the general approach has been to impose gradients upon potentially responsive cells or tissues *in vitro*, as it is, so far, difficult or impossible to detect or modify the hypothetical gradients *in vivo*. For example, in testing whether electric fields might affect the magnitude or orientation of neurite regeneration, trigeminal ganglia were exposed *in vitro* to weak electric currents (Sisken and Smith, 1975). Relative to the controls, experimental preparations showed enhanced neuronal survival and extension of neurites and an accumulation of neurons and nonneuronal cells near the cathode. Similar experiments with dorsal root ganglia (Jaffe and Poo, 1979) again show that neurites can extend more rapidly from the cathodal face of the ganglia. In experiments such as these, however, effects of nonneuronal cells on neurite extension (as discussed above), changes in ionic distribution, and electrophoresis of extracellular components are often difficult to rule out. In addition, the results of such experiments have often been difficult to confirm [for example, compare the contradictory results of S. Ingvar (1920) and D. Ingvar (1947) who, using similar experimental procedures, found that neurites did or did not, respectively, orient in response to electric fields]. Nevertheless, electric currents have recently been reported to occur in developing embryos. Jaffe and Stern (1979) have described currents with densities of ca. 100 μA cm^{-2} in the vicinity of the primitive streak of chicken embryos. Jaffe (1977) hypothesizes that such currents might guide cellular events by causing the electrophoresis of cell surface components. It is not yet clear, however, whether electrical currents persist in embryos to stages when neurogenesis occurs, nor is it known whether such currents might generate sufficient voltage to effect the proposed electrophoretic separations. This model for development presents intriguing possibilities, and deserves further exploration, but the field, at present, should be viewed as a controversial one.

B. CHEMOTROPISM OF NERVE FIBERS

Other attempts have been made to demonstrate the applicability of models of chemotropism by growing nerve fibers. That concentration gradients of diffusible chemicals may exist in embryos has been argued on theoretical grounds (Crick, 1970). However, to demonstrate convincingly and manipulate such gradients *in vivo* has not yet been possible. One approach to this problem (Charlwood *et al.*, 1972) has been to examine the responses of sensory ganglia *in vitro* to the localized application of NGF, which is released by diffusion from the tip of a capillary tube. After 24 hours in culture, neurites are longest on the side of the ganglion nearest the capillary, and some fibers may even penetrate the orifice of the tube. In such an experiment, however, the steepness of the gradient in concentration of NGF, which will change in time as the source of NGF is depleted, is unknown. In addition, the potential interference by nonneuronal cells in this assay (as above; also, see Varon, this volume), the possibility that distant nerve cells may be shielded from the source of NGF by the bulk of the ganglion, and the potential differential survival of neurons that might randomly extend neurites toward a source of NGF (see Campenot, 1977), make a clear interpretation of this experiment impossible.

Perhaps the most careful analysis of the possibility that nerve fibers of sensory neurons might be chemotactic to NGF is that of Letourneau (1978). Isolated neurons were grown in an agar matrix (Strassman *et al.*, 1973) and exposed to a diffusion gradient of NGF. Since the NGF was labeled with ^{125}I, the slope of its gradient could be monitored by measuring the distribution of radioactivity in slices of the agar. In different experiments with labeled β NGF or 7S NGF, gradients of varying steepness were established, in the presence or absence of unlabeled NGF distributed evenly in the matrix. In addition, the effects of gradients of insulin or ovalbumin were also examined. When the positions of nerve fiber tips relative to their cell bodies were determined, a consistent but slight preponderance in displacement of the tips toward the source of NGF was detected (about 60 versus 50% in the absence of an NGF gradient). It is puzzling why, if sensory neurons are indeed chemotactic to NGF, a higher degree of orientation was not observed. It may be that the slope of the NGF gradient in this experiment was inhomogeneous, spatially or temporally, at a cellular level. Alternatively, only part of the population of sensory neurons may be responsive in this way to NGF, other cells perhaps being responsive to other factors. Whether NGF is a chemotactic agent *in vivo* remains unclear, although some evidence for its presence in potential target organs has been obtained (Johnson *et al.*, 1971; Harper and Thoenen, 1980).

That other factors distinct from NGF may be chemotactic agents for certain classes of neurons is suggested by numerous experiments. The typical protocol of these experiments is to culture fragments of nerve tissue near explants of target or nontarget tissue (Chamley *et al.*, 1973; Chamley and Dowel, 1975; Ebendal and Jacobson, 1977; Pollack and Liebig, 1977), or on opposite sides of filters (Coughlin, 1975). The apparent orientation of nerve fibers toward target tissues, after several days of culture, can be striking. In the case of the transfilter culture experiments (Coughlin, 1975), the pattern of neurite extension by the parasympathetic submandibular ganglion of the mouse seems to correspond with the pattern of branching of the salivary epithelium on the other side of the filter, without obvious contact between the two tissues. Nevertheless, these experiments suffer from many of the objections to the work with local application of NGF. As before, it is difficult to rule out possible factors other than chemotaxis, such as the influence of nonneuronal cells, the possible differential survival of randomly extended pioneer fibers and, in addition, local changes in properties of the substratum. Still, these kinds of experiments have prompted much speculation that specific chemotaxis of nerve fibers may operate in embryos. If potential chemotactic agents retain their activity after attempts at purification, then more rigorous experiments should be attempted, perhaps similar to those of Letourneau (1978).

C. REGULATION BY CELLULAR INTERACTIONS

One mechanism by which guidance of nerve fiber extension may occur is through the interactions of fibers among themselves and with other cell types. The interactions among fibers that lead to fasciculation of lateral surfaces may be especially important (Dunn, 1971). In an analysis of the factors that regulate the degree of fasciculation among nerve fibers extending in plasma clots from an explant, Nakai

(1960) found that the consistency of the clot was critical. When the clot was firm, fibers tended to remain separate from one another. However, when the clot was less firm, fasciculation of fibers was prominent (resulting both from the extension of neurites along pioneer fibers and by the zippering of preexisting neurites). This phenomenon also explains the "two-center effect," first described by Weiss (1934), in which the density and parallel organization of neurites seems greatest in the region between two adjacent explants of nerve tissue.

Although Weiss proposed that stress induced within the clot by the presence of the explants might cause alignment of growing nerve fibers (and thus also explain the radial outgrowth of fibers from single explants), Dunn (1971) later showed that enhanced liquefaction of the clot between the explants was the actual cause of the alignment, in that it led to fasciculation of fibers. Indeed, when two explants were cultured in a particularly firm plasma clot, nerve fibers between them did not show enhanced density and were even shorter than normal, even though stress within the clot could still be detected. Dunn attributed the decrease in length of fibers between the explants to the operation of contact inhibition of extension, in which the collision of growth cones with other neurite surfaces leads to the retraction and change in direction of growth cone extension (also, see Harris, 1974 and Heaysman, 1978). He further invoked contact inhibition of extension to explain the even, radial outgrowth of nerve fibers often seen from single explants of nerve tissue (for example, Fig. 7a). In this case, contact inhibition operating among neighboring nerve fibers would permit their growth only toward unpopulated regions, that is, radially outward from explants. Growth would be even, because any fiber that happened to project beyond its neighbors would then grow randomly until the others caught up (Dunn, 1973).

While contact inhibition of extension may explain some aspects of neuronal behavior *in vitro*, other behaviors, such as fasciculation of nerve fibers, seem inconsistent with such a mechanism. In addition,

FIG. 7. Dorsal root ganglia cultured for 3 days using substrata of collagen (a) or partially aligned collagen (b). The shape of the halo of nerve fibers is affected by the alignment of the collagen (the axis of net alignment is indicated by the arrow at the upper right). In (c) a growth cone (gc) at the tip of an axon (ax) is seen in the aligned collagen matrix [cs; as in Fig. (b)]. Even though individual collagen fibers run in many directions, the growth cone nevertheless responds to the overall axis of alignment of the collagen (CAA). The scanning micrograph at the lower right shows apparent points of contact between the microspikes (ms) and collagen fibers. Courtesy of T. Ebendal and Academic Press. (From *Exp. Cell Res.* 98, 159–169, 1976.)

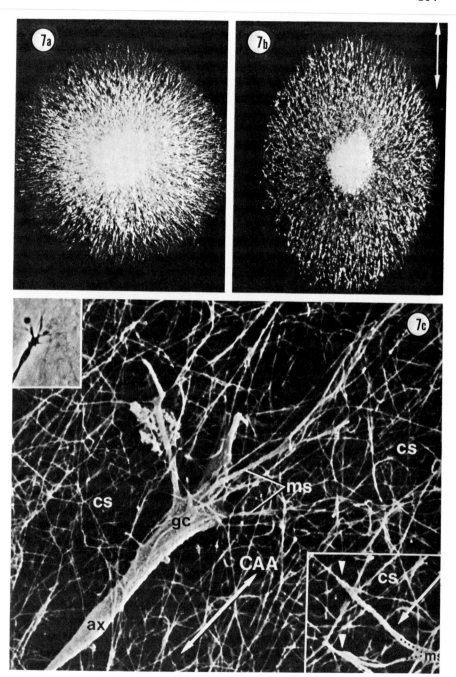

recent studies of the contact relationships among neurites (Wessells *et al.*, 1980) have revealed that growth cones on plastic or glass substrata may interact vigorously with the surfaces of other neurites or growth cones, yet exhibit no inhibition of their motility or extension. When growth cones collide with nonneuronal cells, the motility of each may even be enhanced (Wessells *et al.*, 1980; Ludueña, 1973). Thus, while contact inhibition may be a property of neurites under certain conditions, it seems possible that this inhibition is not of the classic type (Abercrombie *et al.*, 1970) but may more closely resemble type II inhibition (Heaysman, 1978), an event that reflects differential adhesion between cells and substrata.

D. Regulation by the Physical Environment

One way in which guidance of neurites may occur is by the physical alignment of components of the neuronal environment (Weiss, 1961). For example, axons in embryonic or in regenerating newt spinal cord apparently extend within preexisting microchannels along the cell surfaces of the ependymal layer (Norlander and Singer, 1978; Singer *et al.*, 1979). Orientation of neurites can even occur in response to noncellular features of their environment. When neurons are grown on a grooved substratum (Campenot, 1977), or when ganglia are grown in a collagen matrix that has been prepared in such a way that there is a partial alignment of collagen fibrils, then extending neurites will reflect this alignment—a process referred to as "contact guidance" (Fig. 7; Ebendal, 1976). The mechanism by which such an oriented collagen matrix might guide elongating neurites, however, is not certain. One possible mechanism depends upon relative strengths of adhesion by filopodia to nearby collagen fibrils (Fig. 7c, inset; Ebendal, 1976). When a filopodium is extended parallel to a collagen fibril, the potential area of surface contact, and therefore the total strength of adhesion, might be greater than when it is extended perpendicular to a fibril. A growth cone would then respond to the statistical alignment of collagen fibrils by elongating parallel to the axis of the collagen, that is, the axis making the greatest surface area available to the growth cone and thereby permitting the greatest adhesion (see below). A second possible mechanism for guidance by oriented collagen might result from the relative deformability along different axes of the fibrous meshwork. For example, a growth cone exerting sideways tension on an individual collagen fibril might cause a lateral displacement of the fibril toward the cone (also, see Nakai, 1960), and the cone would therefore move "inefficiently" in that direction. Conversely, a cone exerting tension parallel to the axis of a fibril might not cause fibrillar displacement of stretching, and the cone would then be able to elongate more efficiently in the direction of that axis. As with the mechanism proposed by Ebendal, this process would also lead to an apparent orientation of neurites along the axis of net alignment of collagen. Whatever mechanism is correct, it is clear that the physical organization of a neurite's environment can guide its elongation and that adhesive events may be important in this process.

The idea that strength of adhesion of a neuron to its substratum may be an important regulatory factor is supported by many recent experiments. For example, when isolated neurons from ciliary ganglia are cultured in plastic dishes coated with polyornithine and then are exposed to a medium conditioned by heart cells (Helfand et al., 1976), the behavior of filopodial extensions changes rapidly, in that they may then undergo prolonged attachment to the substratum; this is followed by the initiation of growth cone extension (Collins, 1978a). It is interesting that the component of conditioned medium that stimulates neurite extension acts primarily when bound to the substratum (Collins, 1978b). The mechanism of stimulation is unknown but may be mediated by an enhanced adhesion between the cell surface and the substratum. This concept is supported by the experiments of Letourneau (1975a) who found that enhanced cell–substratum adhesion was correlated with an increased frequency and rate of neurite extension by neurons.

Purely adhesive phenomena may also act in the guidance of elongating neurites. When palladium metal is vaporized in patterns upon a culture dish, creating regions of varying adhesiveness to cells, growing neurites will extend along the pathways to which they are most adhesive (Letourneau, 1975b; Helfand et al., 1976). Time-lapse films of such cultures demonstrate that it is the growth cone that is primarily responsive to the pattern, as the cone will rarely grow onto a region of low adhesiveness (even though its filopodia may frequently extend over the surface of such an area).

Finally, in the interference reflection microscope, it is possible to identify regions of growth cones that closely approach the substratum and are presumed to represent sites of adhesion (Fig. 8; also, see Abercrombie et al., 1977). These regions are seen under most portions of the cone, especially near the front edge, and also appear along filopodia, projecting back toward the cone. When the tip of a fine glass needle is maneuvered with a micromanipulator and swept between the growth cone and the substratum, presumably breaking these sites of adhesion, the behavior of the cone may be substantially altered (Wessells and Nuttall, 1978). For example, if one side of the cone is gently

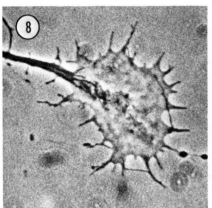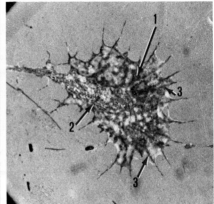

FIG. 8. A living growth cone seen with phase-contrast (left) and interference reflection (right) microscopy (ca. 2 minutes elapsed between the photographs). In the latter, the dark regions are believed to represent areas of close apposition between the lower cell surface and the substratum, whereas in the lighter areas a greater distance separates the two. Note that the dark regions run in broad swaths (1) beneath the more distal portions of the cone, whereas those beneath the thickened, central part of the cone—the extension of the neurite—are in the form of small, discrete points or patches (2). Note also that lines of close apposition (3) occur beneath the leading edge and correspond to sites where parallel arrays of microfilaments would be expected to extend from filopodia into the cortical cytoplasm. (Sensory ganglion cell cultured on a poly-L-ornithine-coated glass substratum in the presence of NGF.) Courtesy of P. C. Letourneau. (From *Exp. Cell Res.* 124, 127–138, 1979.)

lifted, the cone will subsequently extend in the direction of the other side. If only the central part of the cone is lifted, that part will become quiescent and the neurite will branch. These observations may be related to those of Bray (1979), who examined the control of orientation of neurite elongation as a function of mechanical tension exerted on the growth cone. A common observation is that neurites behave as taut fibers, apparently as a consequence of adhesion and extension of the growth cone. Bray made the interesting discovery that when central regions of neurites were laterally deflected, the growth cones at their tips responded by reorienting their extension along and away from the new axis of mechanical tension. This implies that the adhesive interactions of the growth cone with its immediate environment may act coordinately with the cytoskeletal or contractile network within the growth cone and neurite in the regulation of the pattern of neurite extension.

On the basis of these kinds of experiments, and since the adhesiveness of cellular and extracellular components of embryos may well vary both temporally and spatially (Gustafson and Wolpert, 1967; Garber and Moscona, 1972; Roth, 1973; Philips *et al.*, 1977; Marchase, 1977), we suspect that adhesive phenomena in embryos are likely to be important in the guidance of nerve fibers toward their ultimate targets.

VI. Conclusions

We recognize that the distinction we have drawn between the properties of a neuron and of its extracellular environment is in a sense an artificial one and that the activities of a neuron can modify the characteristics of its environment (and vice versa). Nevertheless, we have used this distinction to provide a simple framework within which to discuss what must be a complex and dynamic interplay between the neuron and its environment during their development. Thus, in preparing this review, we were impressed with the accuracy of Harrison's (1910) original speculations on the significance of the growth cone and its two modes of regulation (intracellular and extracellular "forces"). Since then, much of the work on the properties of growth cones and neurites has both confirmed and expanded upon his suggestions. We expect that new areas of research in this field, perhaps focusing on such aspects as the changes in neuronal cell surfaces during development, the regulation of the primary initiation of fiber extension, the organization of intracellular structure, and the interactions of nerve fibers with their environment will continue to be fruitful.

ACKNOWLEDGMENTS

We thank the many people who sent us reprints or preprints of their work. We are especially grateful to F. Collins, T. Ebendal, J. W. Hinds, P. L. Hinds, P. C. Letourneau, and A. Roberts for permission to use their illustrations, and to K. F. Barald, P. C. Letourneau, R. Roth, and K. W. Tosney for helpful comments on the manuscript. While preparing this review, R.N.J. was supported by a scholarship from the National Sciences and Engineering Research Council of Canada. Work from N.K.W.'s laboratory was supported by NIH research grant HD-04708 and by NSF grant PCM-8011704.

REFERENCES

Abercrombie, M., Heaysman, J. E. M., and Pegrum, S. M. (1970). *Exp. Cell Res.* 62, 389–398.
Abercrombie, M., Dunn, G. A., and Heath, J. P. (1977). *In* "Cell and Tissue Interactions" (J. W. Lash and M. M. Burger, eds.). Raven, New York.
Begg, D. A., Rodewalt, R., and Rebhun, L. I. (1978). *J. Cell Biol.* 79, 846–852.
Begg, D. A., and Rebhun, L. I. (1979). *J. Cell Biol.* 83, 241–248.
Binder, L. I., and Rosenbaum, J. L. (1978). *J. Cell Biol.* 79, 500–515.

Birks, R. I., Mackey, M. C., and Weldon, P. R. (1972). *J. Neurocytol.* **1**, 311–340.

Bray, D. (1970). *Proc. Natl. Acad. Sci. U.S.A.* **65**, 905–910.

Bray, D. (1973a). *J. Cell Biol.* **56**, 702–712.

Bray, D. (1973b). *Nature (London)* **244**, 93–96.

Bray, D. (1979). *J. Cell Sci.* **37**, 391–410.

Bray, D., and Bunge, M. B. (1973). In "Locomotion of Tissue Cells." *Ciba Found. Symp.* **14**, 195–209.

Bray, D., Thomas, C., and Shaw, G. (1978). *Proc. Natl. Acad. Sci. U.S.A.* **75**, 5226–5229.

Brenner, S. L., and Korn, E. D. (1980). *J. Biol. Chem.* **255**, 841–844.

Buckley, I. K., and Raju, T. R. (1976). *J. Micros.* **107**, 129–149.

Bunge, M. B. (1973). *J. Cell Biol.* **56**, 713–735.

Bunge, M. B. (1977). *J. Neurocytol.* **6**, 407–439.

Bunge, R. P., and Bunge, M. B. (1978). *J. Cell Biol.* **78**, 943–950.

Burnham, P. A., and Varon, S. (1974). *Neurobiology* **4**, 57–70.

Burstein, D. E., and Greene, L. A. (1978). *Proc. Natl. Acad. Sci. U.S.A.* **75**, 6059–6063.

Cajal, S. R. (1890). *Anat. Anz.* **5**, 609–613, 631–639.

Campenot, R. B. (1977). *Proc. Natl. Acad. Sci. U.S.A.* **74**, 4516–4519.

Chalazonitis, A., and Fischbach, G. D. (1980). *Dev. Biol.* **78**, 173–183.

Chalfie, M., and Thomson, J. N. (1979). *J. Cell Biol.* **82**, 278–289.

Chamley, J. H., and Dowel, J. J. (1975). *Exp. Cell Res.* **90**, 1–7.

Chamley, J. H., Goller, I., and Burnstock, G. (1973). *Dev. Biol.* **31**, 362–379.

Charlwood, K. A., Lamont, D. M., and Banks, B. E. C. (1972). In "Nerve Growth Factor and Its Antiserum" (E. Zaimis and J. Knight, eds.), pp. 102–107. Oxford Univ. Press (Athlone), London and New York.

Chen, W.-T. (1977). *J. Cell Biol.* **75**, 411a.

Cohen, A. M. (1977). *Proc. Natl. Acad. Sci. U.S.A.* **74**, 2899–2903.

Cohen, A. M., and Konigsberg, I. R. (1975). *Dev. Biol.* **46**, 262–280.

Cohen, J., Mares, V., and Lodin, Z. (1973). *J. Neurochem.* **20**, 651–657.

Cohn, R. H., Banerjee, S. D., Shelton, E. R., and Bernfield, M. R. (1972). *Proc. Natl. Acad. Sci. U.S.A.* **69**, 2865–2869.

Collins, F. (1978a). *Dev. Biol.* **65**, 50–57.

Collins, F. (1978b). *Proc. Natl. Acad. Sci. U.S.A.* **75**, 5210–5213.

Condeelis, J. S., and Taylor, D. L. (1977). *J. Cell Biol.* **74**, 901–927.

Cone, C. D., and Cone, C. M. (1976). *Science* **192**, 155–157.

Cone, C. D., and Cone, C. M. (1978). *Exp. Neurol.* **60**, 41–55.

Conrad, M. J., and Singer, S. J. (1979). *Proc. Natl. Acad. Sci. U.S.A.* **76**, 5202–5206.

Constantine-Paton, M., and Capranica, R. R. (1976). *J. Comp. Neurol.* **170**, 17–32.

Coughlin, M. D. (1975). *Dev. Biol.* **43**, 140–158.

Crick, F. (1970). *Nature (London)* **225**, 420–422.

Daniels, M. P. (1973). *J. Cell Biol.* **58**, 463–470.

Daniels, M. (1975). *Ann. N. Y. Acad. Sci.* **253**, 535–544.

Day, W. A. (1980). *J. Ultr. Res.* **70**, 1–7.

De Laat, S. W., Van Der Saag, P. T., Nelemans, S. A., and Shinitzky, M. (1978). *Biochim. Biophys. Acta* **509**, 188–193.

Droz, B., Koenig, H. L., and Giamberardino, L. D. (1973). *Brain Res.* **60**, 93–127.

Droz, B., Rambourg, A., and Koenig, H. L. (1975). *Brain Res.* **93**, 1–13.

Dunn, G. A. (1971). *J. Comp. Neurol.* **143**, 491–508.

Dunn, G. A. (1973). In "Locomotion of Tissue Cells." *Ciba Found. Symp.* **14**, 211–223.

Ebendal, T. (1976). *Exp. Cell Res.* **98**, 159–169.

Ebendal, T., and Jacobson, C.-O. (1977). *Exp. Cell Res.* **105**, 379–387.

Eccles, J. C. (1970). *Proc. Natl. Acad. Sci. U.S.A.* **66**, 294–301.

Edds, K. T. (1977). *J. Cell Biol.* **73**, 479–491.

Estensen, R. D., Rosenberg, M., and Sheridan, J. D. (1971). *Science* 173, 356–358.
Fernandez, H. L., Huneeus, F. C., and Davison, P. F. (1970). *J. Neurobiol.* 1, 395–409.
Fernandez, H. L., Burton, P. R., and Samson, F. E. (1971). *J. Cell Biol.* 51, 176–192.
Fischbach, G. D., and Nelson, P. G. (1977). *In* "Handbook of Physiology I: The Nervous System" (E. R. Kandel, ed.). Waverly Press, Baltimore.
Flanagan, M. D., and Lin, S. (1980). *J. Biol. Chem.* 255, 835–838.
Garber, B. B., and Moscona, A. A. (1972). *Dev. Biol.* 27, 235–243.
Gilbert, D. S. (1975). *J. Physiol.* 253, 257–301.
Gilbert, D. S., Newby, B. J., and Anderton, B. H. (1975). *Nature (London)* 256, 586–589.
Goodman, C. S., O'Shea, M., McCaman, R., and Spitzer, N. C. (1979). *Science* 204, 1219–1222.
Goodman, C. S., Pearson, K. G., and Spitzer, N. C. (1980). *Proc. Natl. Acad. Sci. U.S.A.* 77, 1676–1680.
Grafstein, B. (1977). *In* "Handbook of Physiology I: The Nervous System" (E. R. Kandel, ed.). Waverly Press, Baltimore.
Gray, E. G. (1973). *Brain Res.* 62, 329–335.
Greenberg, J. H., and Schrier, B. K. (1977). *Dev. Biol.* 61, 86–93.
Greene, L. A., and Tischler, A. S. (1976). *Proc. Natl. Acad. Sci. U.S.A.* 73, 2424–2428.
Gustafson, T., and Wolpert, L. (1967). *Biol. Rev.* 42, 442–498.
Haimo, L. T., Telzer, B. R., and Rosenbaum, J. L. (1979) *Proc. Natl. Acad. Sci. U.S.A.* 76, 5759–5763.
Hamberger, A., Blomstrand, C., and Lehninger, A. L. (1970). *J. Cell Biol.* 45, 221–234.
Harper, G. P., and Thoenen, H. (1980). *J. Neurochem.* 34, 5–16.
Harris, A. (1974). *In* "Cell Communication" (R. P. Cox, ed.), pp. 147–185. Wiley, New York.
Harrison, R. G. (1907). *Anat. Rec.* 1, 116–118.
Harrison, R. G. (1910). *J. Exp. Zool.* 9, 787–848.
Hasty, D. L., and Hay, E. D. (1978). *J. Cell Biol.* 78, 756–768.
Heacock, A. M., and Agranoff, B. W. (1977). *Science* 198, 64–66.
Heath, J. P., and Dunn, G. A. (1978). *J. Cell Sci.* 29, 197–212.
Heaysman, J. E. M. (1978). *Int. Rev. Cytol.* 55, 49–66.
Helfand, S. L., Smith, G. A., and Wessells, N. K. (1976). *Dev. Biol.* 50, 541–547.
Helfand, S. L., Riopelle, R. J., and Wessells, N. K. (1978). *Exp. Cell Res.* 113, 39–45.
Heslop, J. P. (1975). *Adv. Comp. Physiol. Biochem.* 6, 75–163.
Heuser, J. E., and Kirschner, M. W. (1980). *J. Cell Biol.* 86, 212–234.
Hibbard, E. (1965). *Exp. Neurol.* 13, 289–301.
Hier, D. B., Arnason, B. G. W., and Young, M. (1972). *Proc. Natl. Acad. Sci. U.S.A.* 69, 2268–2272.
Hinds, J. W., and Hinds, P. L. (1972). *J. Neurocytol.* 1, 169–187.
Hinds, J. W., and Hinds, P. L. (1974). *Dev. Biol.* 37, 381–416.
Hitchcock, S. E. (1977). *J. Cell Biol.* 74, 1–15.
Hughes, A. (1953). *J. Anat.* 87, 150–162.
Ingvar, D. (1947). *Acta Physiol. Scand.* 13, 150–154.
Ingvar, S. (1920). *Proc. Soc. Exp. Biol. Med.* 17, 198–199.
Isenberg, G., and Small, J. V. (1978). *Cytobiology* 16, 326–344.
Isenberg, G., Rieske, E., and Kreutzberg, G. W. (1977). *Cytobiology* 15, 382–389.
Jacobson, M. (1978). "Developmental Neurobiology." Plenum, New York.
Jaffe, L. F. (1977). *Nature (London)* 265, 600–602.
Jaffe, L. F., and Poo, M.-M. (1979). *J. Exp. Zool.* 209, 115–128.
Jaffe, L. F., and Stern, C. D. (1979). *Science* 206, 569–571.
Johnson, D. G., Gordon, P., and Kopin, I. J. (1971). *J. Neurochem.* 18, 2355–2362.
Khan, M. A., and Ochs, S. (1974). *Brain Res.* 81, 413–426.

Kim, H., Binder, L. I., and Rosenbaum, J. L. (1979). *J. Cell Biol.* 80, 266-276.

Kimmel, C. B., and Model, P. G. (1978). *In* "Neurobiology of the Mauthner Cell" (D. Faber and H. Korn, eds.), pp. 183-220. Raven, New York.

Kletzien, R. F., Perdue, J. F., and Springer, A. (1972). *J. Biol. Chem.* 247, 2964-2966.

Koda, L. Y., and Partlow, L. M. (1976). *J. Neurobiol.* 7, 157-172.

Krasne, F. B., and Lee, S.-H. (1977). *Science* 198, 517-519.

Kuczmarski, E. R., and Rosenbaum, J. L. (1976). *J. Cell Biol.* 70, 247a.

Kuczmarski, E. R., and Rosenbaum, J. L. (1979). *J. Cell Biol.* 80, 356-371.

Landis, S. C. (1977). *Soc. Neurosci. Abstr.* 3, 525.

Landis, S. C. (1978). *J. Cell Biol.* 78, R8-R14.

Landmesser, L. (1978). *J. Physiol.* 284, 391-414.

Lapham, L. W. (1968). *Science* 159, 310-312.

Lasek, R. L., and Dower, W. J. (1971). *Science* 172, 278-280.

Lasek, R. J., Gainer, H., and Barker, J. L. (1977). *J. Cell Biol.* 74, 501-523.

Lazarides, E. (1980). *Nature (London)* 283, 249-256.

Leestma, J. E., and Freeman, S. S. (1977). *J. Neurobiol.* 8, 453-467.

Letourneau, P. C. (1975a). *Dev. Biol.* 44, 77-91.

Letourneau, P. C. (1975b). *Dev. Biol.* 44, 92-101.

Letourneau, P. C. (1978). *Dev. Biol.* 66, 183-196.

Letourneau, P. C. (1979). *Exp. Cell Res.* 124, 127-138.

Levi-Montalcini, R. (1966). *Harvey Lect.* 60, 217-259.

Lin, S., and Snyder, S. E. (1977). *J. Biol. Chem.* 252, 5464-5471.

Lopresti, V., Macagno, E. R., and Levinthal, C. (1973). *Proc. Natl. Acad. Sci. U.S.A.* 70, 433-437.

Ludueña, M. A. (1973a). *Dev. Biol.* 33, 268-284.

Ludueña, M. A. (1973b). *Dev. Biol.* 33, 470-476.

Ludueña, M. A., and Wessells, N. K. (1973). *Dev. Biol.* 30, 427-440.

Marchase, R. B. (1977). *J. Cell Biol.* 75, 237-257.

Maupin-Szamier, P., and Pollard, T. D. (1978). *J. Cell Biol.* 77, 837-852.

McCarthy, K. D., and Partlow, L. M. (1976). *Brain Res.* 114, 391-414.

Meinertzhagen, I. A. (1975). *In* "Cell Patterning." *Ciba Found. Symp.* 2a, 265-282.

Metuzals, J., and Izzard, C. S. (1969). *J. Cell Biol.* 43, 456-505.

Mizel, S. B., and Bamburg, J. R. (1975). *Neurobiology* 5, 283-290.

Mizel, S. B., and Bamburg, J. R. (1976). *Dev. Biol.* 49, 20-28.

Mooseker, M. S. (1976). *J. Cell Biol.* 71, 417-433.

Mori, H., Komiya, Y., and Kurokawa, M. (1979). *J. Cell Biol.* 82, 174-184.

Nakai, J. (1956). *Am. J. Anat.* 99, 81-129.

Nakai, J. (1960). *Z. Zellforsch. Mikrosk. Anat.* 52, 427-449.

Nicolson, G. L. (1976). *Biochim. Biophys. Acta* 457, 57-108.

Nordlander, R. H., and Singer, M. (1978). *J. Comp. Neurol.* 180, 349-374.

Nunez, J., Fellous, A., Francon, J., and Lennon, M. (1979). *Proc. Natl. Acad. Sci. U.S.A.* 76, 86-90.

Nuttall, R. P., and Wessells, N. K. (1979). *Exp. Cell Res.* 119, 163-174.

Ochs, S., and Worth, R. M. (1978). *In* "Physiology and Pathobiology of Axons" (S. G. Waxman, ed.), pp. 251-264. Raven, New York.

Palay, S. L., and Chan-Palay, V. (1977). *In* "Handbook of Physiology I: The Nervous System" (E. R. Kandel, ed.). Waverly Press, Baltimore.

Paula-Barbosa, M., and Gray, E. G. (1974). *J. Neurocytol.* 3, 471-486.

Pettmann, B., Louis, J. C., and Sensenbrenner, M. (1979). *Nature (London)* 281, 378-380.

Pfenninger, K. H. (1978). *Annu. Rev. Neurosci.* 1, 445-471.

Pfenninger, K. H., and Bunge, R. P. (1974). *J. Cell Biol.* **63**, 180–196.
Pfenninger, K. H., and Maylié-Pfenninger, M.-F. (1978). *In* "Neuronal Information Transfer" (A. Karrlin, V. M. Tennyson, and H. J. Vogel, eds.). Academic Press, New York.
Philips, H. M., Wiseman, L. L., and Steinberg, M. S. (1977). *Dev. Biol.* **57**, 150–159.
Pollack, E. D., and Liebig, V. (1977). *Science* **197**, 899–900.
Porter, K. R., Byers, H. R., and Ellisman, M. H. (1979). *Neurosci. Study Progr.* **4**, 703–722.
Rieske, E., and Kreutzberg, G. W. (1978). *Brain Res.* **148**, 478–483.
Roberts, A. (1976). *Brain Res.* **118**, 526–530.
Roth, S. (1973). *Q. Rev. Biol.* **48**, 541–563.
Rothman, T. P., Gerschon, M. D., and Holtzer, H. (1978). *Dev. Biol.* **65**, 322–341.
Rutishauser, U., Gall, W. E., and Edelman, G. M. (1978). *J. Cell Biol.* **79**, 382–393.
Schlaepfer, W. W. (1974). *Brain Res.* **69**, 203–215.
Schlaepfer, W. W. (1978). *J. Cell Biol.* **76**, 50–56.
Shaw, G., and Bray, D. (1977). *Exp. Cell Res.* **104**, 55–62.
Sherbany, A. A., Ambron, R. T., and Schwartz, J. H. (1979). *Science* **203**, 78–81.
Singer, M., Nordlander, R. H., and Egar, M. (1979). *J. Comp. Neurol.* **185**, 1–22.
Sisken, B. F., and Smith, S. D. (1975). *J. Embryol. Exp. Morphol.* **33**, 29–41.
Skutelsky, E., and Danon, D. (1976). *J. Cell Biol.* **71**, 232–241.
Smith, J., Fauquet, M., Ziller, C., and Le Douarin, N. M. (1979). *Nature (London)* **282**, 853–855.
Solomon, F. (1979). *Cell* **16**, 165–169.
Speidel, C. C. (1941). *Harvey Lect.* **36**, 126–158.
Spiegelman, B. M., Lopata, M. A., and Kirschner, M. W. (1979). *Cell* **16**, 253–263.
Spitzer, N. C., and Lamborghini, J. E. (1976). *Proc. Natl. Acad. Sci. U.S.A.* **73**, 1641–1645.
Spooner, B. S., Ludueña, M. A., and Wessells, N. K. (1974). *Tissue Cell* **6**, 399–409.
Stoeckel, K., Schwab, M., and Thoenen, H. (1975). *Brain Res.* **89**, 1–14.
Strassman, R. J., Letourneau, P. C., and Wessells, N. K. (1973). *Exp. Cell Res.* **81**, 482–487.
Tennyson, V. M. (1965). *J. Comp. Neurol.* **124**, 267–318.
Tennyson, V. M. (1970). *J. Cell Biol.* **44**, 62–79.
Tilney, L. G., Kiehart, D. P., Sardet, C., and Tilney, M. (1978). *J. Cell Biol.* **77**, 536–550.
Tosney, K. W. (1978). *Dev. Biol.* **62**, 317–333.
Tsukita, S., and Ishikawa, H. (1980). *J. Cell Biol.* **84**, 513–530.
Tuttle, J. B., Suszkiw, J. B., and Ard, M. (1980). *Brain Res.* **183**, 161–180.
Varon, S. S., and Bunge, R. P. (1978). *Annu. Rev. Neurosci.* **1**, 327–361.
Varon, S., Raiborn, C., and Burnham, P. A. (1974). *Neurobiology* **4**, 317–327.
Vasiliev, J. M., Gelfand, I. M., Dommina, O. Y., Komm, S. G., and Olshevskaja, L. V. (1970). *J. Embryol. Exp. Morphol.* **24**, 625–640.
Weiss, P. (1934). *J. Exp. Zool.* **68**, 393–448.
Weiss, P. (1961). *Exp. Cell Res. Suppl.* **8**, 260–281.
Weiss, P., and Hiscoe, H. B. (1948). *J. Exp. Zool.* **107**, 315–395.
Wessells, N. K., and Nuttall, R. P. (1978). *Exp. Cell Res.* **115**, 111–122.
Wessells, N. K., Spooner, B. S., Ash, J. F., Bradley, M. O., Ludueña, M. A., Taylor, E. L., Wrenn, J. T., and Yamada, K. M. (1971). *Science* **171**, 135–143.
Wessells, N. K., Ludueña, M. A., Letourneau, P. C., Wrenn, J. T., and Spooner, B. S. (1974). *Tissue Cell* **6**, 757–776.
Wessells, N. K., Nuttall, R. P., Wrenn, J. T., and Johnson, S. (1976). *Proc. Natl. Acad. Sci. U.S.A.* **73**, 4100–4104.

Wessells, N. K., Johnson, S. R., and Nuttall, R. P. (1978). *Exp. Cell Res.* 117, 335–345.

Wessells, N. K., Geiduschek, J. M., Letourneau, P. C., Nuttall, R. P., and Lueduña-Anderson, M. (1980). *J. Neurocytol.* (in press).

Wolosewick, J. J., and Porter, K. R. (1979). *J. Cell Biol.* 82, 114–139.

Yamada, K. M., and Wessells, N. K. (1971). *Exp. Cell Res.* 66, 346–352.

Yamada, K. M., and Wessells, N. K. (1973). *Dev. Biol.* 31, 413–420.

Yamada, K. M., Spooner, B. S., and Wessells, N. K. (1970). *Proc. Natl. Acad. Sci. U.S.A.* 66, 1206–1212.

Yamada, K. M., Spooner, B. S., and Wessells, N. K. (1971). *J. Cell Biol.* 49, 614–635.

Zenker, W., and Hohberg, E. (1973). *J. Neurocytol.* 2, 143–148.

CHAPTER 6

NERVE GROWTH FACTORS
AND CONTROL OF NERVE GROWTH

Silvio Varon and Ruben Adler

DEPARTMENT OF BIOLOGY, SCHOOL OF MEDICINE
UNIVERSITY OF CALIFORNIA, SAN DIEGO
LA JOLLA, CALIFORNIA

I. Introduction

Neurons are highly evolved cells with specialized functions and even more specialized individual performances. The evolution from an early somatic cell to a mature neuron must involve a sophisticated sequence of extrinsic signals, about which we still know nearly nothing. Among such signals must be those imposing on the evolving cell: (1) restriction to neuronal programs, (2) permanent postmitotic status, (3) neurite elongation, (4) transmitter choice and activities, (5) selection of a target region, and (6) selection of a target cell.

The growth of neurites is not only a most conspicuous behavior of neurons—one that, unlike many others, can be observed with rela-

CURRENT TOPICS IN
DEVELOPMENTAL BIOLOGY, Vol. 16

tively unsophisticated techniques—but also constitutes an essential element in the organization and operation of the communication network that is the principal physiological function of a nervous system. It is, therefore, not surprising that a molecule be labeled "nerve growth factor" (NGF) if its injection into an organism is followed by a dramatic outgrowth of nerve fibers. This was indeed the case of NGF, the existence of which was first recognized through the phenomenal production of nerve fibers that its injection (or, in earlier work, the grafting of tumors later found to release NGF) elicited from spinal and sympathetic ganglia of the chick embryo (Levi-Montalcini and Hamburger, 1951; Cohen and Levi-Montalcini, 1956). NGF and the research efforts centered around it have earned an outstanding place in modern neurobiology. This is not only because of the important functions that NGF serves but also because the 30 years of experience with NGF have inspired and guided productive investigations of several other neurobiological questions.

The time seems ripe to reassess what we have learned about the *regulation of neurite growth* and the *role that so-called nerve growth factors* may play in it. Are there extrinsic factors truly and solely directed toward the promotion, or the inhibition, of neurite growth? Are there other cell behaviors, changes in which might be indirectly reflected in nerve growth responses? As a first approximation, one may view neurite growth as requiring the following components: (1) neuronal survival, (2) production and delivery of building materials, (3) start and stop signals, (4) elongation machinery, (5) permissive terrain, and (6) directional guidance. Inclusion of "neuronal survival" in this list would appear trivial and unnecessary were it not for the fact that survival of neurons is now regarded as being dependent—at least during some portion of their life—on the presence of extrinsic "survival-promoting," or "trophic," agents (see Section II). It seems logical to expect that neuritic growth could be modified by factors acting at *any of the above levels,* and in that sense any such factor could be called "nerve growth-promoting." It seems equally reasonable, however, to put in a special category extrinsic factors, if they exist, that are directly and exclusively concerned with the start and stop signals.

This chapter will concentrate primarily on the first three elements of the list above—the others being covered extensively in another chapter (Johnston and Wessells, this volume). We shall first survey developmental observations indicating the occurrence of target-derived signals that permit a neuron to survive at and beyond a critical time in its life course. We shall then introduce and illustrate the concepts of a trophic drive responsible for both survival and growth-supportive production, and of specifying influences that signal the cell to proceed (or do away) with selected performances. With that conceptual background, we will examine in detail two examples of factors directed toward neurons. The first will be a group of target-derived agents recently shown to promote survival and/or neurite growth in cultured ciliary neurons. The second will be the classic NGF, including the mechanisms through which it may exercise its trophic functions of sensory and sympathetic neurons. We will then briefly survey some signals and mechanisms for neurite growth. Last, we shall move backward into the past of the neuron and discuss whether our concepts regarding the regulation of older cells also apply during earlier phases of their life, before they connect with their target or even acquire a postmitotic neuronal identity.

II. Survival, Trophic, and Specifying Factors

A. DEVELOPMENTAL NEURONAL DEATH

The perception has been built, over the past 30 years, that postmitotic neuronal populations undergo rapid and massive declines in number at specified times during development, which coincide with the time of their synaptic connection to their target cells (cf. Cowan, 1973, and Volume 15, this series; Hollyday and Hamburger, 1976; Varon and Bunge, 1978). Such developmental neuronal cell death has been observed with both peripheral and central neurons and in amphibian, avian, and mammalian species (Hamburger and Levi-Montalcini, 1949; Hamburger, 1958, 1975; Hughes, 1968; Prestige, 1970; Cowan and Wenger, 1968; Kelly and Cowan, 1972; Rogers and Cowan, 1973; Landmesser and Pilar, 1974a,b; Clarke and Cowan, 1976; Clarke et al., 1976; Dibner et al., 1977; among others).

1. Temporal Aspects

Elaborating on suggestions made by Prestige (1970; see also Cowan, in Kerr, 1975; and Varon and Bunge, 1978) we can distinguish several developmental phases in the life history of a neuron, as summarized in Fig. 1:

Phase 0. Proliferation of neuronal precursors is accompanied by influences and events that restrict genetic programs to a neuronal repertoire (*neuroblasts*).

PHASE	NEURONAL STATUS	IMPORTANT FEATURES
0	GENERATION	Proliferation Acquisition of Neuronality
	↓	
1	TARGET INDEPENDENCE	Postmitosis Neurite Elongation Directional Guidance
	↓	
2	TARGET DEPENDENCE	Target Contact Synaptogenesis Death/Survival
	↓	
3	TARGET SUPPORT	Stable Survival Expansion Maturation
	↓ ↑	
4	TARGET DISCONNECTION	Chromatolytic Reaction Degeneration? Survival? Regeneration?

FIG. 1. Schematic subdivisions in the life history of a nerve cell. Features listed are only those relevant to this chapter. Phase 2 is the critical period when developmental neuronal death occurs.

Phase 1. Mitotic activity has ceased. The *postmitotic neuron is independent* of target cells for its survival. Neurite extension takes place, and target cells are reached.

Phase 2. The neuron has become vulnerable to a lack of interaction with its target; i.e., survival is now *target-dependent.* Developmental neuronal death occurs for 40–75% of the neuronal population when a normal innervation territory is available. It is less dramatic if the territory is increased (e.g., by implantation of a supernumerary limb) or, conversely, it may extend to the entire population if the target territory has been removed.

Phase 3. The surviving neurons are stable (i.e., their number remains fixed) as long as their target connections are retained.

Phase 4. A new vulnerability may appear if the connection to a target is subsequently interrupted (e.g., target death or axotomy). The consequences of such a disconnection are initially displayed by a chromatolytic somal reaction (Grafstein, 1975). This reaction may vary in intensity and ultimately lead to cell death or, alternatively, recovery and even regenerative events.

Only marginal speculations have been offered, thus far, on several features of the developmental neuronal death phenomenon. Many concern the transitions from one phase to the next, for example: What controls the outcome of target disconnection at later stages (3–4)?

How are target-dependent neurons selected for either death or survival (2-3)? What is involved in the acquisition of target dependence (1-2)? Conversely, one may ask when and how the target cells acquire or alter their competence to ensure neuronal survival, and how strict the specificity of such a competence vis-à-vis their innervating neuron is. Going further back in time to events preceding the neuronal death event, one may raise additional questions about requirements and regulatory controls for survival and phenotypic expression in phase-1 and even phase-0 cells. Some of these questions will be discussed, purely at a speculative level, later in this chapter (particularly in Section VI).

2. Target-Derived Survival Factors and NGF

The survival of a definitive neuronal population at the time of its synapsing with its target cells (shift from phase 2 to phase 3) has been taken to indicate that the *target supplies a signal* to the innervating cell, which is received by the presynaptic nerve terminal, transmitted retrogradely along the axon, and used by the neuronal soma to sustain life-essential activities (Prestige, 1970; Cowan, 1973; Landmesser and Pilar, 1978). It has been generally assumed, though not established, that the target-derived signal is in the form of a *soluble trophic factor* which is internalized by endocytosis and carried by retrograde axonal transport to the neuronal soma. Some concrete evidence of such a target-derived factor has been recently reported with regard to ciliary ganglionic neurons and will be reviewed in some detail in Section III. An important question yet to be addressed is: Are target-derived factors exclusively concerned with the survival of corresponding neurons, or do they also play a role in other neuronal performances, such as further enlargement and neuritic expansion, functional maturation, and/or execution of functional tasks?

Much inspiration for the concept of target-derived developmental factors has come from the existence of NGF and the evolving perception of what roles it may play in earlier or later stages of neuronal development (see Section IV). Several features of NGF fit the attributes postulated for a target-derived trophic factor, even though no specific demonstration has been provided for its involvement in developmental neuronal death phenomena. In addition, NGF is not only a survival factor but also promotes cell hypertrophy, neurite elongation, and the expression of functional properties (Levi-Montalcini, 1966; Levi-Montalcini and Angeletti, 1968; Varon, 1975a; Chun and Patterson, 1977; Varon and Bunge, 1978). Do these effects result from separate and independent actions of NGF on the responsive neurons, or do they represent different consequences of a single mode of action by the factor? And, in either case, should one expect a corresponding multiplicity of responses from neurons interacting with other, target-derived trophic factors?

B. CONCEPTUAL FRAMEWORK

1. Trophic and Specifying Influences

In the general literature on intercellular regulations, trophic influences have been variably invoked in conjunction with cell death or maintenance, proliferation, hypertrophy, elongation and directional guidance, acquisition of functions, and the magnitude of functional activities. To facilitate the analysis of such disparate events, a distinction has been proposed between trophic and specifying influences of a given factor or hormone (Varon, 1977a; Varon and Bunge, 1978; Varon and Somjen, 1979). The proposition is illustrated in Fig. 2.

The basic concept underlying this proposition is that the balance between anabolic and catabolic activities of a cell determines whether the cell will be in a deficit, a steady-state, or an overproduction condition. *Trophic* influences are defined as directed toward the quantitative regulation of the anabolic machinery of the cell, thus being equally responsible for decline and death (inadequate trophic supply), survival and maintenance (moderate supply), or overproduction (high supply) for growth or secretory activities. What particular programs will benefit from the trophic drive are, in turn, determined by *specifying* influences which instruct the cell for a particular growth modality or the production of selected sets of proteins. This is not to rule out the possibility that a given trophic agent may also be instrumental in regulating the expression of selected programs or be involved in a *tactic* role in guiding or attracting elongating neuronal processes (cf. Section, V,B).

2. Trophic and Specifying Factors

The distinction between trophic and specifying influences has proved useful in the investigation of agents directed toward survival and neuritic production by ciliary ganglionic neurons (see Section III) and in a conceptual analysis of the effects and mode of action of NGF (see Section IV). In addition, it may help in the evaluation of several recent observations concerning cultured cholinergic neurons:

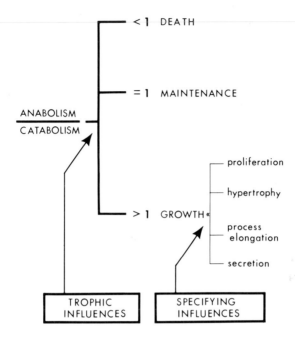

F<small>IG</small>. 2. Conceptual framework for a distinction between trophic and specifying influences. (Adapted from Varon, 1977a.)

1. Explant cultures of *submandibular ganglia*, from embryonic mouse, display dramatic neuritic outgrowth in the presence, but not the absence, of salivary gland epithelium (i.e., its natural target tissue), even when the two tissues are on opposite sides of a filter (Coughlin, 1975b). This suggests that soluble agents are released by the target tissue to act as trophic factors and/or as neurite-promoting agents (a form of specifying factors—see Sections III and V) on the submandibular ganglionic neurons, much as *in vivo* studies had already intimated (Coughlin, 1975a).

2. Dissociated *sympathetic neurons*, from perinatal rat, can be cultured with few if any accompanying nonneuronal cells, provided NGF is present, and they continue to produce tyrosine hydroxylase in accordance with their noradrenergic characteristics *in vivo* (Mains and Patterson, 1973). However, they can be forced to produce considerable amounts of choline acetyltransferase (CAT), in addition to or as a substitute for tyrosine hydroxylase, when cultured with various types of nonneuronal cells (Ko *et al.*, 1976; Patterson *et al.*, 1975) or with medium conditioned over heart muscle (or other) cell cultures (Patterson *et al.*, 1975). Apparently, a shift from noradrenergic to

cholinergic transmitter synthesis (and the use of the latter in functional cholinergic synapses) is elicited by a nondialyzable agent present in the conditioned media. Such an agent would be a typical example of a specifying factor. The response to it will not occur unless a trophic agent is also present, in this case the same NGF that presides over survival, neurite growth, and tyrosine hydroxylase production in the original sympathetic neurons (Chun and Patterson, 1977). It should be noted that neural crest precursors, destined to become either cholinergic or noradrenergic neurons, can be specified for the opposite transmitter mode *in vivo* by transplanting their source neural crest segment to the locality from which the other neurons would have originated (Le Douarin, 1977).

3. Dissociated *spinal cord* cell cultures, from perinatal mouse, increase severalfold their CAT content in the presence of mouse skeletal muscle cells (Giller *et al.*, 1973) or when supplied with medium conditioned over them (Giller *et al.*, 1977). In the latter case, it was not clearly demonstrated whether the nondialyzable constituent of the conditioned medium favored survival of cholinergic neurons (trophic factor), stimulated specifically their CAT production (trophic and/or specifying factors), or induced other neurons to become cholinergic (specifying factor). Attempts to duplicate these observations with spinal cord cell cultures from chick embryo have been, thus far, inconclusive (Popiela *et al.*, 1978).

4. Coaggregate cultures of chick embryo neural retina with optic tectum or telencephalon cells display a substantial increase in CAT activity above the expected contributions from either population cultured separately (Adler and Teitelman, 1974; Adler *et al.*, 1976; Ramirez and Seeds, 1977). These observations suggest trophic and/or specifying influences by one population over the other, but no attempts have yet been made to recognize an involvement of soluble, nondialyzable agents in this interactive system.

It is important to stress that most of the information available and, therefore, the applicability of trophic and specifying concepts have been confined to neurons from phase 2 onward (see Fig. 1). In particular, specifying influences are defined with regard to the *expression* of programs already part of the restricted repertoire of a neuron, rather than in terms of the acquisition of such a repertoire (i.e., the more classic concept of determination). It is, however, possible that the concept of both trophic and specifying influences—as defined in Fig. 2—does apply to earlier behaviors of a neuron (phase 1) or of a neuronal precursor (phase 0), as will be discussed in Section VI.

III. Factors Directed toward
Ciliary Ganglionic Neurons

The NGF model (see Section IV) proposes that (1) trophic factors directed toward neurons, i.e., neuronotrophic factors, may be obtained in pure form, (2) neurons may derive trophic factors from their target cells (including, potentially, other neurons) and from their glial partners, (3) extensive use of culture approaches is essential for the recognition, isolation, and analysis of neuronotrophic agents, (4) neuronotrophic factors can be effectively applied *in vivo* and *in vitro* and in both developmental and regenerative situations. It should be possible to search for and investigate new neuronotrophic factors, following the guidelines provided by the NGF studies. A successful outcome of such a search requires (1) a judicious selection of target neurons, (2) a convenient *in vitro* assay, and (3) an adequate source of trophic activity. Chick embryo ciliary ganglia contain two populations of neurons, both of which are cholinergic and cholinoceptive. They innervate the choroid, ciliary body, and iris muscle cells in the eye. Their development and developmental relations to their target cells have been analyzed in great detail (cf. Landmesser and Pilar, 1978). Between embryonic days 8 and 14, at the time when intraocular synapses are formed, about 50% of the neurons undergo the classic developmental death phenomenon which is enhanced by prior removal of the eye (Landmesser and Pilar, 1974 a,b) and reduced by preimplantation of an additional eye primordium (Narayanan and Narayanan, 1978). Thus, it should be reasonable to assume that the intraocular targets of ciliary ganglionic neurons provide them with a critical ciliary neuronotrophic factor (CNTF) at this and subsequent stages of their development.

When 8-day chick embryo ciliary ganglia (CG) are dissociated and seeded in monolayer cultures, no neurons survive in 24-hour cultures with media containing only fetal calf or horse serum. Like their counterparts *in vivo,* the CG neurons appear to require special extrinsic factors in order to survive. Survival of cultured neurons has, indeed, been achieved by supplying the cultures with (1) skeletal muscle cells (Nishi and Berg, 1977), (2) medium preconditioned over heart cell cultures (Helfand *et al.,* 1976, 1978), or (3) extracts of whole chick embryo (Tuttle, 1977).

A comparative analysis of the effects of chick embryo extract and serum (EE–S) and heart-conditioned medium and serum (HCM–S) was carried out with CG cells cultured for 24 hours on either collagen-coated or polyornithine (PORN)-coated tissue culture plastic (Varon *et al.*, 1979). Figure 3 illustrates some of the results. On collagen, both supplements allowed survival of the same (maximal) number of CG neurons, but only EE–S elicited neurite extension from them. On PORN, HCM–S supported profuse neuritic outgrowth as well as maximal survival, but EE–S allowed only half-maximal survival and little or no neurite production. These observations suggested that the two source materials contain either different CNTFs or the same CNTF accompanied by different neurite-promoting agents.

A. TROPHIC FACTORS (CNTFs)

A working bioassay for survival-promoting (trophic) activity, based on 24-hour cultures of CG neurons, permits one to define one trophic unit (TU) as the activity present in 1 ml of final medium capable of supporting half-maximal survival (Adler *et al.*, 1979b). The assay was used to analyze the regional distribution of CNTF activity in extracts from different chick embryo portions: body and head carcass, viscera (including heart), brain, and eyes (Adler *et al.*, 1979b). As shown in Table I, extracts from 12-day whole embryos contain 8000 TU/embryo, of which approximately 60% resides in the carcass, 30% in the eye, and only 6% in the viscera. The low activity of viscera was of a magnitude comparable to that of HCM–S obtainable from 8-day chick embryo heart cultures (about 200 TU/embryo).

1. Eye CNTF

The 12-day chick embryo eye revealed a disproportionate amount of CNTF activity relative to its size, with a severalfold higher specific activity (over 800 TU/mg) than the rest of the embryo (100 TU/mg). The eye was further subdissected (Adler *et al.*, 1979b) into neural retina, cornea, lens and vitreous, iris, and a fraction (CPE) containing choroid, ciliary body, and pigmented retinal epithelium, as well as the sclera. Bioassay of their extracts demonstrated that the last fraction accounted for nearly 90% of the total eye activity, with a specific activity of 2400 TU/mg (see Table I).

FIG. 3. Comparative behaviors of ciliary ganglionic neurons in different culture systems. Dissociated cells, from a common pool, were seeded with EE and serum on collagen (1) or PORN (2), or with heart-conditioned medium and serum on collagen (3) or PORN (4), or with serum only on either substratum (0). Culture time: 24 hours. Note lack of survival in (0), and occurrence of neurites only in (1) and (4). (For details, see Varon *et al.*, 1979.)

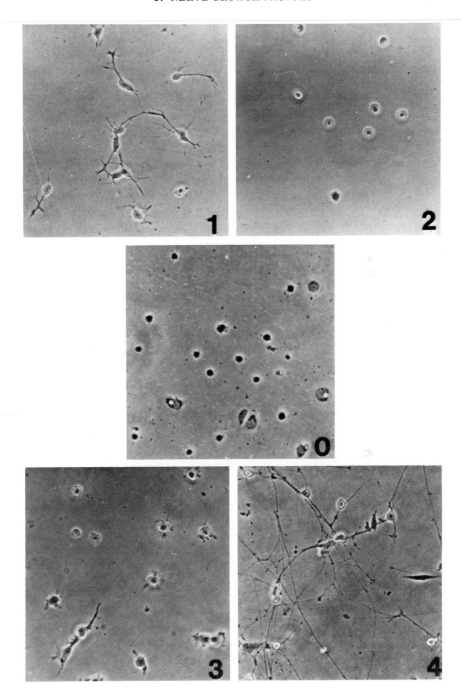

TABLE I

Distribution of Trophic (CNTF) Activity in Embryonic Chick Tissues

Tissue	TU/embryo[a]	Protein/embryo (mg)[a]	TU/mg protein
12-day chick embryo			
Whole embryo	8000 (100)	55.0 (100)	145
Brain	200 (3)	4.8 (9)	42
Viscera	500 (6)	7.2 (13)	69
Carcass	4800 (60)	40.1 (73)	120
Eye	2500 (31)	2.9 (5)	862
12-day chick eye			
Whole eye	2250 (100)	2.80 (100)	804
Neural retina	125 (6)	1.27 (45)	98
Cornea	8 (<1)	0.06 (2)	133
Lens and vitreous	100 (4)	0.58 (21)	172
Iris	33 (2)	0.06 (2)	550
Sclera plus (choroid and ciliary body plus pigment epithelium)	1985 (88)	0.82 (30)	2,421
15-day chick eye			
Whole eye	7100 (100)	4.70 (100)	1,511
Choroid plus pigment epithelium (CPE)	5650 (80)	0.38 (8)	14,870
Remainder	1450 (20)	4.32 (92)	336

[a] Percentages are shown in parentheses.

The same bioassay showed (Landa et al., 1980) that total CNTF activity of the chick embryo eye increased linearly between embryonic days 6 and 11 and again between days 15 and 18. An additional and much sharper rise in both total and specific activities occurred between days 12 and 14, bringing them by day 15 to about 7000 TU/embryo and 1500 TU/mg, respectively. Further refinement of the subdissection technique yielded a sclera-free CPE fraction. As shown in Table I, the CPE fraction continued to be responsible for 80% of the total CNTF activity of 15-day eyes, with a specific activity (nearly 15,000 TU/mg) about 45-fold higher than in the rest of the eye (340 TU/mg).

Work is in progress for the isolation of the eye CNTF. A preliminary study (Manthorpe et al., 1980) indicates that the CPE activity is in the form of a macromolecule with a molecular size of about 35,000 and an isoelectric point of about pH 5.0. The factor can be partially purified by ion-exchange and differential ultrafiltration but tends to lose activity in solution at 4°C (though not on storage at −76°C or in the lyophilized state). To date, batches of at least 100,000 TUs (2 mg protein) can be routinely obtained by a 10-hour procedure. The potency of the CNTF product (20 ng/TU or better) is of the same magnitude as the best NGF preparations (10 ng per biological unit), despite the activity loss incurred and the considerable heterogeneity still revealed by gel electrophoresis.

Altogether, the studies just described offer *direct* evidence for a target-derived neuronotrophic factor. The tissues innervated by CG neurons contain uniquely high levels of a soluble macromolecule, the trophic activity of which is directed toward the very same CG neurons. In addition, the activity of the target tissues increases markedly at a time when developmental cell death for the CG neuronal population is subsiding, raising questions about the role of target-derived factors beyond the promotion of neuronal survival. Finally, experimental approaches are opened up for investigating what extrinsic influences may regulate the levels of trophic activity within the target tissues themselves.

2. Other CNTF Sources

CNTF activity for ciliary ganglionic neurons is also present in chick tissues other than the eye, namely, embryonic carcass and viscera (Adler *et al.*, 1979b), cultured skeletal muscle cells (Nishi and Berg, 1977), and cultured heart cells (Helfand *et al.*, 1976, 1978). These observations indicate that cultured CG neurons can gain trophic support also from peripheral tissues they never innervate *in vivo*. It is possible to interpret such observations as indicating that *cholinergic* neurons in different parts of the nervous system may be sensitive to a more ubiquitous cholinergic neuronotrophic factor. Alternatively, target-derived cholinergic factors from different sources may be sufficiently similar to one another to be able to act on different cholinergic neurons (cf. Adler *et al.*, 1979b). It should be noted that the CAT activity is also at least preserved by chick embryo CNTF extracts (Varon *et al.*, 1979) and increases markedly with time in CG cultures supported by HCM (Nishi and Berg, 1979).

B. Neurite-Promoting Factors

1. Heart-Conditioned Medium

As already mentioned, HCM elicits a profuse outgrowth of neurites from CG neurons on PORN-coated dishes but not on collagen, even though it exerts its trophic activity on both substrata (Varon *et al.,* 1979). Collins (1978a,b) has recently reported that preincubation of PORN dishes with HCM for increasing time confers on the dishes an increasing competence to initiate CG neurite outgrowth. At the same time, this pretreatment reduces the neurite-promoting ability of the HCM itself. Thus, the original HCM appears to contain a material that binds to PORN and is responsible for neurite initiation.

These observations were confirmed and extended in another study (Adler and Varon, 1980). The HCM activities for CG neurons have been experimentally separated into (1) a *trophic factor* (HCM–CNTF) which is retained in solution by HCM preexposed for 24 hours to PORN, collagen, or sulfonated (tissue culture) plastic surfaces, and (2) a *neurite-promoting factor* (HCM–NPF) which adsorbs entirely (over 24 hours) to PORN but not to the other surfaces. The trophic factor present in NPF-deprived HCM supports CG survival on collagen, PORN, or NPF–PORN (i.e., PORN preexposed for 24 hours to original HCM) and yields similar titers of about 4 TU/embryo in the standard bioassay (see Section III,A). Neurites, of course, are only present on NPF–PORN. The NPF, adsorbed on PORN, elicits neurite development but fails to support 24-hour survival of CG neurons. Survival of the neurons requires the addition of trophic factors to the medium. Besides HCM, embryo extract (EE) is also competent to provide trophic support. Neurons supported by EE grow neurites on NPF–PORN but not on untreated PORN. Thus, it appears that the HCM-derived factor fits the definition of a *neurite-specifying* agent and that trophic agents may be unable to elicit neuritic growth without such a specifying partner.

2. *EE and Serum*

Unlike HCM, EE did not deposit neurite-promoting material during a 24-hour incubation over collagen, PORN, or sulfonated plastic (Adler and Varon, 1980). Nevertheless, EE in combination with serum (EE–S) supports neurite extension as well as survival on collagen (Varon *et al.,* 1979). The question, then, arises whether in the latter situation trophic and neurite-specifying activities are provided by the same molecule or by distinct agents as found for the HCM. Preliminary experiments have raised the possibility that in the EE–S collagen system a surface-bound NPF may be provided by the *serum* itself (Adler and Varon, unpublished). If confirmed, this finding might also explain a previous observation that horse serum was considerably more effective than fetal calf serum in supporting neuritic growth in conjunction with EE (Varon *et al.,* 1979).

Sera, of course, have long been viewed as providing adhesion-promoting agents besides trophic and hormonal ones (cf. Paul *et al.,* 1978; Bottenstein *et al.,* 1980). Several investigators have also been concerned with substratum-attached materials (SAMs) deposited on culture surfaces from serum (e.g., Culp and Buniel, 1976) or from microexudates released by cultured cells (e.g., Weiss *et al.,* 1975). It will be of interest to examine the possible relevance of such materials to the promotion of neuritic growth in the terms described here for CG neuronal cultures (see also Section, V,A). One may further speculate that neurite growth *in vivo* may be similarly specified by agents presented by the surface of the cells (glial or other) that comprise the natural terrains on which neurite elongation takes place. It is, for example, a common observation in tissue culture that neurons and neuritic growth fare better on the surface of nonneuronal cells than on cell-free substrata (cf. Varon and Saier, 1975). Furthermore, different neurite-promoting surfaces may be involved in defining preferential pathways for guiding growing axons to their destination. Indeed, selective extension of neurites along radial glial fibers *in vivo* has been described as an important migration mechanism in developing cortical tissues (Rakic, 1971, 1972; see also Varon and Somjen, 1979).

IV. The NGF

NGF designates a particular protein that elicits neurite outgrowth and other responses selectively from sympathetic and spinal sensory ganglionic neurons. Several reviews have covered the investigations of the NGF phenomenon, now spanning a quarter-century (Levi-Montalcini, 1966, 1976; Varon, 1975a; Hendry, 1976; Bradshaw and Young, 1976; Mobley *et al.,* 1977; Varon and Bunge, 1978; Yu *et al.,* 1978a). The fortunate abundance of NGF in the submaxillary gland of adult male mice (as yet not entirely explained) has permitted purification of the NGF protein and an extensive characterization of its molecular properties. However, the natural sources of NGF *in vivo* and at different stages of development have not yet been firmly identified. Considerable evidence points to two such possible sources, namely, peripheral tissues innervated by sensory and sympathetic neurons (Angeletti and Vigneti, 1971; Johnson *et al.,* 1971; Hendry and Iversen, 1973; Young *et al.,* 1976), and peripheral glial cells (Burnham *et al.,* 1972; Varon *et al.,* 1974a–d; Ebendal and Jacobson, 1975; see also Schwartz *et al.,* 1977; Murphy *et al.,* 1977; Perez-Polo *et al.,* 1977). Interestingly, C1300 mouse neuroblastoma cell cultures have also been reported to produce and release NGF (Murphy *et al.,* 1975).

A. ROLES OF NGF

1. NGF as a Target-Derived Factor

The following points support the view that NGF may be a target-derived factor in the sense defined in preceding sections, even though no involvement in a traditional developmental cell death phenomenon has been explicity reported:

1. NGF is required for the *survival*, as well as the growth and function, of the responsive neurons. This has been amply demonstrated *in vitro* (e.g., Levi-Montalcini and Angeletti, 1968; Burnham *et al.*, 1972; Greene, 1977; Bottenstein *et al.*, 1980). *In vivo*, a corresponding demonstration would require abolition of all the NGF sources, and, indeed, systemic treatments with anti-NGF immune sera can lead to nearly total immunosympathectomy (Levi-Montalcini and Booker, 1960b; Steiner and Schonbaum, 1972; Zaimis and Knight, 1972).

2. A *temporal pattern* in neuronal responsiveness to exogenous NGF has been reported for both spinal and sympathetic neurons (see Section VI), much like the temporal restrictions of target dependence in traditional cell death systems (cf. Fig. 1).

3. *Most body tissues* (and fluids) contain NGF in very minute amounts. Given the widespread distribution of spinal and sympathetic ganglionic nerve endings, this ubiquitous presence of NGF is in keeping with its occurrence in target tissues for such neurons. In fact, there have been suggestions that NGF distribution in peripheral tissues may grossly parallel sympathetic innervation (Angeletti and Vigneti, 1971).

4. NGF *can be taken up and retrogradely* transported to sympathetic or spinal ganglia from their respective innervation territories. Administration of ^{125}I-labeled NGF into the eye (cf. Hendry, 1976), a selected dermatome (Stoeckel and Thoenen, 1975), or chick embryo leg (Brunso-Bechtold and Hamburger, 1978) has permitted demonstration of the specificity of the uptake and tracing of the factor along the nerve fibers and into the ganglionic somata.

5. NGF also plays a role for *injured* neurons, a situation analogous to phase 4 (cf. Fig. 1). Target-derived signals can be prevented from reaching sympathetic neurons by axotomy or colchicine treatment—the latter presumably acting by interfering with retrograde axonal transport (Grafstein, 1975; Purves, 1975, 1976). Characteristic neuronal responses to either treatment are chromatolysis and loss of presynaptic and satellite glial contacts. All the above consequences are

prevented by topical treatment of the affected ganglion with NGF
(Purves and Njå, 1976; West and Bunge, 1976).

2. NGF as a Trophic Factor

As already mentioned, NGF is required for survival of its target
neurons, both *in vivo* and *in vitro*. In addition, however, NGF also
causes hypertrophy, neurite elongation, and a rise in transmitter-syn-
thesizing enzymes in the responsive neurons. The conceptual framework presented in previous sections (cf. Section II,B and Fig. 2) provides an interpretation of this multiplicity of NGF effects, namely,
that separate influences *specify* each of these consequences while
NGF supplies in all cases the required *trophic* drive. It is important to
stress that the same trophic agent could well lead to different consequences if it is applied:

1. To cells with different programs (e.g., sensory and sympathetic
neurons).
2. To the same cell at different developmental ages (either because
of different specifying influences or of different endogenous levels of
trophic factor already acting on the cell).
3. To different parts of the same cell if they contain different
cellular machineries (see Section V,B).
4. In different amounts to the same cell (e.g., if different cell
machineries require different "trophic thresholds" for their mobilization).

We do not know how much NGF is naturally available to responsive neurons *in vivo* at different developmental stages. It is possible
that NGF availability increases with age (from the same or from additional sources) and that the exogenous supply of high levels of NGF
merely compresses in a brief time the consequences that a gradual increase in NGF would elicit progressively during the natural course of
developmental events. Nor do we know, at present, what other extrinsic influences act upon NGF-responsive neurons at different developmental stages (see also Section VI). Be it as it may, the same NGF has
been reported to (1) increase cell numbers at early embryonic
stages—although it remains debated whether this reflects promotion
of neuroblast proliferation (Levi-Montalcini and Hamburger, 1951;
Levi-Montalcini and Booker, 1960a) or rescue of neurons from their
expected developmental cell death (cf. Hendry, 1976); (2) promote
neuronal hypertrophy and neurite growth if applied at more advanced
stages; (3) cause hypertrophy and hyperfunction in postnatal animals;

or (4) support survival, axonal regrowth, and functional recovery after axotomy in the adult.

In vitro, there is clear evidence that different levels of NGF do result in and are required for the expression of different behaviors. Chun and Patterson (1977) have reported a detailed study on dissociated rat sympathetic neurons in monolayer culture. Increasing NGF concentrations led to increasing neuronal hypertrophy and neuritic growth. However, it took at least 2- to 10-fold higher NGF levels to elicit an elevation in tyrosine hydroxylase activity. As already mentioned (Section II,B,2), NGF was equally required, and at equally high concentrations, for an increase in CAT when the specifying factor for this type of transmitter behavior was also provided (Chun and Patterson, 1977).

3. Does NGF Have Other Functions besides the Trophic One?

Neuritic growth may involve NGF not only as a source of trophic drive (with the resulting support of axonal supplies) but possibly also in some more direct, specifying terms and even in some tactic (directional) terms. Both of these possibilities will be discussed in Section V,B. An even more puzzling phenomenon is the ability of NGF to act on a pheochromocytoma clonal line, PC12, and convert the growing tumoral cells into mitotically stable neuronal elements (Greene and Tischler, 1976; Greene, 1978). This, and other related observations, will be taken up in Section VI.

B. THE PROBLEMS OF A UNIFIED VIEW OF NGF ACTION

1. Delivery Routes and Molecular Targets

The effects of NGF include both membrane and nuclear manifestations (cf. Varon, 1975a; Yu *et al.,* 1978a). Since nuclear events may represent consequences of membrane events, and vice versa, there has been considerable debate as to whether NGF acts solely at membrane sites, solely at the level of the cell nucleus, or at both. The debate is encouraged by equally convincing evidence for a physical association of NGF with extracellular (surface) and intracellular locations. NGF is known to bind to high-affinity sites on the surface of ganglionic cell membranes, which could represent true receptors for its action (Banerjee *et al.,* 1973, 1976; Herrup and Shooter, 1973; Frazier *et al.,* 1974; Levi-Montalcini *et al.,* 1974; Sutter *et al.,* 1979). However, NGF can also be internalized both *in vivo* (cf. Hendry, 1976) and *in vitro* (Norr and Varon, 1975)—as is postulated for target-derived trophic agents—and the surface binding sites could be viewed as preferential gates for this process. In addition, NGF has also been reported to bind directly to nuclear materials (Andres *et al.*, 1977).

Figure 4 presents a schematic view of the various locations of NGF, on or in the cells, that may be involved in the action of the factor. Exogenous NGF (x) must first associate with a surface membrane binding site (solid bar on the membrane). As already stated, this location (Fig. 4A) may be viewed as the only site of action (cf. Frazier *et al.*, 1973a; Levi-Montalcini *et al.*, 1974). In this case, the NGF–receptor complex will cause membrane alterations which will lead to intracellular changes (arrows), such as the release of "second messages," and these in turn will elicit via various cellular machineries all the ultimate consequences usually manifest to the observer. Alternatively, the surface-bound NGF may be internalized by an endocytotic sequence (Fig. 4A–C). This could end with fusion of the endocytotic vesicle to the nuclear membrane (Fig. 4D), and either the exposure of bound NGF to intranuclear targets or the release of free NGF for interaction with them. The hypothesis of a direct action of NGF on nuclear targets must imply such a transport sequence, since it is difficult to conceive how NGF could appear in a free form in the

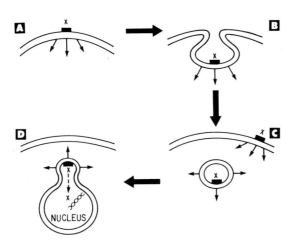

Fig. 4. Schematic view of several cellular locations that NGF many occupy. (A) NGF (x) binds to a receptor (solid bar) on the surface membrane and triggers changes (small arrows) inside the cell. (B and C) Endocytotic internalization of receptor-bound NGF. (D) The endocytotic vesicle has fused with the nuclear membrane, and the NGF may be released from the receptor into the nucleoplasm.

cytoplasm short of traversing the plasma or the vesicular membrane by means that no one has yet proposed.

There is, however, a way to accommodate both the surface receptor view and the internalized NGF view. This is based on the consideration that, throughout the several locations taken by NGF, the factor would always be on the same surface receptor facing away from the cell cytoplasm and therefore would continue to have the opportunity to alter the underlying membrane and cause cytoplasmic changes (Fig. 4, arrows), regardless of whether the NGF–receptor–membrane complex was situated at the periphery of the cell (Fig. 4A), inside the cell (Fig. 4B and C), or as part of the nuclear membrane (Fig. 4D).

2. Initial Effects and Ultimate Manifestations

We have repeatedly mentioned the diverse and multiple manifestations of NGF action displayed by NGF-responsive cells. All such manifestations require several hours or even days before they can be perceived by the observer. A principle of parsimony would dictate that the initial steps of NGF action be common to most if not all of the ultimate manifestations. In other words, the various long-term consequences we generally perceive are likely to be the end products of separate sequences of cellular events, *involving both common and separate cellular machineries,* and it is reasonable to postulate that such sequences branch out from common, earlier events toward which the action of the factor is truly directed.

This concept is illustrated schematically in Fig. 5. The association of NGF with a cellular receptor (R_1), whatever its location, triggers a series of rapid events (-x-x-) leading to the alteration of a key property of the cell (1). Changes in this property directly affect the behavior of more than one additional property (2 and 3), which in turn control a variety of other events (branched lines). Some of these events may lead to longer-term consequences independently of other activities, while other events may converge before eliciting their ostensive expression. Extrinsic influences (arrows), other than the trophic one provided by NGF, may select the branches the cell will follow (e.g., specifying factors). Alternatively, they may exert additional regulation or even control of certain critical steps required for the ultimate expression of one or more cell behaviors. The likelihood that such controls occur has been explicitly raised in a recent study on cultured spinal ganglionic neurons, in which neuronal survival could be ensured, in the absence of serum, by a defined combination of NGF, insulin, progesterone, transferrin, and putrescine (Bottenstein *et al.,* 1980). While NGF was essential for the survival of all the neurons, omission of any

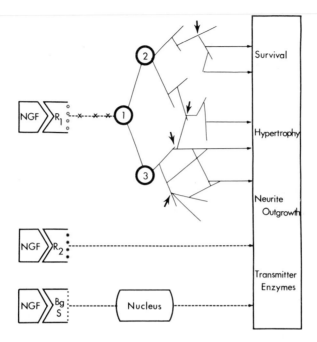

FIG. 5. Diagrammatic representations of sequential events linking the presentation of NGF (left) with its long-term consequences (right). See text for details.

one of the other ingredient reduced the number of survivors by about half. The scheme in Fig. 5 assigns to NGF the role of a "master" factor (controlling all cell machineries), in contrast to the other agents which act as pleiotypic factors (Hershko *et al.*, 1971) and control only discrete groups of cellular events.

The debate on whether NGF acts on a single molecular target (e.g., surface receptors) or on more than one is only briefly illustrated in Fig. 5 by the two broken lines. One may conceive that neuronal surfaces present to NGF two different classes of receptors (R_1 and R_2) each responsible for different sequential events (such as those illustrated for R_1) and eventually different outward manifestations. Alternatively, NGF could associate with a nonreceptor binding site (Bg. S), so as to be transported to a nuclear site of action.

C. MECHANISMS OF TROPHIC ACTION

If one accepts the premise that all (or at least most) of the NGF effects originate from the initial interaction between NGF and its membrane receptors, one would like to discover *short-latency responses* (on the order of minutes) likely to occur at the very beginning of the sequence of events diagrammed in Fig. 5. One would also like to identify among these responses some that could invest *key cell properties* pervasive enough to affect a whole variety of cellular machineries and behaviors. The following is a brief survey of some recent studies that may have brought these two objectives within reach.

1. Cyclic AMP Responses

Cyclic nucleotides, in particular cyclic AMP, have received considerable attention in several cell systems as potential second messages linking the interaction between a hormone or factor and its surface receptor with the more complex cellular responses elicited by that agent. Early suggestions of a cyclic AMP involvement in the mode of action of NGF (Roisen *et al.*, 1972; Hier *et al.*, 1972) have been refuted by other investigators (Frazier *et al.*, 1973b). More recently, however, several investigators have reported an NGF-induced cyclic AMP elevation in rat superior cervical ganglia (Nikodijevich *et al.*, 1975), chick dorsal root ganglia (Narumi and Fujita, 1978; Skaper *et al.*, 1979), PC12 cells (Schubert and Whitlock, 1977), and adrenal medulla (Otten *et al.*, 1978). In all cases, the cyclic AMP rise occurred a few minutes after NGF presentation and subsided within 30 minutes. Thus, this cyclic AMP response may constitute at least one of the short-latency responses being sought and could, in principle, represent an early "crossroad" event in the scheme in Fig. 5. The possibility has been debated (cf. Yu *et al.*, 1978a) that, in fact, a cyclic AMP response may lead, via activation of cytoplasmic protein kinases, to several nuclear events already reported, for example, phosphorylation of a specific nuclear protein (Yu *et al.*, 1978b) and transcription-dependent selective induction of tyrosine hydroxylase (McDonnell *et al.*, 1977) and ornithine decarboxylase (Hatanaka *et al.*, 1978).

2. Permeation Responses

Dorsal root ganglia (DRGs) from 8-day chick embryos are traditional NGF targets. Early studies had shown that RNA and protein labeling (from exogenous radiouridine and radioleucine, respectively) occurred at higher levels in DRGs incubated in the presence than in the absence of exogenous NGF (Angeletti *et al.*, 1965). These differences were subsequently found to be mainly due to a decline in labeling competence by NGF-deprived ganglia (Partlow and Larrabee, 1971; Burnham *et al.*, 1974). Horii and Varon (1975) incubated DRG cell suspensions for various times without NGF and then supplied them with NGF and a pulse of [³H]uridine. They found that for about 6 hours the NGF-deprived cells remained competent to recover their full RNA-labeling capability within minutes of the delayed NGF presentation. Longer NGF deprivations led to progressively reduced recoveries and (after 15–20 hours) to irreversible damage and death. Six-hour NGF-deprived cells, thus, provided for the first time an experimental system in which quantitative RNA-labeling changes were induced within minutes of NGF presentation, while identical cell aliquots (not receiving the factor) served as rigorous controls.

This experimental system further allowed the demonstration that the action of NGF resulted in an increase in the *net uptake* of [3H]uridine, rather than an increased rate of RNA synthesis (Horii and Varon, 1975, 1977). The ability of DRG cells to accumulate acid-soluble radioactivity during a radiouridine pulse displayed the same changes as their ability to label RNA: Both abilities declined progressively in NGF-deprived cells and were fully restored on 6-hour-delayed NGF presentation, the magnitude of this restoration depending on the dose of NGF. This uridine response not only occurred within minutes of the NGF administration but also was not blocked by actinomycin or cycloheximide and was therefore independent of transcriptional and translational events. Here, then, was another example of a short-latency response affecting membrane properties without involving macromolecular syntheses.

DRG cells are able to synthesize uridine and, therefore, do not depend on its exogenous supply. Of much greater significance to the cell would be the control by NGF of exogenous substrates essential to the cell, such as glucose or certain amino acids (Horii and Varon, 1977). Skaper and Varon (1979a) demonstrated that the uptake of 2-deoxyglucose (2DG, a nonmetabolizable glucose analog using the same transport system) displayed the same temporal behavior and the same dependence on time and dose of NGF administration as had been found for uridine. Despite the cellular heterogeneity of the preparation (which comprises neurons, Schwann cells, and capsule-derived fibroblasts), it was possible to distinguish operationally two components within the total 2DG accumulation. One component is NGF-dependent and presumably occurs only in the NGF-sensitive DRG neurons. The other component is NGF-independent and presumably reflects uptake activity of the other DRG cells. Transport kinetics were quite different for these two components of 2DG uptake. Furthermore, it was possible to show that the time required for full recovery of the first component upon presentation of NGF (5–10 minutes in the previous studies) depended mainly on the concentration of NGF and could be reduced to *less than 1 minute* with suprasaturating levels of the factor. This indicated that recovery time reflected the time needed for equilibration of exogenous NGF with its surface receptors, rather than a relatively slow development of sequential cellular changes aimed at restoring transport competence.

3. Sodium Responses

The ability of NGF to control various, though not all, transport systems would be exercised most parsimoniously if what NGF regulates is a single feature common to all affected transport systems. Skaper and Varon (1979b) showed that all transport systems found sensitive to NGF (nucleosides, glucose, and α-aminoisobutyric acid, but not leucine) also required the presence of sodium ions in the outside medium and were disrupted by pretreatment with ouabain and veratridine, as well as dinitrophenol. The findings indicated that such transport systems were sodium-coupled, with transmembrane sodium gradients providing the driving force (cf. Varon and Wilbrandt, 1966; Crane, 1977; Hopfer, 1977).

The question then arises whether the Na^+ gradient and the NGF requirements of the same transport systems are independent of each other or whether, conversely, one is subordinate to the other. An investigation of the uptake and release of $^{22}Na^+$ by DRG cells (Skaper and Varon, 1979c) revealed two critical features:

1. In the absence of NGF, the cells developed a progressive propensity to take up and retain Na^+ from the medium. As a result, cells deprived of NGF for 6 hours accumulated Na^+ at about sixfold higher levels than NGF-supported cells.

2. NGF-deprived cells preloaded with $^{22}Na^+$ and then provided with NGF released their excess sodium content. The release occurred within the same time frame (a few minutes) as that already observed for the uridine and glucose responses (as well as the cyclic AMP response).

The best explanation for these observations is that NGF controls a *sodium extrusion mechanism* in its responsive neurons (Skaper and Varon, 1980a). This would in turn result in a regulation of transmembrane sodium gradient and of sodium gradient-dependent transport systems. A similar *sodium response* to NGF has subsequently been observed with undissociated DRGs and sympathetic ganglia from chick embryo, the two traditional targets of NGF action (Skaper and Varon, 1980b).

4. An Ionic Model

It is possible that the Na^+ response constitutes only a passive adjustment of Na^+ to alterations imposed on other ions (e.g., Ca^{2+}) and that the true objective of NGF action is the regulation of the latter rather than a control of sodium pumps. Thus, the question of *how*

NGF exerts its control on Na^+ extrusion remains unresolved for the present time. However, the unpredicted recognition of an ionic involvement in the action of NGF opens up entirely new perspectives and invites the speculation (Varon and Skaper, 1980) that *ionic events* occupy a key position in the sequence of cellular changes induced by NGF—i.e., positions 1 or 2 in the scheme in Fig. 5. In an even more speculative spirit, one may wonder whether neuronotrophic factors in general exert all their influence on target neurons via a strict modulation of transmembrane ionic potential and/or intracellular ionic environments.

The ability to maintain selective ionic permeabilities and asymmetric ionic distributions (hence an electrochemical gradient) across the plasma membrane is a universal property of animal cells. Thus, it appears reasonable to assume that selected intracellular ionic environments and/or transmembrane potentials are critical for cell survival. Regulation of the uptake of essential exogenous substrates via Na^+ gradients—as illustrated by the effects of NGF on DRG cells—may be but one example of how neuronotrophic factors may control cell survival and cell growth by an ionic mechanism. Other possibilities to be considered are, for example, regulation of cyclic nucleotide systems (another short-latency effect of NGF on ganglionic cells) and selective activation–inhibition of cytoplasmic or nuclear enzymes. There is a vast literature on the importance of ionic regulation on a variety of cell performances, including developmental events, which cannot be appropriately reviewed here (cf. Jaffe and Nuccitelli, 1977). A few pertinent examples are the regulation of DNA synthesis (Cone and Cone, 1976, 1978), cyclic nucleotides (e.g., Rebhun, 1977; Rasmussen and Goodman, 1977), cell elongation (see Section V), and the shift of neurotransmitter modalities (Walicke and Patterson, 1978).

V. Neurite Growth

It has been firmly established by a number of investigators that neurite growth involves the operation of an elongation machinery, the *growth cone*. The growth cone is supplied with new materials from the proximal neurite (and ultimately the cell soma) via effective anterograde transport and is engaged in adhesive interactions with the substratum on which it operates (cf. Johnston and Wessells, this volume). This assumes, of course, that the genetic program for neuritic growth has already been included in the "restricted" repertoire of the differentiated cell, as is by definition the case for a neuron (see also Section VI).

Considerable knowledge has been gained on the morphology and dynamics of growth cones, including changes in plasma membrane and cytoskeletal structures involved in their operation. Much, however, remains to be learned about what regulates growth cone activities, i.e., the start and stop signals and the molecular mechanisms through which they are translated into action. One may also have to distinguish among different situations, namely, (1) *initiation* of neuritic growth from cells never previously engaged in such activities, (2) *elongation* of neurites already initiated, and (3) *reinitiation* of neuritic growth, from either the soma or a neuritic stump, in cells that have already performed such tasks in their previous history (e.g., axotomized neurons *in vivo*, excised ganglia, or dissociated neurons *in vitro*). We shall limit ourselves to a brief discussion of triggers and mechanisms controlling neuritic growth that are particularly pertinent to other problems explored in this chapter.

A. Adhesion as an Extrinsic Trigger

There appears to be no doubt that the activity of growth cones is strongly influenced by the relative adhesiveness of their surface membrane toward the substratum on which they operate (cf. Varon, 1975b; Varon and Saier, 1975; Sidman and Wessells, 1975). Adhesive interactions are likely to play a key role in the initiation (Collins, 1978a) as well as the subsequent elongation of neurites. It remains to be ascertained, however, to what extent adhesive interactions constitute a true trigger for neuritic growth (in the terms defined as neurite-specifying influences) or instead represent permissive conditions for it, i.e., *allow* neuritic expression under appropriate trophic and specifying influences. Growth cone–substratum interactions can conceivably be varied by altering either the adhesiveness of the substratum or that of the growth cone (or the cell) membrane.

1. Substratum Adhesiveness

In vitro studies have shown that growth cone activity prefers highly adhesive substrata (Letourneau and Wessells, 1974; Letourneau, 1975). Substratum adhesiveness can be altered experimentally by three main approaches:

1. The first approach is to choose different *solid surfaces* (e.g., regular or sulfonated plastic or glass) or to *coat* them with defined materials (e.g., collagen, polylysine, or PORN). Surface coating, however, is not an all-or-none operation. The concentration of polycat-

ion used in the coating process influences adhesiveness and stability of the coat and may also have adverse effects on the cells (cf. Varon, 1979). Collagen can be applied in a variety of ways, resulting in different cell behaviors (cf. Varon, 1977b; Adler et al., 1979a).

2. Another approach is to allow the *deposition* of biological materials, from media supplements or from other cultured cells, before the population under study is to be seeded. These SAMs have been studied mainly in terms of their ability to support the proliferation of nonneural cells (e.g., Grinnell, 1978; Weiss et al., 1975; Culp and Buniel, 1976). Schubert (1976) has analyzed the microexudates generated by several clonal cells and has reported that those from neuronal, glial, and muscle cells are uniquely characterized by the predominance of a 55,000-dalton protein. We have already discussed (Section III,B) the SAM derived from heart-conditioned media (Collins, 1978a,b; Adler and Varon, 1980) as a special case of neurite-promoting material.

3. A third approach is to use *live cells* as a preattached culture substrate. While this approach may come closest to an *in situ* situation, it makes it difficult to distinguish between adhesive interactions and influences mediated by soluble agents released from the cell layer (nutrients and factors). In fact, in various cell systems, preformed cell layers have been used either to supply needed materials (the feeder layer concept) or to study the selective adhesiveness of one cell population to another (cf. Glaser, 1978). The distinction between adhesive and supportive influences by preformed cell layers was particularly evident in studies on mouse DRG neurons dissociated by trypsin and purified by differential attachment on tissue culture plastic (Varon et al., 1973). Layers of other cells, regardless of their source, were required for attachment of these neurons even in the presence of NGF. Only ganglionic nonneuronal (glial) cells, the physiological partners of DRG neurons, displayed the additional capacity of substituting for the required exogenous NGF their own supply of NGF-like materials (Varon et al., 1974 a–d).

2. Membrane Adhesiveness

Surface properties of the plasma membrane, hence presumably of the growth cone, are altered by a variety of cellular activities (Moscona, 1974). Agents that affect such cellular activities may also cause changes in membrane adhesion and, therefore, in the adhesive interactions between membrane and substratum. Few studies have addressed specifically this question in the context of neurite regulation. A variety of treatments have led to neuritic outgrowth from neuroblastoma clonal cell cultures (cf. McMorris *et al.*, 1973; Schubert *et al.*, 1973). Several such treatments, for example, serum withdrawal and the use of certain conditioned media, may have operated via alterations of the culture substratum. However, other treatments such as the use of hypertonic media (Ross *et al.*, 1975), 5-bromodeoxyuridine (Lasher and Cahn, 1969; Bishoff and Holzer, 1970; Brown, 1971), and antimitotic agents (e.g., Carlin *et al.*, 1974) may well act via changes imposed on the cell membrane itself (Schubert *et al.*, 1971; see also Varon and Saier, 1975). Among the several consequences of NGF treatment, the possibility has been noted that this factor also results in increased adhesiveness of the responsive cells (Weston, 1972; Varon *et al.*, 1974a,b). The ability of NGF to elicit neurite growth from the PC12 pheochromocytoma clonal cells (Greene and Tischler, 1976) and from normal adrenal medullary cells (Unsicker *et al.*, 1978) may or may not involve adhesive changes in the cell membrane (see Section VI).

B. MOLECULAR MECHANISMS

Adhesive interactions with a substratum can initiate a chain of complex events (cf. Varon, 1979) involving both the membrane and the submembrane scaffold of cytoskeletal structures, such as those described in a growth cone. Little is known about the transduction mechanisms that may link surface interactions and submembrane structural changes. It has been reported that cell adhesion changes can lead to changes in permeation properties and cyclic nucleotide activities, among others. It is possible that agents regulating neuritic growth do so by eliciting events similar to those generated by adhesive interactions. In what follows, we shall restrict ourselves to the area of *ionic involvements,* to which several observations have begun to point strongly. For the purpose of discussion, we shall assume that the nerve cell is provided with a fixed trophic drive adequate not only for maintenance but also for growth.

1. The Concept of Continuous Growth

In vivo, axonal elongation ceases either when growth cones have reached their synaptic targets (cf. Rees *et al.,* 1976) or when body dimensions have imposed final distances between the nerve cell soma and its connected target cells. *In vitro,* information on how finite neurite production is has been rarely sought (Bray, 1970), and one may conceive of elongation ceasing either on connection with existing targets or on exhaustion of growth potential (i.e., when the drive from the fixed level of trophic agents provided in the medium is balanced out by the maintenance needs of the "grown" cell). Where a stop signal may be involved, one may ask whether (1) the same mechanism involved in the start response applies in reverse in the stop response, and (2) whether the stop signal is merely the removal of the start signal or a different signal eliciting opposite reactions.

Paul Weiss (1968) has proposed the concept of continuous growth, according to which axoplasmic flow (i.e., the supply of materials from soma to nerve ending) does not differ under conditions where further elongation takes place or those where elongation has ceased. In the first case, the material reaching the distal portion of an axon would be used to build new axonal length, while in the other case it would be degraded and dissipated (at least in part via retrograde axonal transport). The continuous growth concept can easily be equated to a constant trophic drive supplying a constant flow of materials down the axon. The concept, however, also implicates a start–stop mechanism *at the axonal ending* for deciding the fate of the arriving materials. Lasek and Hoffman (1976) have suggested that the decision may lie with certain Ca^{2+}-sensitive proteases which accompany the axoplasmic flow in an inactive form and which may be activated to degrade axoskeletal material by an influx (or a local mobilization) of Ca^{2+} at the nerve fiber terminal. Possible involvements of divalent cations in neurite growth have been repeatedly noted (cf. Varon and Saier, 1975). In particular, Schlaepfer and Bunge (1973) have reported that, in culture, the breakdown of amputated neurites may result from an influx of Ca^{2+} and could be long delayed in Ca^{2+}-free medium.

2. Electric and Ionic Currents

Several observations on plant and animal cells (reviewed in Jaffe and Nuccitelli, 1977) have indicated that cells may be the object of natural transcellular ionic currents and that elongation in such cells occurs in cell regions that are the site of entry of an ionic current. In fact, it appears that the local entry of a current may both precede and predict the onset of elongation at this site. The current involves Ca^{2+} which may move in first and cause a larger, steady influx of other ions (and a corresponding efflux of ions from the other pole of the cell). Based on his studies on fucoid eggs, Jaffe (1968) proposed that the local trigger for polarized cell growth was the entry of such an ionic current. This would make the site of entry electropositive relative to the rest of the cell and possibly guide electrophoretic migration of molecules along the membrane or of intracellular vesicles toward the membrane in the direction of the growing pole. Such a model mechanism could apply to the growth of a neurite. One may speculate that inward currents at the growth cone would sustain neurite elongation, or even that inward currents at localized regions of the somal membrane could determine there the initiation of neurite growth.

The imposition of electric fields on cultured explants has been used to examine electrical influences on neuritic growth. An increase in neuritic outgrowth toward the cathode and a reduction or suppression of it toward the anode have been reported with explants of chick embryo medulla (Marsh and Beams, 1946) and spinal ganglia (see Jaffe and Nuccitelli, 1977). More recently, Sisken and Smith (1975) applied direct electric currents to explant cultures of chick embryo trigeminal ganglia and observed a preferential migration of neurons, neurites, and nonneuronal elements toward the catode. In addition, there was increased survival of neurons within the explant and a faster than expected rate of neuritic outgrowth, not unlike that reported in dorsal root ganglia under the influence of NGF. These authors suggested that such effects could reflect a gradient of cations (especially Ca^{2+}) moving outward from cathode to explant, rather than a direct electrical stimulation of the ganglionic cells by the imposed fields (see also Pilla, 1974). In a subsequent study (Sisken and Lafferty, 1978), explant cultures of DRGs were reported to display similar neuronal and neuritic behaviors whether subjected to minute amounts of electric current or under treatment with NGF.

3. Localized Triggers of Ionic Changes

Neurite growth is an exquisitely polarized phenomenon. Thus, both initiation and elongation signals must be applied to discrete, narrowly restricted portions of the cell membrane. In the case of initiation, the localized nature of the response implies in turn either a membrane mosaic of receptive patches or a signal source that is itself localized. Direct application of an electrical stimulus could be such a localized signal, as would be the use of a mechanical stimulus (Wessells and Nuttall, 1978), and both could operate via ionic perturbations. A substratum-bound adhesive material could also work in the same manner, given a sufficient microheterogeneity of its distribution on the surface.

Soluble agents, on the other hand, could act in a neurite-promoting capacity only if (1) the agent is available in the form of a concentration gradient or bound to the substratum itself, or (2) sites receptive to the agent are uniquely or more extensively present on the growing region of the membrane. NGF, as we have already discussed (Section IV,C), may exert its trophic action via the regulation of ionic mechanisms.

Also mentioned (Section IV,A) has been the concept that a neurono-trophic agent may elicit different responses via the same mechanism if it acts on differently specialized portions of a cell. Thus, it appears possible that interactions of NGF with receptors located on a growth cone, leading to local changes in ionic mechanisms, could exert a neurite-stimulating action separate from the overall trophic activity of NGF.

Two recent reports appear pertinent to the questions raised in the preceding paragraph:

1. Campenot (1977) has cultured dissociated rat sympathetic neurons in a multichamber vessel in which the soma and distal neurites were separated by fluid-impermeable barriers. Extension of neurites into the distal compartment required the presence of NGF in both somal and distal compartments. Subsequent removal of somal NGF had no effect on the survival of either the soma or the distal neurites. However, removal of NGF from the distal compartment caused eventual degeneration of the distal neurites. These findings confirm the ability of sympathetic neurons to receive their NGF at the nerve endings but also indicate that neuritic growth requires *local* action of NGF besides general trophic support.

2. Letourneau (1978) cultured chick spinal sensory neurons in a semisolid agar matrix in which a concentration gradient of NGF had been established. He described evidence for a preferential orientation and an enhanced extension of neurites up the gradient of NGF. These data confirm the receptivity to NGF of growth cones of elongating neurites and are at least compatible with the concept of a local neurite-promoting effect of NGF. In addition, they demonstrate the capability of gradients of a trophic factor to serve as *tactic* signals. Levi-Montalcini (1976) had already reported that intracerebral injections of NGF into newborn rodents caused aberrant growth of sympathetic nerve fibers into the central nervous system. One may also note, in this context, that gradients of auxins, i.e., growth-promoting factors for plant cells, have been strongly implicated in the generation of inward currents of ions into the growing pole of such cells (cf. Jaffe and Nuccitelli, 1977).

VI. Trophic and Neurite-Promoting Influences in Early Development

Most culture systems applied to the study of trophic and neurite-promoting mechanisms have used neurons near or beyond the developmental stages at which they acquire their target dependence. This temporal parameter should always be kept in mind before attempting to generalize interpretations of the experimental data. What is valid for a neuron at a given developmental stage is not necessarily valid for the same cell at a temporally close but biologically different phase of its history—as witnessed by the several transitions occurring in this period (between phases 1 and 2, or phases 2 and 3). For example, most neurons used in these culture studies already bear neurites at the time of their collection, and one important effect of dissociation procedures is precisely the loss of such neurites (cf. Varon, 1970, 1975b, 1977b; Varon and Saier, 1975). Neuritic growth in culture, consequently, represents in most cases a *regenerative* phenomenon, the regulation of which may or may not be identical to that operating during phase 1, when *de novo* neurite development normally starts. This distinction becomes even more important when the neurons under consideration have already developed synaptic contacts with their target. One should also keep a distinction between target-dependent neurons that have not yet undergone the death or survival selection ("unrescued" cells, comprising both future casualties and future survivors) and those collected after cell death has taken place ("rescued" neurons only).

With the above considerations in mind, we shall address in this section the following question: Do the trophic and neurite-specifying concepts, thus far developed only for target-dependent neurons, also apply to neurons in phase 1 and 2 or, even earlier, to nerve cell precursors? Because only scanty information is available on specific requirements of early neuronal elements, this section will necessarily involve more speculation and questions than it will provide answers.

A. PHASE-1 NEURONS

As already mentioned (Section II,A), early neurons seem to be able to survive and differentiate in the absence of trophic and/or specifying influences from their future targets. This conclusion is firmly supported by the study of several neuronal populations, the target territories of which were removed well before the neurites could have reached them (cf. Hamburger, 1975; Prestige, 1970; Cowan, 1973; Landmesser and Pilar, 1978). At least three alternative explanations can be proposed for the target-independent survival (and growth) of these early neurons: (1) The neurons require the *same* trophic factor now as they will after encountering the target (target-derived trophic factor, or $T'TF$), but they receive it from different sources; (2) they are already under strict control of a trophic factor, but a *different* one; or

(3) their survival *does not* require extrinsic trophic support, but merely adequate availability of nutrients. One should note that, since neurite growth is taking place during this phase, it could also be asked whether neurite-specifying influences are being applied at this time in the same manner as they have been perceived to operate later on (see Section V). The few experimental data concerning these problems come exclusively from NGF studies, but they will hopefully inspire investigations with the more recently identified CNTF.

From the beginning of the NGF investigations, it was pointed out by Levi-Montalcini that the susceptibility to exogenous NGF by NGF-related neurons has a temporal dimension (cf. Levi-Montalcini, 1966; Levi-Montalcini and Angeletti, 1968). Chick embryo DRG neurons are reputed to respond to exogenous NGF only after embryonic day 6 (E6) and to stop requiring it after E14–17, both *in vivo* (Levi-Montalcini and Hamburger, 1951; Winick and Greenberg, 1965) and *in vitro* (Luduena, 1973; Letourneau, 1975). Chick sympathetic neurons also begin to respond to exogenous NGF after 6–8 days of development (Levi-Montalcini and Hamburger, 1951; Winick and Greenberg, 1965; Partlow and Larrabee, 1971). More recently, Coughlin *et al.*, (1977, 1978) examined explant cultures of mouse superior cervical ganglia (SCG) of different prenatal ages. They reported that early (E13–15) sympathetic neurons did not require exogenous NGF for their survival over the next few days, and were even capable of expressing in its absence "differentiated" behaviors such as neurite production and tyrosine hydroxylase activity. A strict requirement for exogenous NGF develops in the next few days (*in vivo* or *in vitro*), so that late fetal (E18) or newborn cells will die if NGF is not provided in the medium.

At first glance, all the above observations seem to indicate that NGF becomes relevant to sensory and sympathetic neurons only at some stage in development, before which these neurons either need no trophic factors or depend on a trophic factor that is not NGF. However, several questions will have to be fully resolved before a clear and firm interpretation can be made.

1. Alternative Sources of Trophic Factors

Both the *in vivo* and the *in vitro* experiments described above concern "intact" ganglia, in which the neurons remain closely associated with nonneuronal ganglionic elements. It is of particular interest that neonatal mouse DRG neurons do not appear to require exogenous NGF in *explant* cultures, while they depend on it for their survival and neurite growth in *dissociated cell* cultures (Varon *et al.*, 1973). It has been demonstrated that nonneuronal cells from DRGs of fetal and perinatal mice (as well as of embryonic chick and neonatal rat) are able to produce and deliver to their neurons a trophic agent that is immunochemically as well as biologically analogous to NGF (Burnham et al., 1972; Varon et al., 1974a–d; Varon and Bunge, 1978). NGF-like active agents have also been detected from other glial sources (Longo and Penhoet, 1974; Arnason et al., 1974; Ebendal and Jacobson, 1975; Murphy et al., 1977; Schwartz et al., 1977; Perez-Polo et al., 1977), as well as from normal and neoplastic fibroblasts (cf. Young et al., 1976). It would be of obvious interest to determine whether NGF is also produced by nonneuronal cells at developmental stages corresponding to the phase 1 of neuronal life. Equally interesting, though even more speculative, is the possibility that the early neurons supply themselves with a needed trophic factor, as suggested by the ability of neoplastic sympathetic neurons (C1300 clones) to generate NGF (Murphy et al., 1975).

If NGF were already available to early ganglionic neurons from nonneuronal cells in situ or in explant cultures, the mode of its delivery would be particularly critical. Coughlin et al. (1977, 1978) found that antiserum against NGF interfered only modestly, if at all, with the neuronal behavior in NGF-free explant cultures of early SCGs. Thus, one would have to postulate that the putative transfer of NGF from endogenous source to recipient neurons is inaccessible to the antibody.

2. Sensitivity versus Need for Exogenous Trophic Factors

Whether or not target-independent neurons receive trophic factors from other sources, an important question is whether they are already sensitive to target-derived factors. Exogenous administration of NGF, while not needed by early mouse SCG cultures, did elicit marked increases in both neurite growth and tyrosine hydroxylase activity (Coughlin et al., 1977, 1978). One must conclude, then, that receptors for NGF are already present several days before the requirement for exogenous NGF becomes apparent—a necessary, though not sufficient, point for a thesis that endogenous NGF is already involved at the earlier times. Interestingly, these same early SCG neurons responded to the presence of target tissues (submaxillary gland) in their culture, again with an increase in both neurite and enzyme activities, and this stimulation by target tissue was not abolished by anti-NGF antiserum (Coughlin et al., 1977, 1978). Thus, receptors may also be already present for target-derived factors different from the traditional NGF.

B. PHASE-2 NEURONS

1. Acquisition of Target Dependence

At or about the time of interaction with target tissues, the neuron must acquire a special kind of vulnerability, which makes it target-dependent for its future survival. The nature of the switch from target independence to target dependence is likely to reflect the survival requirements in the preceding phase. If a neuron is already receiving T'TF from other sources, the new target dependence will merely reflect a *quantitative* change, either an exacerbation of the need for T'TF or a decrease in the availability of it from previous sources. Alternatively, if the early neuron required either a different trophic factor or no trophic factor at all, the transition to target dependence will constitute a *qualitative* event, requiring the operation of an extrinsic signal to turn on the mechanism responsible for the new state. The neuron will thus become specified for the new trophic dependence. The turning on of a program for a selective trophic dependence should be regarded as the acquisition of a new differentiated trait, much as is the ability to grow neurites or the selection of transmitter programs. In favor of this second view may be the ultrastructural changes reported to occur (on interaction with the target tissues) in *all* ciliary ganglionic neurons, regardless of whether they will later survive or become casualties (Pilar and Landmesser, 1976).

A somewhat different view of the transition to target dependence can be proposed by extending to neurons the concept of "determination for cell death," introduced by Saunders and co-workers after their studies with the limb bud (Saunders and Fallon, 1966; Fallon and Saunders, 1968). These authors have proposed that cells can be *determined* for cell death much as they can be determined to become nerve or cartilage cells. Cell death, in other words, represent the phenotypic expression of a particular genetic program. Based on this view, the transition from phase 1 to 2 involves the reception of a signal that turns on a "death machinery." The activity of this machinery would inevitably bring the cell to its death unless a saving factor is provided to switch it off or to switch on compensating processes.

2. Selection of Survivors and/or Casualties

Nothing is known about the mechanisms by which phase-2 neurons are sorted out for survival or, conversely, death. One possibility is that the choice reflects an already implicit distinction between two subpopulations of phase-2 neurons. While there is no evidence for such a lack of uniformity, one cannot dismiss subtle distinctions pertaining to, for example, qualitative or quantitative differences in receptors for the target-derived trophic factor. A more likely view is that the selection process concerns the interaction between target cells and a uniformly competent population of neurite terminals.

   a. *Differentiation of Target Cells as Trophic Sources.* As debated for the neuronal population, the question can be raised whether all target cells are *equally* competent as trophic sources for the incoming neurites. The only information currently available on temporal changes in the concentration of a trophic factor in target tissues derives from the recent studies on intraocular CNTF (Adler *et al.*, 1979b; Landa *et al.*, 1980). As already discussed (Section III,A), these studies found a temporal correlation between CNTF activity in the intraocular target tissues and the rescue from developmental death of the CG neurons that innervate them. It is not known, thus far, whether the CNTF increase is due to activation of preexisting factor or greater production of it, or whether it reflects a uniform increase in all target cells rather than a progressive recruitment of active targets. In any case, it is not illogical to assume that the developmental correlation found for the ciliary system, between trophic requirements of the neurons and the capability of target tissues to provide trophic support to them, may also be found wherever a target-derived trophic factor is related to the prevention of neuronal cell death. Nor is it likely that such a correlation is due to mere chance. Landa *et al.* (1980) have suggested that the neurites invading the target tissue could themselves provide target cells with the signal that triggers an increase in trophic activity. It is also conceivable that both the target dependence of the neurons and the trophic competence of the target tissue are regulated by a common exogenous signal.

   b. *Delivery Mechanisms.* If both the target cells and the neurite terminals are uniformly competent, the selection process must rely on a *selective mode* of delivery of trophic factor from one to the other. As previously discussed, the prevailing view is that the only neurons from a target-dependent population that will survive are those that make correct contact with the target so as to receive an adequate trophic supply. It is considered unlikely that freely diffusible trophic agents could explain either the selective rescue that generally takes place or the restricted effects that partial target ablation has on the survival of corresponding neurons (e.g., Clarke *et al.*, 1976). Thus, some form of special contact between neurite terminal and target cell must be involved, which would channel the released trophic agent directly to the recipient neuritic terminal. Selectivity could be determined by the availability of a restricted number of contact sites [or, as proposed by Landmesser and Pilar (1978), of synaptic sites] through which the target cells could dispense trophic factor to the neurites. It is at the very least unlikely that such contact sites involve direct transfer from cytoplasm to cytoplasm (e.g., via a gap junction), since trophic factors are presumed to be macromolecules like NGF (cf. Varon, 1975a) or the newly investigated CNTF (Manthorpe *et al.*, 1980). However, the transfer of macromolecules from one cell to another with minimal "spillage" is no longer unknown (cf. Lasek *et al.*, 1977; Varon and Somjen, 1979) and presumably involves exocytotic release from the donor into a confined space and rapid capture from it by the recipient cell element.

C. Neurons in Phases 3 and 4

All available information indicates that, after adequate contact with the target cells is established, neurons continue to depend upon trophic support from the target. Cellular responses of neurons that lose preexisting connections, however, are completely different from those of phase-2 cells that fail to develop them. One most obvious difference is the speed of the neuronal response to peripheral deprivation. In his experiments with amphibian motor neurons, Prestige (1970) observed that early extirpation of the limb (i.e., at phase 2) causes immediate neuronal degeneration; in contrast, removal of the limb from older animals, the motor neurons of which have already been in phase 3 for a long time, leads to neuronal death only after a 4- to 6-month lag. Similarly, with chick embryo ciliary ganglia, phase-2 neurons that fail to achieve effective contact with the intraocular targets die in less than 6 days, while disconnection from the eye in the postnatal animal triggers a slow reaction, at the end of which only some of the neurons die (Landmesser and Pilar, 1978). A second very important characteristic of phase 4, then, is that peripheral deprivation does not always result in the death of the deprived neurons. At least in the peripheral nervous system, late disconnection from the main target tissue may well lead to neuronal recovery and even to the regeneration of fully functional synapses with the same target (e.g., Olson and Mamfors, 1970; Guth, 1974).

Does this mean that maturation in phase 3 is accompanied by a decrease in neuronal requirements for trophic factors, or by an increasing independence from target tissue as a source of them? The question could be debated in the same terms as that concerning the early target independence of phase-1 neurons (Section VI,A). And, indeed, several studies have suggested that, as development advances,

NGF-dependent neurons lose or reduce their needs for *exogenous* NGF (levi-Montalcini and Angeletti, 1968; Lazarus *et al.*, 1976; Chun and Patterson, 1977). However, there are several indications that NGF remains important even for older neurons. DRG neurons in the adult animal, while apparently indifferent to exogenous NGF, retain the capability for taking it up at their terminals and transporting it retrogradely to their somata (Stoeckel and Thoenen, 1975). More to the point, the effects of surgical or functional axotomy on adult sympathetic neurons can be prevented by topical administration of NGF (Purves and Njå, 1976; West and Bunge, 1976). In this case, the consequences of axotomy can be reasonably interpreted to reflect a decreased supply of target-derived NGF in the face of a maintained need for it. Thus, increased independence from the target tissue may represent a shift in NGF sources from target tissue to glial cells, the development of which also advances with age (cf. Lasek *et al.*, 1977; Varon and Bunge, 1978; Varon and Somjen, 1979). Based on the competence of the new sources, and the degree of their loss on axotomy (which also reduces the number of available Schwann cells), the remaining supply of trophic factors may be insufficient for survival in some cases, or support survival and even regeneration in some others.

The question, obviously, has not yet been entirely resolved. Nevertheless, the possibility of rescuing damaged, mature neurons by the exogenous supply of appropriate trophic factors has such implications for regenerative events in the nervous system that no effort should be spared to advance our understanding of this problem.

D. Neuronal Precursors (Phase 0)

A brief incursion into the *mare ignotum* of neuronal precursors appears warranted, indeed demanded, by recent findings of NGF effects on certain pheochromocytoma clonal cells, PC12 (Greene and Tischler, 1976), and on chromaffin cells from normal adrenal medulla (Unsicker *et al.*, 1978). Such findings raise important questions regarding several problems debated in this chapter, among them the roles of NGF (Section IV), the promotion of neuritic growth (Section V), and the susceptibility of early cells to trophic and specifying factors (Section VI).

Chromaffin cells of the adrenal medulla, and the pheochromocytoma tumoral cells originating from them, are derivatives of the neural crest, just as spinal sensory and sympathetic neurons are. It is not, therefore, particularly surprising that chromaffin cells can be responsive to NGF. However, it has not been established whether chromaffin cells represent one additional developmental step beyond sympathetic neurons—as suggested by their ability to process further norepinephrine to epinephrine—or whether they are to be viewed as separate derivatives of a precursor shared with sympathetic neurons. This point has important bearings on the interpretation of the findings to be discussed in this section.

1. Normal Chromaffin Cells

In vivo, chromaffin cells have been reported to accumulate intravenously injected ^{125}I-labeled NGF and to respond to it with a selective induction of catecholamine-synthesizing enzymes, tyrosine hydroxylase and dopamine β-hydroxylase (DBH) (Otten et al., 1977). When transplanted outside their adrenal locale, these cells also display the ability to grow processes (Olson and Malmfors, 1970; Unsicker and Chamley, 1977). Unsicker et al. (1978) now report that, in monolayer cultures, chromaffin cells do not require exogenous NGF for survival but respond to its presence with both an increase in tyrosine hydroxylase and an outgrowth of neuritelike processes. Neurite outgrowth, but not tyrosine hydroxylase induction, is prevented by the additional presence of the soluble glucocorticoid dexamethasone. It appears, then, that chromaffin cells (1) are intrinsically programmed for neurite production, (2) are prevented from expressing the neurite program by glucocorticoid hormones originating from the adrenal cortex, and (3) are stimulated to express it (in the absence of glucocorticoids) by NGF either exogenously supplied (Unsicker et al., 1978) or locally produced (Unsicker and Chamley, 1977). It is possible that their basal production of tyrosine hydroxylase and DBH also reflects a local availability of NGF, as it is enhanced by exogenous NGF and not affected, or even potentiated, by glucocorticoids. In all these respects, then, chromaffin cells resemble sympathetic neurons. Two special points must be noted. One is that chromaffin cells can only derive NGF from the bloodstream or the local producers. The other point is that exogenous NGF is not required for their survival in vitro, nor has it been reported to elicit hyperplastic or hypertrophic responses in vivo. It is very tempting to see an analogy between these behaviors and those of NGF-related neurons either in phase 1 (when they would not yet be target-dependent) or in phase 3 (when they may have escaped target dependence).

2. The Pheochromocytoma PC12 Clonal Line

Greene and Tischler (1976) have isolated a rat pheochromocytoma clonal line which, when cultured with NGF, develops neuronal characteristics, namely, (1) postmitotic status, (2) neurite production, (3) electrical excitability (Dichter et al., 1977), and (4) the ability to form cholinergic synapses on skeletal muscle cells (Schubert et al., 1977). Like normal chromaffin cells, PC12 cells do not require NGF for survival provided serum is present; however, if serum is withdrawn, the cultured cell will die unless NGF is present (Greene, 1978). In serum-free medium, NGF supports survival in a manner that is not transcription-dependent, while it elicits neuritic outgrowth in a manner that is (Greene, 1978). In serum-containing media, and in the absence of NGF, PC12 cells produce both tyrosine hydroxylase and CAT, and increases in both enzymes depend on cell density (Greene and Rein, 1977; Schubert et al., 1977). CAT production was enhanced by NGF (Greene and Rein, 1977), as well as by coculture with a variety of heterologous cells or by the use of conditioned media derived from them (Schubert et al., 1977).

It is clear from the above, and several other, observations that the PC12 cell population will become an increasingly powerful tool for the study of many questions pertaining to NGF, and also to other modes of extrinsic cell regulation.

3. Comments and Conclusions

A review of the many questions raised by the PC12 and the chromaffin cell observations would be inappropriate to the dimensions of this chapter. However, a list of some such questions appears to be a fit conclusion to the many issues raised here.

1. *Roles of NGF (and, by extrapolation, of other neuronotrophic factors).* We have sustained the view that NGF is concerned with the survival and growth of responsive neurons, i.e., a trophic function, and may solicit selective expressions of differentiated properties only in a secondary or a localized manner (Sections IV and V). Can these concepts accommodate the ability of NGF to convert chromaffin cells to neuronlike elements, or perhaps to unmask neuronal properties already intrinsic to such cells?

2. *Mode of action of NGF.* We have speculated that an ionic mechanism of action could fit both the trophic functions by NGF (Section IV,C) and putative neurite-triggering actions exerted by it in localized regions of the nerve cell (Section V,B). Do similar ionic perturbations accompany the actions of NGF on PC12 and primary chromaffin cells?

2. *Trophic dependence of early neurons or their precursors.* We have debated the question whether the same trophic factors that rescue target-dependent neurons from developmental cell death are also involved in earlier stages of their cell history, or whether different (or no) trophic factors are required at such times. Which factors are

needed by PC12 and/or chromaffin cells for their survival and growth before exogenous NGF is brought into the picture?

These, and many others, are tantalizing questions, indeed. It seems that, both conceptually and experimentally, we are at the threshold of vast and portentous advances in our understanding of cellular neurobiology. From phase 0 to phase 4, one stretches through the development of nerve cells into questions of the regulation of their mature behavior and the control of their regenerative capability. The future is wide open and promises rich rewards to the increasingly numerous investigators in these areas.

ACKNOWLEDGMENTS

This work was supported by USPHS grant NS-07606 from the National Institute of Neurological and Communicative Disorders and Stroke.

REFERENCES

Adler, R., and Teitelman, G., (1974). *Dev. Biol.* **39**, 317–321.

Adler, R., and Varon, S. (1980). *Brain Res.* **188**, 437–448.

Adler, R., Teitelman, G., and Suburo, A. (1976). *Dev. Biol.* **50**, 48–57.

Adler, R., Manthorpe, M., and Varon, S. (1979a). *Dev. Biol.* **69**, 424–435.

Adler, R., Landa, K. B., Manthorpe, M., and Varon, S. (1979b). *Science* **204**, 1434–1436

Andres, R., Jeng, I., and Bradshaw, R. (1977). *Proc. Natl. Acad. Sci. U.S.A.* **74**, 2785–2789.

Angeletti, P., and Vigneti, E. (1971). *Brain Res.* **33**, 601–604.

Angeletti, P. U., Gandini-Attardi, D., Toschi, G., Salvi, M. L., and Levi-Montalcini, R. (1965). *Biochim. Biophys. Acta* **25**, 111–120.

Arnason, G., Oger, J., Pantazis, N., and Young, M. (1974). *J. Clin. Invest.* **53**, 2a.

Banerjee, S. P., Snyder, S. H., Cuatrecasas, P., and Greene, L. A. (1973). *Proc. Natl. Acad. Sci. U.S.A.* **70**, 2519–2523.

Banerjee, S., Cuatrecasas, P., and Snyder, S. (1976). *J. Biol. Chem.* **251**, 5680–5685.

Bischoff, R., and Holtzer, H. (1970). *J. Cell Biol.* **44**, 134–150.

Bottenstein, J. E., Skaper, S. D., Varon, S., and Sato, G. (1980). *Exp. Cell Res.* **125**, 183–190.

Bradshaw, R. A., and Young, M. (1976). *Biochem. Pharmacol.* **25**, 1445–1449.

Bray, D. (1970). *Proc. Natl. Acad. Sci. U.S.A.* **65**, 905–910.

Brown, J. C. (1971). *Exp. Cell Res.* **69**, 440–443.

Brunso-Bechtold, J. K., and Hamburger, V. (1978). *Soc. Neurosci. Abstr.* **4**, 31.

Burnham, P., Raiborn, C., and Varon, S. (1972). *Proc. Natl. Acad. Sci. U.S.A.* **69**, 3556–3560.

Burnham, P. A., Silva, J., and Varon, S. (1974). *J. Neurochem.* **23**, 689–697.

Campenot, R. B. (1977). *Proc. Natl. Acad. Sci. U.S.A.* **74**, 4516–4519.

Carlin, S. C., Rosenberg, R. N., VandeVenter, L., and Friedkin, M. (1974). *Mol. Pharmacol.* **10**, 194–203.

Chun, L., and Patterson, P. (1977). *J. Cell Biol.* **75**, 694–718.

Clarke, P. G. H., and Cowan, W. M. (1976). *J. Comp. Neurol.* **167**, 143-164.
Clarke, P. G. H., Rogers, L. A., and Cowan, W. M. (1976). *J. Comp. Neurol.* **167**, 125-142.
Cohen, S., and Levi-Montalcini, R. (1956). *Proc. Natl. Acad. Sci. U.S.A.* **42**, 571-574.
Collins, F. (1978a). *Dev. Biol.* **65**, 50-57.
Collins, F. (1978b). *Proc. Natl. Acad. Sci. U.S.A.* **75**, 5210-5213.
Cone, C. D., and Cone, C. M. (1976). *Science* **192**, 155-157.
Cone, C. D., and Cone, C. M. (1978). *Exp. Neurol.* **60**, 41-55.
Coughlin, M. (1975a). *Dev. Biol.* **43**, 123-139.
Coughlin, M. (1975b). *Dev. Biol.* **43**, 140-158.
Coughlin, M., Boyer, D., and Black, I. (1977). *Proc. Natl. Acad. Sci. U.S.A.* **74**, 3438-3442.
Coughlin, M., Dibner, M. D., Boyer, D. M., and Black, I. (1978). *Dev. Biol.* **66**, 513-528.
Cowan, W. M. (1973). *In* "Development and Aging in the Nervous System" (M. Rockstein and M. L. Sussman, eds.), pp. 19-41. Academic Press, New York.
Cowan, M., and Wenger, E. (1968). *J. Exp. Zool.* **168**, 105-124.
Crane, R. K. (1977). *Rev. Physiol. Biochem. Pharmacol.* **78**, 99-160.
Culp, L., and Buniel, J. (1976). *J. Cell Physiol.* **88**, 89-106.
Dibner, M., Mytilineou, C., and Black, I. (1977). *Brain Res.* **123**, 301-310.
Dichter, M. A., Tischler, A. S., and Greene, L. A. (1977). *Nature (London)* **268**, 501-504.
Ebendal, T., and Jacobson, C. O. (1975). *Zoon* **3**, 169-172.
Fallon, J. F., and Saunders, Jr., J. W. (1968). *Dev. Biol.* **18**, 553-570.
Frazier, W., Boyd, L., and Bradshaw, R. (1973a). *Proc. Natl. Acad. Sci. U.S.A.* **70**, 2931-2935.
Frazier, W., Ohlendorf, C., Boyd, L., Aloe, L., Johnson, E., Ferrendelli, J., and Bradshaw, R. (1973b). *Proc. Natl. Acad. Sci. U.S.A.* **70**, 2448-2452.
Frazier, W., Boyd, L., Szutowicz, A., Pulliam, M. W., and Bradshaw, R. (1974). *Biochem. Biophys. Res. Commun.* **57**, 1096-1103.
Giller, E. L., Schrier, B. K., Shainberg, A., Fisk, R., and Nelson, P. (1973). *Science* **182**, 588-589.
Giller, E. L., Neale, J. H., Bullock, P. N., Schrier, B. K., and Nelson, P. G. (1977). *J. Cell Biol.* **74**, 16-29.
Glaser, L. (1978). *Rev. Physiol. Pharmacol.* **83**, 89-122.
Grafstein, B. (1975). *Exp. Neurol.* **48**, 32-51.
Greene, L. A. (1977). *Dev. Biol.* **58**, 96-105.
Greene, L. A. (1978). *J. Cell Biol.* **78**, 747-755.
Greene, L. A., and Rein, G. (1977). *J. Neurochem.* **29**, 141-150.
Greene, L., and Tischler, A. (1976). *Proc. Natl. Acad. Sci. U.S.A.* **73**, 2424-2428.
Grinnell, F. (1978). *Int. Rev. Cytol.* **58**, 65-114.
Guth, L. (1974). *Exp. Neurol.* **45**, 606-654.
Hamburger, V. (1958). *Am. J. Anat.* **102**, 365-410.
Hamburger, V. (1975). *J. Comp. Neurol.* **160**, 535-546.
Hamburger, V., and Levi-Montalcini, R. (1949). *J. Exp. Zool.* **111**, 457-501.
Hatanaka, H., Thoenen, H., and Otten, U. (1978). *FEBS Lett.* **92**, 313-316.
Helfand, S. L., Smith, G. A., and Wessells, N. (1976). *Dev. Biol.* **50**, 541-547.
Helfand, S. L., Riopelle, R. J., and Wessells, N. K. (1978). *Exp. Cell Res.* **113**, 39-45.
Hendry, J. A. (1976). *Rev. Neurosci.* **2**, 149-193.
Hendry, I., and Iversen, L. (1973). *Nature (London)* **243**, 500-504.
Herrup, K., and Shooter, E. M. (1973). *Proc. Natl. Acad. Sci. U.S.A.* **70**, 3884-3888.
Hershko, A., Mamont, P., Shields, R., and Tomkins, G. (1971). *Nature (London) New Biol.* **232**, 206-211.

Hier, D., Arnason, B., and Young, M. (1972). *Proc. Natl. Acad. Sci. U.S.A.* **69**, 2268-2272.

Hollyday, M., and Hamburger, V. (1976). *J. Comp. Neurol.* **170**, 311-320.

Hopfer, U. (1977). *J. Supramol. Struct.* **7**, 1-13.

Horii, Z., and Varon, S. (1975). *J. Neurosci. Res.* **1**, 361-375.

Horii, Z., and Varon, S. (1977). *Brain Res.* **124**, 121-133.

Hughes, A., F. (1968). "Aspects of Neural Ontogeny." Academic Press, New York.

Jaffe, L. (1968). *Adv. Morphogen.* **7**, 295-328.

Jaffe, L., and Nuccitelli, R. (1977). *Annu. Rev. Biophys. Bioeng.* **6**, 445-476.

Johnson, D. G., Gordon, P., and Kopin, I. J. (1971). *J. Neurochem.* **18**, 2355-2362.

Kelly, J. P., and Cowan, W. M. (1972). *Brain Res.* **42**, 263-288.

Kerr, F. W. L. (1975). *Exp. Neurol.* **48**, 16-31.

Ko, C.-P., Burton, H., and Bunge, R. P. (1976). *Brain Res.* **117**, 437-485.

Landa, K. B., Adler, R., Manthorpe, M., and Varon, S. (1980). *Dev. Biol.* **74**, 401-408.

Landmesser, L., and Pilar, G. (1974a). *J. Physiol. London* **241**, 715-736.

Landmesser, L., and Pilar, G. (1974b). *J. Physiol. London* **241**, 737-749.

Landmesser, L., and Pilar, G. (1978). *Fed. Proc.* **37**, 2016-2022.

Lasek, R., and Hoffman, P. (1976). *Cold Spring Harbor Conf. Cell Prolif.* **3**, 1021-1049.

Lasek, R. J., Gainer, H., and Barker, J. L. (1977). *J. Cell. Biol.* **74**, 501-523.

Lasher, R., and Cahn, R. D. (1969). *Dev. Biol.* **19**, 415-435.

Lazarus, K., Bradshaw, A., West, N., and Bunge, R. (1976). *Brain Res.* **113**, 159-164.

LeDouarin, N. M. (1977). *In* "Cell Interactions in Differentiation" (M. Karkinen-Jääskeläinen, ed.), pp. 171-190. Academic Press, New York.

Letourneau, P. C. (1975). *Dev. Biol.* **44**, 77-91.

Letourneau, P. C. (1978). *Dev. Biol.* **66**, 183-196.

Letourneau, P. C., and Wessells, N. (1974). *J. Cell. Biol.* **61**, 56-69.

Levi-Montalcini, R. (1966). *Harvey Lect.* **60**, 217-259.

Levi-Montalcini, R. (1976). *In* "Perspectives in Brain Research, Progress in Brain Research" (M. A. Corner and D. F. Swaab, eds.), Vol. 45, pp. 235-258. Elsevier/North-Holland, Amsterdam.

Levi-Montalcini, R., and Angeletti, P. (1968). *Physiol. Rev.* **48**, 534-569.

Levi-Montalcini, R., and Booker, B. (1960a). *Proc. Natl. Acad. Sci. U.S.A.* **46**, 373-383.

Levi-Montalcini, R., and Booker, B. (1960b). *Proc. Natl. Acad. Sci. U.S.A.* **46**, 384-391.

Levi-Montalcini, R., and Hamburger, V. (1951). *J. Exp. Zool.* **116**, 321-362.

Levi-Montalcini, R., Revoltella, R., and Calissano, P. (1974). *In* "Recent Progress in Hormone Research" (R. Greep, ed.), Vol. 30, pp. 635-669. Academic Press, New York.

Longo, A., and Penhoet, E. (1974). *Proc. Natl. Acad. Sci. U.S.A.* **71**, 2347-2349.

Luduena, M. A. (1973). *Dev. Biol.* **33**, 470-476.

McDonnell, P. C., Tolson, N., and Guroff, G. (1977). *J. Biol. Chem.* **252**, 5859-5863.

McMorris, F. A., Nelson, P. G., and Ruddle, F.H., eds. (1973). *Neurosci. Res. Prog. Bull.* **11** (5).

Mains, R. E., and Patterson, P. H. (1973). *J. Cell. Biol.* **59**, 329-366.

Manthorpe, M., Skaper, S., Adler, R., Landa, K., and Varon, S. (1980). *J. Neurochem.* **34**, 69-75.

Marsh, G., and Beams, H. W. (1946). *J. Cell Comp. Physiol.* **27**, 139-157.

Mobley, W. C., Server, A. C., Ishii, D. N., Riopelle, R. J., and Shooter, E. M. (1977). *New Engl. J. Med.* **297**, 1096-1104; 1149-1158; 1211-1218.

Moscona, A. A., ed. (1974). "The Cell Surface in Development." Wiley, New York.

Murphy, R. A., Pantazis, N. J., Arnason, B. G. W., and Young, M. (1975). *Proc. Natl. Sci. U.S.A.* **72**, 1895-1898.

Murphy, R. A., Oger, J., Saide, J. D., Blanchard, M. H., Arnason, B. G. W., Hogan, C., Pantazis, N. J., and Young, M. (1977). *J. Cell. Biol.* 72, 769–773.

Narayanan, C. H., and Narayanan, Y. (1978). *J. Embryol. Exp. Morphol.* 47, 137–148.

Narumi, S., and Fujita, T. (1978). *Neuropharmacology* 17, 73–76.

Nikodijevic, B., Nikodijevic, O., Yu, M-Y. W., Pollard, H., and Guroff, G. (1975). *Proc. Natl. Acad. Sci. U.S.A.* 72, 4769–4771.

Nishi, R., and Berg, D. K. (1977). *Proc. Natl. Acad. Sci. U.S.A.* 74, 5171–5175.

Nishi, R., and Berg, D. K. (1979). *Nature (London)* 277, 232–234.

Norr, S., and Varon, S. (1975). *Neurobiology* 5, 101–118.

Olson, L., and Malmfors, T. (1970). *Acta Physiol. Scand. Suppl.* 348, 1–112.

Otten, U., Schwab, M., Gagnon, C., and Thoenen, H. (1977). *Brain Res.* 133, 291–303.

Otten, U., Hatanaka, H., and Thoenen, H. (1978). *Brain Res.* 140, 385–389.

Partlow, L., and Larrabee, M. (1971). *J. Neurochem.* 18, 2101–2118.

Patterson, P. H., Reichardt, L. F., and Chun, L. L. Y. (1975). *Cold Spring Harbor Symp. Quant. Biol.* 40, 389–397.

Paul, D., Ristow, H. J., Rupniak, H. T., and Messmer, T. O. (1978). *In* "Molecular Control of Proliferation and Differentiation" (J. Papaconstantinou and W. J. Rutter, eds.), pp. 65–79. Academic Press, New York.

Perez-Polo, J. R., Hall, K., Livingston, K., and Westlund, K. (1977). *Life Sci.* 21, 1535–1544.

Pilar, G., and Landmesser, L. (1976). *J. Cell. Biol.* 68, 339–356.

Pilla, A. A. (1974). *Ann. N.Y. Acad. Sci.* 238, 149–170.

Popiela, H., Manthorpe, M., Adler, R., and Varon, S. (1978). *Trans. Am. Soc. Neurochem.* 9, 49.

Prestige, M. C. (1970). *In* "The Neurosciences Second Study Program" (F. O. Schmitt, ed.-in-chief), pp. 73–82. Rockefeller Univ. Press, New York.

Purves, D. (1975). *J. Physiol. (London)* 252, 429–463.

Purves, D. (1976). *J. Physiol. (London)* 259, 159–175.

Purves, D., and Njå, A. (1976). *Nature (London)* 260, 535–536.

Rakic, P. (1971). *J. Comp. Neurol.* 141, 283–312.

Rakic, P. (1972). *J. Comp. Neurol.* 145, 61–84.

Ramirez, G., and Seeds, N. (1977). *Dev. Biol.* 60, 153–162.

Rasmussen, H., and Goodman, D. (1977). *Physiol. Rev.* 57, 421–509.

Rebhun, L. (1977). *Int. Rev. Cytol.* 49, 1–54.

Rees, R., Bunge, M., and Bunge, R. (1976). *J. Cell. Biol.* 68, 240–263.

Rogers, L., and Cowan, M. (1973). *J. Comp. Neurol.* 147, 291–320.

Roisen, F. J., Murphy, R. A., and Braden, W. G. (1972). *J. Neurobiol.* 3, 347–368.

Ross, J., Olmsted, J. B., and Rosenbaum, J. L. (1975). *Tissue Cell* 7, 107–136.

Saunders, Jr., J. W., and Fallon, J. F. (1966). *In* "Major Problems in Developmental Biology" (M. Locke, ed.), pp. 289–314. Academic Press, New York.

Schlaepfer, W. W., and Bunge, R. P. (1973). *J. Cell Biol.* 59, 456–470.

Schubert, D. (1976). *Exp. Cell Res.* 102, 329–340.

Schubert, D., and Whitlock, C. (1977). *Proc. Natl. Acad. Sci. U.S.A.* 74, 4055–4058.

Schubert, D., Humphreys, S., de Vitry, F., and Jacob, F. (1971). *Dev. Biol.* 25, 514–546.

Schubert, D., Harris, A. J., Heinemann, S., Kidokoro, Y., Patrick, J., and Steinbach, J. H. (1973). *In* "Tissue Culture of the Nervous System" (G. Sato, ed.), pp. 55–86. Plenum, New York.

Schubert, D., Heineman, S., and Kidokoro, Y. (1977). *Proc. Natl. Acad. Sci. U.S.A.* 74, 2579–2583.

Schwartz, J., Chuang, D. M., and Costa, E. (1977). *Brain Res.* 137, 369–375.

Sidman, R. L., and Wessells, N. K. (1975). *Exp. Neurol.* 48 (3), part 2, 237–251.

Sisken, B. F., and Lafferty, J. F. (1978). *Bioelectrochem. Bioenerget.* **5**, 459–472.
Sisken, B. F., and Smith, S. D. (1975). *J. Embryol. Exp. Morphol.* **33**, 29–41.
Skaper, S. D., and Varon, S. (1979a). *Brain Res.* **163**, 89–100.
Skaper, S. D., and Varon, S. (1979b). *Brain Res.* **172**, 303–313.
Skaper, S. D., and Varon, S. (1979c). *Biochem. Biophys. Res. Commun.* **88**, 563–568.
Skaper, S. D., and Varon, S. (1980a). *J. Neurochem.* **34**, 1654–1660.
Skaper, S. D., and Varon, S. (1980b). *Brain Res.* (in press).
Skaper, S. D., Bottenstein, J. E., and Varon, S. (1979). *J. Neurochem.* **32**, 1845–1851.
Steiner, G., and Schonbaum, E., eds. (1972). "Immunosympathectomy." Elsevier, Amsterdam.
Stoeckel, K., and Thoenen, H. (1975). *Brain Res.* **85**, 337–342.
Sutter, A., Riopelle, R. J., Harris-Warrick, R. M., and Shooter, E. M. (1979). *J. Biol. Chem.* **254**, 5972–5982.
Tuttle, J. B. (1977). *Soc. Neurosci. Abstr.* **3**, 529.
Unsicker, K., and Chamley, J. H. (1977). *Cell Tissue Res.* **177**, 247–268.
Unsicker, K., Krisch, B., Otten, U., and Thoenen, H. (1978). *Proc Natl. Acad. Sci. U.S.A.* **75**, 3498–3502.
Varon, S. (1970). *In* "The Neurosciences: Second Study Program" (F. Schmitt, ed.), pp. 83–97. Rockefeller Univ. Press, New York.
Varon, S. (1975a). *Exp. Neurol.* **48**, (3), part 2, 75–92.
Varon, S. (1975b). *Exp. Neurol.* **48**, (3), part 2, 93–134.
Varon, S. (1977a). *Exp. Neurol.* **54**, 1–6.
Varon, S. (1977b). *In* "Cell, Tissue, and Organ Culture in Neurobiology" (S. Fedoroff and L. Hertz, eds.), pp. 237–261. Academic Press, New York.
Varon, S. (1979). *Neurochem. Res.* **4**, 155–173.
Varon, S., and Bunge, R. (1978). *Annu. Rev. Neurosci.* **1**, 327–362.
Varon, S., and Saier, M. (1975). *Exp. Neurol.* **48**, (3), part 2, 135–162.
Varon, S., and Skaper, S. D. (1980). *In* "Tissue Culture in Neurobiology" (E. Giacobini, A. Vernadakis, and A. Shahar, eds.), pp. 333–347. Raven, New York.
Varon, S., and Somjen, G. (1979). *Neurosci. Res. Progr. Bull.* (in press).
Varon, S., and Wilbrandt, W. (1966). *In* "Intracellular Transport" (K. Brehme-Warren, ed.), pp. 119–139. Academic Press, New York.
Varon, S., Raiborn, C., and Tyszka, E. (1973) *Brain Res.* **54**, 51–63.
Varon, S., Raiborn, C., and Burnham, P. A. (1974a). *J. Neurobiol.* **5**, 355–371.
Varon, S., Raiborn, C., and Burnham, P. A. (1974b). *Neurobiology* **4**, 231–252.
Varon, S., Raiborn, C., and Burnham, P. A. (1974c). *Neurobiology* **4**, 317–327.
Varon, S., Raiborn, C., and Norr, S. (1974d). *Exp. Cell Res.* **88**, 247–256.
Varon, S., Manthorpe, M., and Adler, R. (1979). *Brain Res.* **173**, 29–45.
Walicke, P. A., and Patterson, P. H. (1978). *Soc. Neurosci. Abstr.* **4**, 129.
Weiss, L., Poste, G., MacKearnin, A., and Willett, K. (1975). *J. Cell Biol.* **64**, 135–145.
Weiss, P. (1968). "Dynamics of Development: Experiments and Inferences; Selected Papers on Developmental Biology." Academic Press, New York.
Wessells, N. K., and Nuttall, R. P. (1978). *Exp. Cell Res.* **115**, 111–122.
West, N., and Bunge, R. (1976). *Soc. Neurosci. Abstr.* **2**, 1038.
Weston, J. A. (1972). *In* "Cell Interactions" (L. G. Silvestri, ed.), pp. 286–292. Elsevier, Amsterdam.
Winick, M., and Greenberg, R. E. (1965). *Nature (London)* **205**, 180–181.
Young, M., Murphy, R. A., Saide, J. D., Pantazis, N. J., Blanchard, M. H., and Arnason, B. G. W. (1976). *In* "Surface Membrane Receptors" (R. A. Bradshaw, W. A. Frazier, R. C. Merrell and G. I. Gottlieb, eds.), pp. 247–268. Plenum, New York.
Yu, M. W., Lakshmanan, J., and Guroff, G. (1978a). *In* "Essay in Neurochemistry and

Neuropharmacology" (M. B. H. Youdim, W. Lovenberg, D. F. Sharman and J. R. Lagnado, eds.), Vol. 3, pp. 33-48. Wiley, New York.

Yu, M. W., Hori, S., Tolson, N., Huff, K., and Guroff, G. (1978b). *Biochem. Biophys. Res. Commun.* **81**, 941-946.

Zaimis, E., and Knight, J. eds. (1972). "Nerve Growth Factor and its Antiserum. A symposium." Oxford Univ. Press, Athlone, London and New York.

CHAPTER 7

REQUIREMENTS FOR THE FORMATION AND MAINTENANCE OF NEUROMUSCULAR CONNECTIONS

Terje Lømo

INSTITUTE OF NEUROPHYSIOLOGY
UNIVERSITY OF OSLO
OSLO, NORWAY

and

Jan K. S. Jansen

INSTITUTE OF PHYSIOLOGY
UNIVERSITY OF OSLO
OSLO, NORWAY

I. Introduction

In this chapter we shall review recent contributions to our understanding of the control of the motor innervation of mammalian skeletal muscle. The field is developing rapidly, but as yet our insight is only fragmentary and it is difficult to predict which of the present features will remain important in the final picture. We will make no attempt to give a comprehensive review, since that would require an en-

CURRENT TOPICS IN
DEVELOPMENTAL BIOLOGY, Vol. 16

tire book rather than a single chapter. Unavoidably, we have probably given too much prominence to material of which we have first-hand knowledge at the expense of neglecting other contributions.

The behavior and development of neuromuscular junctions (NMJs) are of more general interest than the understanding of one aspect of the control of muscle. Synaptic mechanisms are in principle similar throughout the nervous system, and there are several developmental analogies as well. The simplicity of the neuromuscular system and the experimental advantages it offers will certainly favor further study of the development of synaptic connections and contribute to our understanding of the "wiring" of the nervous system.

II. Formation of Neuromuscular Junctions

For synapses to form, certain requirements must be fulfilled. Axons must come close to the target cells, the target cells must be receptive to innervation, and the axons and the target cells must be appropriately matched.

A. Axonal Growth

Axonal growth ensures that the axons come sufficiently close to the muscle fiber membrane to interact with it locally to form a functional contact. It is a general property of developing neurons to grow by extending their axons. However, the process is self-limiting and stops when the required number of functional contacts has been formed. Axons that fail to make contact disappear as these neurons die (Hamburger, 1975). Successful axons that have stopped growing can be stimulated to further growth by axonal injury (Guth, 1956), paralysis of the muscle (Duchen and Strich, 1968; Brown and Ironton, 1977; Pestronk and Drachman, 1978), or disturbance of the peripheral tissue (Jones and Tuffery, 1973).

The stimuli that cause axons to start or stop growing are unknown. It is likely that signals to axonal growth are produced by inactive muscles. For example, axonal sprouts arise from the terminal end plate region (terminal sprouting) in muscles paralyzed by botulinum toxin (Duchen and Strich, 1968), tetanus toxin (Duchen and Tonge, 1973), or conduction blocks in the nerve (Brown and Ironton, 1977; Pestronk and Drachman, 1978). Similar sprouts arise in partially denervated muscles (Edds, 1953).

Partial denervation affects not only denervated muscle fibers and motoneurons whose axons have been cut but also the remaining intact motor units. The innervated muscle fibers develop denervation-like properties such as tetrodotoxin (TTX)-resistant action potentials (Cangiano and Lutzemberger, 1977), the remaining axons sprout (Brown and Ironton, 1978), and the afterhyperpolarization following each action potential in the motoneurons is shortened (Huizar et al., 1977). Similar changes, both in the muscle and the motoneurons, are seen after impulse conduction is blocked locally in the motor nerve. Restoration of muscle activity by direct electrical stimulation prevents many and possibly all of these changes. Thus direct stimulation of the muscle prevents hypersensitivity to acetylcholine (ACh) and other denervation-like changes (Lømo and Westgaard, 1975), as well as terminal sprouting induced by partial denervation or botulinum toxin (Brown et al., 1977; Ironton et al., 1978). Stimulation of the nerve distal to a local conduction block prevents both hypersensitivity to ACh in the muscle and shortening of the afterhyperpolarization in the motoneurons (Lømo and Rosenthal, 1972; Czéh et al., 1978). On the other hand, stimulation proximal to the cuff, which does not restore muscle activity but activates the motoneurons at least antidromically, has no such effect (Czéh et al., 1978).

These results may be interpreted as follows. Muscle inactivity, however caused, alters the properties of the muscle fiber and generates signals that affect not only the motoneurons which in some situations still innervate the fibers but also adjacent muscle fibers and their motoneurons. Some of the effects may be due to substances released by inactive muscle fibers (Section II,C,2). In addition there may be retrograde effects dependent on intimate contact between inactive muscle fibers and their nerve terminals. The effects on the properties of the motoneurons in the spinal cord may be secondary to the sprouting caused by local interactions in the periphery or to altered flow of trophic substances from the muscle to the motoneuron (Czéh et al., 1978). In either case it appears that signals controlled by the level of muscular activity act retrogradely from the muscle to the motoneuron, and such signals may be important vehicles for the control of axonal growth.

In partially denervated muscles, sprouts arise not only from unmyelinated terminals (terminal sprouting) but also from nodes of Ranvier (collateral sprouting) (Brown and Ironton, 1978). Direct electrical stimulation of the muscle blocks terminal sprouting but not collateral sprouting (Ironton et al., 1978). This suggests that collateral sprouting may be influenced by factors not directly related to muscle activity. One possibility is that breakdown products from degenerat-

256 TERJE LØMO AND JAN K. S. JANSEN

ing axons or local reactions to these products have stimulatory effects on the remaining intact axons within the nerve trunk. Several recent observations suggest that "products of nerve degeneration" cause denervation-like changes in the muscle (Gordon et al., 1976; Lømo and Westgaard, 1976; Brown et al., 1978a; Cangiano et al., 1978; Cangiano and Lutzemberger, 1977). Local reactions to a foreign body on the muscle (e.g., a small piece of thread) do not only cause hypersensitivity to ACh in the muscle (Jones and Vrbová, 1974) but stimulate terminal sprouting as well, although the muscle is innervated and presumably active (Jones and Tuffery, 1973).

Axonal sprouting is not necessarily confined to the peripheral region whose function has been disturbed. Recently it has been shown in the frog that section of axons on one side can lead to the sprouting of corresponding axons on the other side (Rotshenker and McMahan, 1976). The mechanism is unclear. One possibility is that axotomized motoneurons release signals within the spinal cord which affect adjacent motoneurons (Rotshenker, 1978).

In conclusion it appears that stimuli to axonal growth arise form inactive muscles and from local pathological conditions after axotomy, nerve degeneration, or inflammation.

B. Axonal Guidance

To achieve its purpose axonal growth must be directed so that the axons reach their appropriate target. It has recently become clear that axons find their correct place in the periphery at very early stages in development before individual muscles separate out from the muscle mass (Landmesser, 1978). This occurs before the period of motoneuron death. Apparently establishment of the correct connections does not involve a period of widely distributed axonal outgrowth with subsequent elimination of inappropriate contacts.

Many experiments and much speculation have been presented to uncover the factors responsible for the directed growth of axons. In higher vertebrates the possibility that a muscle has an absolute specific affinity only for its appropriate nerve has been ruled out by the demonstration that seemingly any motor nerve can innervate any skeletal muscle both in development (Hollyday et al., 1977; Morris, 1978) and in adults after nerve crossing (see Purves, 1976a). Thus somatic motor axons have been shown to be able to form functional junctions with all inappropriate muscles that have been tested. The interpretation that the rat EDL becomes preferentially reinnervated by its appropriate axons because the muscle has a greater affinity for this nerve (Hoh, 1975) has recently been questioned (Riley, 1978).

What is it then that causes the directed growth of axons? A simple view is that they grow out according to the position of their motoneurons in the spinal cord and for this reason take up a fixed position in the extending nerve trunks. Parallel developments in the tissues through which the trunks grow will cause the trunks to branch, possibly by presenting the axons with mechanical obstacles. Axons lying to the left will therefore always take a left turn and so end up in a particular place in the developing muscle mass. Such a scheme places the emphasis on temporal and positional influences during development rather than on differential chemical affinities. It has received support from recent experiments by Lewis (Lewis, 1978). By grafting limb buds in chicks he produced wings with two elbows in series and found that the nerve trunks branched normally at the first elbow. One of the trunks branched into three with a middle branch (the median nerve) directed distally toward its normal destination in the hand. On meeting the second elbow, however, the median nerve behaved like the first trunk and branched into three, with fibers lying medially turning medially to innervate forelimb muscles they do not normally innervate. Only the middle fibers continued distally toward the hand. This suggests that the fibers branch according to their relative position in the nerve trunk. By peeling the branching fibers away from the main trunk it was shown that their relative position in the trunk was maintained proximally at least as far as the brachial plexus. The importance of morphogenetic factors in the ordering of nerve muscle connections in development has recently been emphasized (Horder, 1978).

In lower animals the situation seems more complex. Functional recovery after nerve injuries is far better than in higher vertebrates. Many observations indicate that regenerating axons find their appropriate target, even after extensive rerouting, by cues whose nature is unknown (Grimm, 1971; Cass and Mark, 1975). The factors that guide axons in regeneration and in normal development may well be different. Regenerating axons tend to follow previously established paths and have been shown (Letinsky et al., 1976) to grow with amazing precision along the remains of the original synaptic gutter on the surface of the muscle fibers. Similar cues are not available for growing axons in normal development.

C. Role of Muscle Activity in Synapse Formation

1. Experimental Preparation

It has recently become clear that muscle activity plays important roles in the formation of NMJs. It affects axonal growth (Section

II,A), receptivity to innervation, the appearance of junctional acetylcholinesterase (AChE), and the maintenance of synaptic connections.
The evidence we have obtained is based largely on experiments with
an adult preparation where the precise role of muscle activity is easier
to study than in normal development. It is therefore necessary to
describe this preparation briefly and to present the reasons for believ-
ing that it is a good model for what occurs in normal development.

The preparation is obtained by transplanting a foreign nerve (the
superficial fibular nerve) onto the soleus muscle in adult rats, allowing
the nerve 3 weeks to grow over the soleus, and then cutting the
original nerve to denervate the soleus so that it will accept innervation
by the foreign nerve (Frank et al., 1975). Transmission from the
foreign nerve is first detected 2½–3 days after the original nerve is
cut, at the time when full hypersensitivity to ACh is just developing
in the denervated muscle. The new junctions form ectopically, and
their development recapitulates many of the processes seen in normal
development. Thus preceding the onset of transmission the fibers
have a generalized sensitivity to ACh. When transmission is first
detected, the sensitivity at the synaptic site is already higher than
that resulting from denervation alone, presumably because the nerve
terminals quickly cause ACh receptors to accumulate in the postsynaptic membrane (Lømo and Slater, 1980a). Evoked muscle impulse activity begins on the third to the fourth day after the soleus nerve is
cut. Soon afterward the denervation hypersensitivity begins to
decline, and further innervation is arrested (Lømo and Slater, 1978).
Many of the junctions develop gradually toward fully mature, permanent junctions with characteristic synaptic properties (Korneliussen
and Sommerschild, 1976; Cangiano et al., 1980; Lømo and Slater,
1980b). Initially several axons contact each muscle fiber, but within
2–3 weeks many of these are eliminated (Kuffler et al., 1977; Lømo and
Slater, 1980a). All these events mimic those observed in normal development.

2. Axonal Sprouting (see also Section II,A)

It has been suggested that terminal sprouting may be caused by
some highly local stimulus, possibly the ACh receptor itself, appearing in the membrane of inactive muscle fibers (Pestronk and
Drachman, 1978; Brown et al., 1978b). This is unlikely in a situation
where a foreign nerve makes ectopic junctions with a denervated muscle. In this case the transplanted nerve does not appear to grow close
to the muscle fiber membrane, and specialized structures for transmit-

ter release do not develop as long as the muscle keeps its own innervation. Within a few days after the original nerve is cut, however, axonal sprouts and close, specialized nerve–muscle contacts appear (Korneliussen and Sommerschild, 1976). Before the original nerve is cut the distance between the overlying foreign nerve and the muscle is probably too great to be bridged by ACh receptors or other molecules in the plasma membrane. Local products of nerve degeneration are also unlikely factors, because the transplanted nerve is intact. A more likely explanation is that the inactive muscle fibers release a diffusible substance that stimulates the adjacent axons to sprout. More direct evidence for a diffusible factor inducing sprouting has been obtained in sympathetic motoneurons (Olson and Malmfors, 1970).

3. Receptivity in Innervation

During development innervation causes the muscle fiber to become refractory to further innervation. Receptivity to innervation is restored when muscles are paralyzed by denervation (Elsberg, 1917; Fex and Thesleff, 1967), botulinum toxin (Tonge, 1974), or conduction blocks in the nerve (Jansen et al., 1973); Cangiano et al., 1980). A foreign nerve can then form new junctions anywhere along the surface of the muscle fiber. No new junctions form if muscle activity is maintained after the original nerve is cut by direct stimulation of the muscle (Jansen et al., 1973; Lømo and Slater, 1978). The same stimulation also prevents the development of ACh hypersensitivity and other membrane changes usually caused by denervation (Lømo and Westgaard, 1975). The appearance of ACh receptors always precedes the formation of NMJs. The muscle fibers that most quickly become hypersensitive to ACh after denervation are also first innervated by the foreign nerve (Lømo and Slater, 1978). Direct stimulation of denervated muscles does not prevent reinnervation of the denervated end plates which retain their high ACh sensitivity despite the stimulation (Frank et al., 1975).

These results suggest that the formation of NMJs requires the presence of ACh receptors and cannot occur in normally active fibers where these receptors are absent except at the end plates. However, although ACh receptors are clearly necessary for neuromuscular transmission, it is not known whether the ACh receptor itself or other molecules present in inactive muscle fibers are required for the initial nerve–muscle interactions. It appears that the ACh binding part of the ACh receptor is not involved in this process, since NMJs may form in the presence of curare (Cohen, 1972) or α-bungarotoxin (Van

Essen and Jansen, 1974). This does not exclude the possibility that other parts of the ACh receptor molecule are involved.

Recent experiments by McMahan and collaborators have extended this picture (Sanes et al., 1978). In a preparation where the muscle fibers have degenerated but where the basal lamina around each fiber remains as a ghost, they have found that regenerating nerve fibers differentiate into specialized presynaptic terminals when they reach the synaptic gutters still present in the basal lamina. Apparently the basal lamina contains structures capable of inducing nerve terminal differentiation, possibly as a result of earlier contact with the nerve (see Section III,B).

In conclusion it appears that muscle activity induces and maintains properties in the extrajunctional part of the muscle fiber membrane, which prevent local nerve–muscle interactions.

4. Induction of Junctional AChE

During the formation of NMJs both in normal development (Bennett and Pettigrew, 1974) and in experimental situations (Lømo and Slater, 1980b) the appearance of AChE is delayed relative to the appearance of ACh receptors and transmission. With a transplanted foreign nerve, transmission begins $2\frac{1}{2}$–3 days after the original nerve is cut, and 3–4 days later junctional AChE appears. No AChE appears if the foreign nerve is cut at about the time transmission starts. However, if these muscles are stimulated directly for 4 days or more, plaques of intense AChE activity appear underneath the degenerated foreign nerve. These plaques are similar to the plaques induced by an intact nerve. They have the same form and distribution and coincide with discrete spots of high ACh sensitivity which, like postjunctional ACh sensitivities elsewhere, are resistant to the effects of muscle activity (Section II,D). These sites are therefore new postjunctional sites induced by the foreign nerve in the brief interval beginning with the muscle becoming receptive to innervation and ending with degeneration of the fibular nerve a day or so later. Stimulation starting after the foreign nerve has degenerated is equally effective.

In a different type of experiment the soleus is denervated, treated locally with the irreversible cholinesterase (ChE) inhibitor diisopropyl fluorophosphate (DFP), and stimulated directly for the following 7 days. At this time intensely stained, distinctly outlined ChE plaques are seen in the stimulated muscles at the original end plate band, while no histochemically detectable ChE can be demonstrated in the nonstimulated control muscles (Lømo, unpublished).

These results show that muscle activity causes the accumulation of junctional AChE. They are in agreement with a recent report that AChE fails to appear at NMJs formed *in vitro* in the presence of curare, an effect that can be overcome by direct stimulation of the muscle (Rubin *et al.*, 1978). Moreover, we have shown that, when all impulse conduction is blocked locally both in the original nerve and in the transplanted foreign nerve, the foreign nerve, nevertheless, forms new ectopic junctions with the muscle. In this case, the muscle remains inactive throughout the period of synapse formation and little AChE appears at the new NMJs. AChE is also much reduced at the old NMJs (Cangiano *et al.*, 1980). These results indicate that the muscle fiber is an important source of junctional AChE and that muscle activity controls its appearance. In addition, junctional AChE may come from the nerve (Skau and Brimijoin, 1978) and be subject to influences that are independent of muscle activity (Fernandez and Inestrosa, 1976; Younkin *et al.*, 1978).

What is the mechanism behind the effect of muscle activity on junctional AChE? It is helpful to compare junctional AChE and extrajunctional ACh receptors in different experimental situations. Soon after the onset of transmission and evoked muscle activity extrajunctional ACh sensitivity disappears, while at the same time junctional AChE activity appears at the new synapses (Lømo and Slater, 1978; Lømo and Slater, 1980b). If ordinary muscles are paralyzed by denervation or a nerve impulse conduction block, extrajunctional ACh sensitivity develops, while junctinal AChE activity is much reduced (Butler *et al.*, 1978; Cangiano *et al.*, 1980). Direct stimulation of denervated muscles abolishes extrajunctional ACh sensitivity and causes the reappearance of strong ChE activity at denervated end plates treated with DFP at the time of denervation (Lømo and Rosenthal, 1972; Lømo, unpublished). Apparently muscle activity has reciprocal effects on junctional AChE activity and extrajunctional ACh sensitivity. Morphological work suggests that AChE is synthesized in the junctional region of the muscle fiber and released into the synaptic cleft by exocytosis (Wake, 1976). There is also evidence that impulse activity can cause the release of AChE from nerves (Skau and Brimijoin, 1978; Chubb *et al.*, 1976) and adrenal chromaffin cells (Chubb and Smith, 1975). It seems likely therefore that muscle activity turns on synthesis and release of junctional AChE in addition to its better known effect of turning off synthesis of extrajunctional ACh receptors (Hall and Reiness, 1977).

Muscle activity has important regulatory effects only if it is imposed on the muscle at not too infrequent intervals. Spontaneous fibrillatory activity has little effect on ACh sensitivity, probably because it is cyclic, having long-lasting periods of inactivity alternating with briefer self-terminating periods of activity (Purves and Sakmann, 1974). Infrequent trains of electrical stimuli are without effect on ACh sensitivity (Lømo and Westgaard, 1975), and little AChE appears at ectopic synapses forming between impulse-blocked nerves and paralyzed but spontaneously fibrillating muscle fibers (Cangiano et al., 1980).

D. SYNAPTIC PROCESSES INDEPENDENT OF MUSCLE ACTIVITY

Ongoing muscle activity does not appear to influence local membrane interactions leading to the establishment of individual nerve-muscle contacts with their machinery for impulse transmission. For example, all impulse conduction may be blocked in the transplanted foreign nerve and yet its terminals will form functional NMJs (but with little AChE) with the completely paralyzed soleus muscle. Conversely, the foreign nerve may be left intact and the muscle stimulated directly starting 2 days after the original nerve is cut. This is before transmission and substantial hypersensitivity to ACh appears, and yet the foreign nerve goes on to make numerous new NMJs with the virogously stimulated soleus muscle (Lømo and Slater, 1978). These results show that synapse formation readily occurs both in the presence and in the absence of imposed muscle activity.

That stimulation prevents synapse formation when it begins on the day the original nerve is cut but not when it starts 2 days later may seem surprising. However, when stimulation starts on day 2, the hypersensitivity to ACh, which is just appearing at this time, continues to rise in spite of the stimulation and is fully developed on day 3. Only then does it begin to decline and is completely abolished after a further 4–5 days of stimulation (Lømo and Slater, 1978). Consequently a transient hypersensitivity develops during stimulation. It is elicited by inactivity which for the rat soleus must exceed a critical duration of 1–2 days. The appearance of hypersensitivity is delayed, as is its disappearance when activity is resumed. These delays could be at the level of synthesis of the receptor, during its incorporation into the membrane, or at any earlier step in the process that links activity and receptor synthesis. A regulating factor within the fiber whose concentration changes slowly when the level of activity is altered could constitute such a link. Whatever the underlying molecular mechanisms, it appears that the transient muscular response is sufficient both to stimulate sprouting from adjacent axons and to make the muscle fiber briefly receptive to innervation. Furthermore, since NMJs form during the stimulation, their formation must be intrinsically resistant to the effects of muscle activity.

E. POSSIBLE NEUROTROPHIC EFFECTS

Only a few points will be mentioned here. It is possible that substances moved by axonal transport from the neuron to the muscle contribute to the normal control of muscle membrane properties which in turn affect synapse formation. For example, the nerve induces local postjunctional specializations, and these effects do not appear to depend on muscle activity (Sections II,D and III,A). They could instead depend on substances released by the nerve or on intimate contact with the nerve.

A separate issue is whether similar neural influences can affect membrane properties such as ACh sensitivity along the rest of the fiber. In many experimental situations extrajunctional membrane properties change in a way that cannot be accounted for by changes in muscle activity alone. The difference could be due to a loss of neurotrophic influences (Miledi, 1963; Harris and Thesleff, 1972; Pestronk et al., 1976; Gilliatt et al., 1978) or to the appearance of pathological factors such as inflammation and products of nerve degeneration, arising because of the experimental intervention (Gordon et al., 1976; Lømo and Westgaard, 1976; Brown et al., 1978a; Cangiano and Lutzemberger, 1977). Both interpretations have their proponents. The distinction is important, and the problem needs to be resolved.

Several observations indicate that nerves release substances that influence ACh sensitivity or receptor distribution (Younkin et al., 1978; Christian et al., 1978), AChE activity (Lentz, 1974; Younkin et al., 1978), or other membrane properties when applied to muscles in culture. The role of such factors on junctional and extrajunctional membranes in normal regulation remains to be established. However, this is a promising approach in identifying possible neurotrophic substances.

III. Development of the Synaptic Machinery

During the formation of NMJs the ACh receptors at the sites of nerve–muscle contact acquire different properties from extrajunctional ACh receptors. The receptors become more densely packed and acquire a slower turnover rate (Berg and Hall, 1975) and resistance to the effects of muscle activity (Lømo and Rosenthal, 1972). In addition, the open time of the ionic channels associated with the receptors becomes shorter (Katz and Miledi, 1972). Other changes also occur. For example AChE accumulates in the synaptic cleft (Section II,C,4), and junctional folds develop. Why do these changes occur only at the end plate and how are they related?

A. JUNCTIONAL ACh RECEPTORS

In rats and chickens at the time of birth or hatching the density of junctional ACh receptors is already high (Bevan and Steinbach, 1977; Burden, 1977), while the open time of their associated ionic gates are still long, as in extrajunctional membranes (Schuetze and Fischbach, 1978). At the same time the receptors have a slow (junctional-like) turnover rate in rats (Berg and Hall, 1975) and a fast (extrajunctional-like) turnover rate in chickens (Burden, 1977). Overall mature characteristics appear only after a further few weeks of development. A similar dissociation of properties during synapse formation has been observed *in vitro* (Schuetze *et al.*, 1978). In rats the nerve can be cut at birth and yet ACh receptors will remain densely packed at the denervated end plates for some time (Slater, 1978). The results suggest that different junctional features may develop independently along different time courses and that the stability of the end plate may not depend on a slow turnover rate of the receptors.

During the formation of NMJs between a transplanted foreign nerve and the denervated soleus we have found that the postjunctional sensitivity to ACh is already higher than background denervation hypersensitivity when transmission is first detected. At the same early time many MEPPs also have a normal or larger than normal amplitude, although the fibers are large and have a relatively low input resistance. Movements of ACh receptors toward sites of developing NMJs have been demonstrated *in vitro* (Anderson *et al.*, 1977). It seems likely that similar movements take place in the present experimental situation and that this may account both for the large amplitude of the MEPPs and the higher than background sensitivity to ACh at the new junctions. It is possible therefore that the extrajunctional ACh receptors present at low densities in inactive fibers provide a pool of receptors ready to be rapidly mobilized and concentrated upon contact with a nerve terminal.

The biochemical stability of the early postjunctional receptors at ectopic end plates is not yet known. Their turnover rate and channel open time may well be extrajunctional-like initially, as they are at normal end plates in newly hatched chickens. Functional stability, however, is induced very early. Thus the foreign nerve can be cut many hours before it would normally form transmitting new junctions and yet it will induce peaks of ACh sensitivity which are resistant to the effects of muscle stimulation and indistinguishable from those observed with an intact fibular nerve. Since the nerve terminals degenerate within less than 24 hours of the cutting of the nerve, this suggests that the nerve requires less than a day to induce sites that are stable and have a high postsynaptic ACh sensitivity. Early resistance to the effects of muscle activity is of course essential for survival of the synapse and represents a fundamental distinction between junctional and extrajunctional membranes.

B. JUNCTIONAL AChE

Electrical stimulation of denervated muscle causes AChE to accumulate both at old and newly formed end plates (Section II,C,4). AChE accumulates at immature junctions even when the synapse-forming nerve is cut very early and thus allowed only a brief period in which to interact with the muscle. What causes the AChE to accumulate precisely at sites of nerve–muscle contact?

One attractive possibility is that structural changes or "traces" are induced in the surface of the muscle fiber immediately underneath the nerve terminal. These traces would bind or cause aggregation of AChE molecules released by active muscle fibers. They are likely to be in the basal lamina, because junctional AChE is normally associated with this structure (McMahan *et al.*, 1978). They persist in inactive muscles and are resistant to the effects of muscle activity, as are the aggregates of ACh receptors ("hot spots") in the immediately underlying plasma membrane. This explains why muscle activity starting long after degeneration of the nerve terminals nonetheless causes AChE to accumulate both at new and old denervated junctions (Section II,C,4).

There are different kinds of AChE of varying molecular complexity. The most complex in mammals is the easily aggregated 16S form found mainly at motor end plates. It is also the more labile form and disappears quickly from denervated end plates (Hall, 1973; Vigny *et al.*, 1976). It is possible that the 16S form is the form of AChE that primarily acts on ACh released from nerve terminals. A 10- to 14-day conduction block in the motor nerve causes a marked reduction in junctional AChE activity as judged by the time course and sensitivity to neostigmine of MEPPs. Despite this reduction in "functional AChE" the end plates stain well for AChE activity by histochemical techniques (Cangiano et al., 1980). The reduction in functional AChE activity could be due to a preferential loss of the more labile 16S form, the remaining forms being less efficient perhaps because they are further from sites of ACh release.

IV. Maintenance of Synaptic Connections

Like the formation of new synapses their maintenance and continual function depend on several factors, presynaptic as well as postsynaptic. Again, these are at present not especially known. However, our present insights provide strong indications for their existence.

The presynaptic factors depend on the structural integrity of the neuron. Most axon terminals degenerate within a few days after their connection with the cell body has been severed. Presumably this is due to interruption of the transport of essential substances from the cell body through the axon. We do not know which substances or which components of the axonal transport system are the critical ones for maintenance of the axon terminals. There is also conflicting evidence on how long motor end plates can remain functional after a blocking of axonal transport with colchicine (Albuquerque et al., 1972; Cangiano and Fried, 1977; see also Purves, 1976). Surprisingly, motor axons differ dramatically in their sensitivity to axonal injury. While the peripheral part of vertebrate motor axons degenerates within a week or so after axonal section, some invertebrate motor axons (Hoy et al., 1967; Van Essen and Jansen, 1977) remain alive and functional for many months. The length of the peripheral isolated segment of the axon is apparently important with regard to how long the terminals survive (Frank, 1974). This could be explained by the presence of a reservoir of essential substances in the longer nerve stumps, but there are certainly alternative possibilities.

The postsynaptic part of the end plate may also be influenced by neural factors transported along the axons. The concept of a trophic influence of nerve on muscle has remained viable in spite of little direct evidence. The recurring question has involved the extent to which the so-called neurotrophic effects on muscle are the consequences of muscle activity or mediated by specific neurotrophic substances. A recent contribution demonstrates that extracts of nerves enriched with transported material enhance the ACh sensitivity and the AChE activity of muscle in organ cultures. Substances with a similar effect on muscle were released by indirect stimulation of nerve–muscle preparations (Younkin *et al.*, 1978). These results are promising indications of transported neural substances with an effect on postsynaptic enzymes essential for synaptic function. However, an adequate understanding of the importance of axonal transport mechanisms in the maintenance of presynaptic as well as postsynaptic aspects of synaptic function requires the development of techniques for controlled prevention of the transport of specific substances.

Our knowledge of the postsynaptic factors required for the maintenance of synapses is equally vague. As will become apparent, however, the postsynaptic factors are interesting in that they provide simpler explanations for many of the striking changes in the pattern of innervation of muscles that occur during development and in certain experimental situations.

The most direct evidence for the importance of the postsynaptic structures, or target tissue, for maintenance of the presynaptic neurons has come from the study of cell death during early development. About half of the original number of motoneurons die during a relatively short period about halfway through embryonic development.

The process affects neurons that have already grown axons to the target muscle (Prestige, 1970, 1976) and probably even established synaptic connections (Landmesser and Pilar, 1976). Deprived of their target tissue by early extirpation of the limb bud all the motoneurons die during the same period (Hamburger, 1958), while an expansion of the target area by implantation of supernumerary limbs permits the maintenance of a greater number of motoneurons (Hollyday and Hamburger, 1976). This shows that specific neurons are not predetermined to die on account of an inherent defect, but rather that their survival depends on the availability of some factor associated with target tissue. The production of this factor, hence the proportion of surviving motoneurons, appears to depend on the state of activity of the muscle. Treatment with agents that bind to the ACh receptor and thereby paralyzes the muscle increases the number of surviving motoneurons by 50% or more (Pittman and Oppenheim, 1978; Laing and Prestige, 1978). There are interesting analogies in this with the requirements for maintenance of synapses at later stages of development.

The critical nature of the factors maintaining functional synapses is clearly demonstrated by the common occurrence of an initial overproduction followed by a subsequent loss of synapses during development. This was first demonstrated by Redfern (1970) in the rat diaphragm. Mammalian skeletal muscle is particularly favorable for the demonstration, since the mature muscle fibers are virtually always innervated only by a single axon. In the newborn animal, on the other hand, all the muscle fibers are innervated by branches of several different motor axons—in the rat soleus, by as many as five axons on the average (Brown et al., 1976). In the soleus all these form perfectly functional synapses, each by itself capable of activating the muscle fiber. This, however, does not seem to be the case for all muscles. In the sternothyroid muscle of the mouse all the muscle fibers are multiply innervated at birth and during the first postnatal week. However, some of the end plate potentials are much smaller than others, and the contraction of each motor unit is relatively no larger than in mature muscle (K. Nicolaysen and P. Hoff, personal communication). This suggests that a substantial fraction of the synapses is unable to activate the muscle fibers fully at birth. Whether the inefficient ones are in the state of being eliminated is not known.

In the rat soleus virtually all polyneuronal innervation is eliminated during the first $2\frac{1}{2}$ weeks postnatally (Brown et al., 1976). A similar time course has been reported for the diaphragm (Redfern, 1970; Bennett and Pettigrew, 1974). In the rabbit, on the other hand, Bixby and Van Essen (1979) found that polyinnervation was eliminated about 1 week earlier in the diaphragm than in the soleus. This may be due to a longitudinal gradient in the timing of the elimination of polyinnervation. They found no difference between proximal and distal muscles of the same limb or between fast and slow muscles.

The dominant part of the neonatal polyneuronal innervation occurs within a muscle. A particular group of motoneurons is not usually involved in the innervation of other muscles (Landmesser, 1978). However, in prenatal rats Harris and Dennis (1977) found that some intercostal muscle fibers were erroneously innervated by axons from the next posterior segment in addition to their appropriate innervation. Most of these mistakes were eliminated just before birth, that is, a week or two before the elimination of polyinnervation within muscles.

The loss of synapses could be caused by the death of entire motoneurons, as tentatively suggested by Bennett and Pettigrew (1974). Alternatively, the elimination of synapses could be due to reduced terminal branching of a constant number of motor axons supplying the muscle. For the rat soleus (Brown et al., 1976) and the sternothyroid muscles of mice (Nicolaysen and Hoff, personal communication) the number of motor units to the muscle remains constant throughout the period of postnatal synapse elimination. For the soleus muscle the relative size of motor units decreases approximately in parallel with the loss of polyinnervation. Hence the elimination is brought about by reduced terminal branching of the motor axons.

The morphological correlate of this process is at present controversial. Korneliussen and Jansen (1976) found a reduced number of terminal axonal profiles in their cross sections of the end plate region corresponding in time to the elimination of functional synapses in the muscle. Since they found no signs of degenerating terminal or preterminal axons over the relevant period, they suggested that the elimination was accompanied by retraction of the terminal branches into the parent axons. This interpretation is supported by the light microscopical observations of silver-stained material by O'Brien et al. (1978). Rosenthal and Taraskevitch (1977), on the other hand, described electron-dense terminal axons in comparable material and attributed them to axonal degeneration. The final resolution of the controversy probably will require a detailed comparison of the ultrastructural picture of such immature axons during various stages of axonal degeneration and their appearance during the elimination process.

While the question of terminal degeneration is still open, all investigations agree that all the terminals of the multiple axons innervating immature muscle fibers are located within a circumscribed end plate region, relatively no larger than the end plate of a single axon on a mature muscle fiber (Bennett and Pettigrew, 1974; Brown et al., 1976; O'Brian et al., 1978). In the end plate region the different axons that supply the end plate arborize, making the total terminal pattern rather like that of a mature terminal. There is no indication that the individual terminals occupy separate areas within the region. It has been appealing to consider these overlapping terminal areas of importance in the elimination process.

However, in other experimental situations separately located end plates can be eliminated as well. This has been demonstrated, for instance, in a rat soleus muscle innervated by foreign nerves (Brown et al., 1976; Kuffler et al., 1977). In these experiments the foreign nerve forms new end plates in the neighborhood where the nerve was implanted outside the original end plate band after the original nerve supply to the muscle was interrupted. At early times (1–2 weeks) after the new end plates have been formed, many muscle fibers are innervated by several axons, and most of the foreign end plates are located in separate regions along the muscle fiber (Frank et al., 1975). This has been demonstrated by functional localization of active end plates, as well as by the distribution of end plate AChE on isolated muscle fibers. After several weeks the majority of the muscle fibers are only innervated by a single foreign axon and there is a significant reduction in the frequency of muscle fibers with multiple esterase sites. This demonstrates an inability to maintain multiple innervation of individual muscle fibers even with end plates at separate sites. The reduction in esterase-stained sites shows that even end plates that have reached a relatively advanced stage of development are not protected from destruction.

However, if the original nerve is allowed to regenerate to the muscle soon after the foreign end plates have formed, it will reinnervate its original end plates, even on muscle fibers with foreign nerve innervation (Frank et al., 1975; Brown et al., 1976). The two end plates are maintained and remain functional on individual muscle fibers for prolonged periods. In these muscles there is usually a distance of several millimeters between the ectopic foreign end plates and the original end plates. The distance is longer than the distance between different foreign end plates on individual muscle fibers. Hence the enhanced survival of end plates could be due to the greater separation between them. Alternatively, the original nerve or the original end plate site could be preferentially maintained. A more advanced stage of maturation (Section V,A) could explain the additional survival of the foreign end plate on the same fibrers. However, when two foreign nerves were implanted at opposite ends of the muscle, many muscle fibers remained innervated by both nerves for prolonged periods (Kuffler et al., 1977). When the two nerves were implanted in the same region of the muscle, usually only a single axon terminal was maintained and remained functional. This suggests that individual muscle fibers can maintain several end plates permanently if the distance between them is sufficiently great. This is commonly seen normally in focally innervated muscle fibers of lower vertebrates, such as the frog sartorius (Katz and Kuffler, 1941).

There appears to be a characteristic morphological correlation with the loss of multiple ectopic innervation in the rat soleus muscle (Kuffler et al., 1980). At late times, when the majority of the muscles were only functionally innervated by single axons, there was still an appreciable number of individual muscle fibers with multiple ectopic end plate esterase sites. Electron microscopically many of these sites had all the hallmarks of ordinary end plates with presynaptic terminals overlying a characteristic folded region of the muscle fiber. At some sites, however, there were no axon terminals, while the esterase-stained region of the muscle fiber was differentiated as the postsynaptic part of the end plate. The "vacant" sites were interpreted as being the remains of end plates that had been functional initially and later eliminated. Apparently transitional stages with only a fraction of the postsynaptic region occupied by presynaptic terminals were also observed. The muscle fibers with such vacant or partially vacant end plate sites always had a fully intact, normal end plate in addition to the vacant site.

If the vacant sites are considered the remains of eliminated end plates, their distribution provides a measure of the distance over which the interaction between end plates can operate. In the foreign nerve region of the rat soleus we found vacant sites as far as 3 mm away from the fully innervated site of the same muscle fiber.

Ultrastructural examination also suggests a possible sequence of morphological stages of functionally inactivated end plates. Initially there is a progressive retraction of the presynaptic terminals from the postsynaptic area. This is completed while the postsynaptic structure of the end plate remains essentially intact. With time, however, even this dedifferentiates; the folds become progressively wider, and ultimately even the AChE decreases beyond histological detection. The whole process probably takes many months.

Even though some of the multiple esterase sites of these muscles were found to be vacant, and obviously nonfunctional, there were still more muscle fibers with fully intact multiple end plates than expected from the functional examination of their innervation. This discrepancy is probably explained by the innervation of separate end plates on the same muscle by terminal branches of the same axon. To see the possible implications of this we have to consider the mechanisms of elimination of functional end plates.

V. Competitive Interaction among End Plates

As mentioned, vertebrate muscle fibers are commonly innervated by several motor axons at early stages of development. The subsequent elimination of innervation in mammalian muscle results in the survival of only one axon to each fiber. The consistency of the end result suggests a competitive interaction among the axons innervating the same muscle fiber as a mechanism for elimination of the redundant innervation. This implies that the elimination is due to the existence of the surviving terminal.

A. ELIMINATION OF END PLATES IN EXPERIMENTAL SITUATIONS

There is some direct support for this in other cases of elimination of end plates. In salamanders Dennis and Yip (1978) induced cross-innervation of a denervated muscle by transplanting a foreign motor nerve. After about 4 weeks the majority of the muscle fibers were innervated by the foreign nerve. By this time the original nerve had regenerated, and as it reinnervated the muscle there was a progressive reduction in the frequency of muscle fibers innervated by the foreign nerve. When the original nerve was prevented from reinnervating the muscle, the fibers remained innervated by the foreign nerve. A comparable development has been described for reinnervation of the frog sartorius by two foreign nerves (Grinnell *et al.*, 1977). A single foreign nerve will innervate virtually the entire muscle. A second nerve, even after a delayed arrival, will superinnervate a substantial number of the fibers at the expense of the first.

Another commonly examined situation providing evidence for competitive interactions among motor axons is seen after partial denervation of a muscle. This leads to sprouting of the remaining axons and to an increase in their peripheral field of innervation. As the original axons regenerate, they reclaim some of their territory at the expense of the sprouted terminals (Cass *et al.*, 1973; Brown and Ironton, 1978; Haimann *et al.*, 1976; Thompson, 1978). A particularly convincing demonstration of this is from the peroneus tertius muscle of mice (Brown and Ironton, 1978). After section of 75% of the motor axons to this muscle the remaining three motor units sprout and innervate all the muscle fibers of the muscle. At later times, as the severed axons regenerate, some muscle fibers are innervated by both regenerated axons and by sprouts from the remaining ones, and some fibers are innervated only by the regenerated axons.

Suppression of the sprouted terminals in some of these situations requires arrival of the reinnervating axons at reasonably early times after formation of the terminals from the sprouting axons. When the regenerating axons are delayed, for instance, by nerve resection instead of a simple nerve section, the terminals from the sprouted axons have apparently established themselves sufficiently to prevent the regenerating axons from reclaiming their territory (Slack, 1978; Thompson, 1978). This could be due to complete occupancy of all available synaptic space by the terminals of the sprouted axons. Alternatively, it could be due to the progressive loss of a factor required for the reestablishment of synapses from the regenerating nerve (Frank *et al.*, 1975). In a comparable experiment Bennett *et al.* (1979) observed that foreign end plates were protected from competitive interaction with end plates of original axons reinnervating the same muscle fibers after a delay.

The morphological correlate of the suppression of foreign innervation has been controversial. Originally, Mark and Marotte (1972) suggested on indirect evidence that the inactivated end plates remained morphologically intact and might be rapidly reactivated by removal of the repressing original end plates. This has not been supported by subsequent more direct approaches to the question. After the foreign end plates had been functionally inactivated in the axolotl muscle, all the remaining end plates were labeled with horseradish peroxidase (HRP) after stimulation of the original nerve (Dennis and Yip, 1978). Similar labeling of the regressing foreign end plates showed that their morphological frequency and distribution corresponded to that determined physiologically (Bennett et al., 1979). Hence the functional repression of foreign end plates in axolotl muscle appears to be associated with withdrawal of the presynaptic terminals. This is entirely in line with the observations of the vacant end plates in the foreign nerve region of rat soleus muscles (Section IV).

The complete functional repression of foreign end plates in the axolotl muscle takes several weeks. During this period there is a progressive reduction in the average amplitude of the foreign nerve end plate potentials. At the same time the amplitude of the spontaneous MEPPs remains unchanged, indicating that the postsynaptic sensitivity to the transmitter remains normal. Hence the repression is due to a reduced number of quanta of transmitter released by each nerve impulse (Dennis and Yip, 1978). Since the quantal content of the end plate potential is correlated with the size of the terminal (Kuno et al., 1971; Bennett and Raftos, 1977), this may be a direct consequence of gradual withdrawal of the presynaptic terminal. This, however, remains to be directly demonstrated.

B. ELIMINATION OF END PLATES IN NORMAL DEVELOPMENT

With this evidence concerning the competitive nature of the elimination of foreign innervation in experimental situations we can return to the physiological elimination of early postnatal polyinnervation. As mentioned, it consists of a progressive reduction in the size of the motor units supplying the muscle. However, the loss of motor terminals is not random. To reach the final stage of one axon terminal for each muscle fiber that would imply an appreciable fraction of the muscle fibers losing all their innervation. This does not seem to happen (Brown et al., 1976). A competitive interaction among the motor terminals innervating a particular muscle fiber is again an attractive possibility. Evidence in this direction has been sought by partial denervation of muscles while their motor units are still in the neonatal expanded size. This procedure should ideally eliminate potential competitors, and the prediction is that the reduction in motor unit size will be prevented. The experiment has been performed on two different muscles, the soleus (Brown *et al.*, 1976; Thompson and Jansen, 1977) and the fourth lumbrical (Betz *et al.*, 1980) of the rat. In the lumbrical muscle the average size of the motor units at birth is 1.5 times their size in the mature muscle after polyneuronal innervation has been eliminated. Single motor units remaining in the muscle after partial denervation at birth retain approximately their original size. Hence the normal reduction in size of motor units can be explained entirely by competitive interaction among motor terminals.

In the soleus muscle the results are more complicated. The average size of the motor units at birth is about five times their normal mature size. In muscles partially denervated just after birth the motor units were on average reduced in size to about 1.8 their mature size in spite of the presence of a large number of denervated muscle fibers in these muscles. Hence neonatal motor units are different from adult motor units which sprout to innervate 4 to 5 times their normal number of muscle fibers in partially denervated muscles (Thompson and Jansen, 1977). Based on the cases with only a few remaining motor units after neonatal partial denervation it was even likely that some muscle fibers lost their last motor terminal during the reduction in motor unit size. This means that competition alone cannot explain the entire reduction in the number of motor units. In addition, there appears to be an inherent tendency of motoneurons to lose terminals during this stage of development.

C. Mechanism of Interactions among End Plates

The motoneuron, the muscle fiber, and the terminal Schwann cells are all intimately involved in establishment of the neuromuscular end plates. In principle all three types of cells could be involved in the process of eliminating the redundant innervation. At present we have no established views on the functional significance of the Schwann cells associated with the end plate. Their possible contribution therefore remains entirely speculative.

As we have seen, synaptic repression is primarily a presynaptic failure eventually leading to the retraction of axon terminals. The focus is therefore on maintenance of the terminals and the question concerns the nature of the required signals. The competitive interaction among terminals is restricted to the group innervating the same muscle fiber. It also involves end plates separately located on the muscle fiber. Therefore the signal must be mediated by the muscle fiber. The simplest view is that it is produced by the muscle fiber. Such a substance has already been required to explain the induction of sprouting and synapse formation, and there is no a priori reason why it cannot serve as a synapse-maintaining substance as well. Produced by the muscle in limiting amounts, the availability of such a substance could determine the number of axon terminals that could remain functional on a muscle fiber. If a nerve fiber depleted the substance over a certain region of the muscle fiber, the limited distance of the competitive interaction among terminals would follow. Being diffusible it could explain the moderate delay in the early postnatal elimination seen after partial denervation (Thompson and Jansen, 1977).

The amount produced by the muscle fiber should depend on the level of activity of the fiber. There is evidence that a paralyzed muscle can entertain more functional end plates than the normally active muscle (Benoit and Changeux, 1978; Thompson et al., 1979), and conversely that the postnatal elimination is accelerated by additional electrical stimulation of muscles (O'Brien et al., 1978).

At present this scheme is entirely speculative. Its merit is probably mainly in summarizing the significant observations requiring an explanation and in restricting the number of postulated factors involved. Other models have been advanced and may be equally valid. O'Brien et al. (1978) suggested that proteolytic enzymes released by muscle fibers during activity might digest and eliminate competing end plates on the same muscle fiber. It is not obvious how the surviving end plate is protected from the proteolytic action in this scheme. In general, it is probably premature to discuss the various possibilities systematically.

Nevertheless, since one or another form of competitive interaction among end plates appears to take place, we may consider the nature of the competitive advantage of the surviving terminal. In the axolotl the original nerve eliminates the foreign end plates (Dennis and Yip, 1978), even though it reinnervates the muscle after the foreign end plates are fully active. This suggests a preference for the original nerve. The preference is apparently not due to more efficient activation of the muscle fibers by the original nerve. In this respect the foreign nerves do equally well. In terms of competition for a substance produced by the muscle fibers it is rather a question of a more efficient uptake of the substance by the original nerve.

However, in other cases a preference for a particular nerve can probably not explain the competitive advantage of the survivor. In mammals there is in general very little evidence for a muscle preferring its original motoneurons (Section II,B). In the rat soleus nerve, for instance, a foreign nerve can prevent reinnervation of the original nerve (Frank et al., 1975). Furthermore, in recent experiments Bixby and Van Essen (1978) have shown that a transplanted foreign nerve to some extent can superinnervate a fully intact and functional muscle. This cross-innervation takes place at the original end plates and leads to suppression of some of the original innervation. This behavior is easily accommodated by the current idea suggesting the importance of the total number of terminals supported by a motorneuron in the outcome of competitive interactions. After partial denervation of a muscle the regenerating axons are able to reclaim some of the muscle fibers innervated by sprouts from remaining axons (Section V,A). This is a question of competition among axons that are all native to the muscle. The axons that have sprouted into new territory have increased their number of terminal branches. In the competition they lose to the regenerating axons in the process of re-forming their first terminals.

A similar phenomenon may occur during the early postnatal elimination of polyinnervation. Just after birth there is a large scatter in the size of the motor units. Some motor units are no larger than their mature size, while others may be 10 times as large. Elimination of the redundant innervation takes place particularly at the expense of the large units, so that in the mature state the motor units are much more uniform in size (Brown et al., 1976). It appears reasonable that there is a limit to the number of terminals that can be supported by a particular motoneuron (Thompson and Jansen, 1977) and from that quite appealing to accept the competitive disadvantage of an overextended peripheral field of innervation.

This again means that the fate of a particular end plate is determined by the total number of end plates belonging to its axon. Hence the outcome of the competition cannot be determined only locally, at the terminal, but also depends on signals probably from the cell body. This perhaps explains why different terminals from the same axon apparently can remain functional on the same muscle fiber. An example of this from ectopically innervated rat soleus was mentioned above, and it also appears to occur commonly in various pathological states (Woolf and Coërs, 1974).

Based on the views advanced here the activity of the muscle determines the availability of the end plate-maintaining factor and does not directly determine the outcome of the competitive interaction among end plates on the same muscle fiber. In an alternative model the "selective stabilization" of a synapse is linked directly to its level of activity (Changeux and Danchin, 1976). However, the concept of stabilization through activity does not in any simple way lead to predictions about the outcome of the competitive interactions we have discussed. Nevertheless, a link between functional activity and maintenance of synapses appears to be required in other situations, and its demonstration remains a major achievement in our understanding of the development of the nervous system.

VI. Conclusions

Only fragments of the mechanisms underlying the establishment and maintenance of neuromuscular connections are known presently. Descriptively the system is much better known. We have emphasized some aspects of neuromuscular development that appear particularly significant in the construction of mechanistic explanations of the processes:

1. Some form of guidance is required to explain how motor axons reach their target. It appears to depend on pathways in the intervening tissue and axonal position in the nerve trunk rather than on chemotactic signals originating in the target. At close range signals from receptive muscles can induce axonal sprouting and differentiation of the terminals.

2. Formation of end plates requires a receptive state in the muscle fiber. This state is associated with the presence of ACh receptors in the muscle membrane. The ACh receptor is necessary for neuromuscular transmission, but it is not known whether it also serves as a necessary recognition factor in the postsynaptic membrane.

3. Synapse formation leads to a series of events postsynaptically. ACh receptors aggregate and begin to turn over more slowly. The open time of the ionic channels becomes shorter, and AChE activity and junctional folds appear. The aggregation and stabilization of ACh receptors in the plasma membrane do not require muscle activity and are intrinsically resistant to the effects of muscle activity. The process is accompanied by similarly persistent changes (traces), probably in the overlying basal lamina, which determine the localization of AChE appearing later. This junctional AChE is produced, at least in part, by the muscle fiber.

4. The receptive state of the muscle fiber and several of the subse-

quent events are directly, although often with an appreciable time lag, dependent on the level of activity in the muscle fiber. Activity prevents ectopic synapse formation but not reinnervation of original end plate sites. Activity reduces the rate of synthesis of ACh receptors, eliminates extrajunctional ACh sensitivity, and causes the appearance of junctional AChE.

5. During development an excess number of motoneurons are formed. Normally a substantial fraction of these neurons die. The number surviving is determined by the size of the available target. Inactivity of the muscle enhances motoneuronal survival. Similarly the number of end plates a neuron can maintain is limited and increases in paralyzed muscle.

6. During initial formation synapses are often formed in excessive numbers. The redundant ones are subsequently eliminated. Physiologically this is seen as a progressive reduction in the quantal content of the end plate potential. Morphologically elimination is probably caused by the retraction of redundant terminals from postsynaptic membranes which initially remain intact. Competitive interaction among axon terminals innervating the same muscle fiber appears to be an important element of the elimination process. There is probably a limitation to the distance over which the competitive process can work along a muscle fiber. Larger than normal motor units appear to be disfavored in the competitive interaction. In addition, part of the normal postnatal loss of end plates appears to be noncompetitive and due to an inherent property of motoneurons at this stage of development.

7. Hypothetical factors produced by the muscle and acting on the motoneurons figure prominently in simple explanations of these neuromuscular interactions.

REFERENCES

Albuquerque, E. X., Warnick, J. E., Tasse, J. R., and Sansone, F. M. (1972). *Exp. Neurol.* **37**, 607–634.
Anderson, M. J., Cohen, M. W., and Zorychta, E. (1977). *J. Physiol.* **268**, 731–756.
Bennett, M. R., and Pettigrew, A. G. (1974). *J. Physiol.* **241**, 515–545.
Bennett, M. R., and Raftos, J. (1977). *J. Physiol.* **265**, 261–295.
Bennett, M. R., McGrath, P. A., and Davey, D. F. (1979). *Brain Res.* **173**, 451–469.
Benoit, P., and Changeux, J.-P. (1978). *Brain Res.* **149**, 89–96.
Berg, D. K., and Hall, Z. W. (1975). *J. Physiol.* **252**, 771–789.
Betz, W. J., Caldwell, J. H., and Ribchester, R. R. (1980). *J. Physiol.* **303**, 265–280.
Bevan, S., and Steinbach, J.H. (1977). *J. Physiol.* **267**, 195–213.
Bixby, J. L., and Van Essen, D. C. (1978). *Soc. Neurosci. Abstr.* **4**, 367.
Bixby, J. L., and Van Essen, D. C. (1979). *Brain Res.* **169**, 275–286.
Brown, M. C., and Ironton, R. (1977). *Nature (London)* **265**, 459–461.

Brown, M. C., and Ironton, R. (1978). *J. Physiol.* 278, 325-348.
Brown, M. C., Jansen, J. K. S., and Van Essen, D. C. (1976). *J. Physiol.* 261, 387-422.
Brown, M. C., Goodwin, G. M., and Ironton, R. (1977). *J. Physiol.* 267, 42-43P.
Brown, M. C., Holland, R. L., and Ironton, R. (1978a). *Nature (London)* 275, 652-654.
Brown, M. C., Holland, R. L., and Ironton, R. (1978b). *J. Physiol.* 282, 7P-8P.
Burden, S. (1977). *Dev. Biol.* 61, 79-85.
Butler, I. J., Drachman, D. B., and Goldberg, A. M. (1978). *J. Physiol.* 274, 583-600.
Cangiano, A., and Fried, J. A. (1977). *J. Physiol.* 265, 63-84.
Cangiano, A., and Lutzemberger, L. (1977). *Science* 196, 542-544.
Cangiano, A., Lømo, T., and Lutzemberger, L. (1978). *Eur. Neurosci. Meet., 2nd Florence, Sept.* (Abstr.)
Cangiano, A., Lømo T., Lutzemberger, L., and Sveen, O. (1980). *Acta Physiol. Scand.* (in press).
Cass, D. T., and Mark, R. F. (1975). *Proc. R. Soc. B* 190, 45-58.
Cass, D.T., Sutton, T. J., and Mark, R. F. (1973). *Nature (London)* 243, 201-203.
Changeux, J.-P., and Danchin, A. (1976). *Nature (London)* 264, 705-712.
Christian, C. N., Daniels, M. P., Sugiyama, H., Vogel, Z., Jacques, L., and Nelson, P. G. (1978). *Proc. Natl. Acad. Sci. U.S.A.* 75, 4011-4015.
Chubb, I. W., and Smith, A. D. (1975). *Proc. R. Soc. B* 191, 263-269.
Chubb, I. W., Goodman, S., and Smith, A. D. (1976). *Neuroscience* 1, 57-62.
Cohen, M. W. (1972). *Brain Res.* 41, 457-463.
Czéh, G., Gallego, R., Kudo, N., and Kuno, M. (1978). *J. Physiol.* 281, 239-252.
Dennis, M. J., and Yip, J. W. (1978). *J. Physiol.* 274, 299-310.
Duchen, L. W., and Strich, S. J. (1968). *Q. J. Exp. Physiol.* 53, 84-89.
Duchen, L. W., and Tonge, D. A. (1973). *J. Physiol.* 228, 151-172.
Edds, M. V. (1953). *Q. Rev. Biol.* 28, 260-276.
Elsberg, C. A. (1917). *Science* 45, 318-320.
Fernandez, H. L., and Inestrosa, N. C. (1976). *Nature (London)* 262, 55-56.
Fex, S., and Thesleff, S. (1967). *Life Sci.* 6, 635-639.
Frank, E. (1974). *J. Physiol.* 242, 371-382.
Frank, E., Jansen, J. K. S., Lømo, T., and Westgaard, R. H. (1975). *J. Physiol.* 247, 725-743.
Gilliatt, R. W., Westgaard, R. H., and Williams, I. R. (1978). *J. Physiol.* 280, 499-514.
Gordon, T., Jones, R., and Vrbová, G. (1976). *Prog. Neurobiol.* 6, 103-136.
Grimm, L. M. (1971). *J. Exp. Zool.* 178, 479-496.
Grinnell, A. D., Rheuben, M. B., and Letinsky, M. S. (1977). *Nature (London)* 265, 368-370.
Guth, L. (1956). *Physiol. Rev.* 36, 441-478.
Haimann, C., Mallart, A., and Zilber-Gachelin, N. F. (1976). *Neurosci. Lett.* 3, 15-20.
Hall, Z. W. (1973). *J. Neurobiol.* 4, 343-361.
Hall, Z. W., and Reiness, C. G. (1977). *Nature (London)* 268, 655-657.
Hamburger, V. (1958). *Am. J. Anat.* 102, 365-410.
Hamburger, V. (1975). *J. Comp. Neurol.* 160, 535-546.
Harris, A. J., and Dennis, M. J. (1977). *Soc. Neurosci. Abstr.* 3, 107.
Harris, J. B., and Thesleff, S. (1972). *Nature (London) New Biol.* 236, 60-61.
Hoh, J. F. Y. (1975). *J. Physiol.* 251, 791-803.
Hollyday, M., and Hamburger, V. (1976). *J. Comp. Neurol.* 170, 311-320.
Hollyday, M., Hamburger, V., and Farris, J. (1977). *Proc. Natl. Acad. Sci. U.S.A.* 74, 3582-3586.
Horder, T. J. (1978). *Zoon* 6, 181-192.
Hoy, R. R., Bittner, G. D., and Kennedy, D. (1967). *Science* 156, 251-257.

280   TERJE LØMO AND JAN K. S. JANSEN

Huizar, P., Kudo, N., Kuno, M., and Miyata, Y. (1977). J. Physiol. 265, 175-191.
Ironton, R., Brown, M. C., and Holland, R. L. (1978). Brain Res. 156, 351-354.
Jansen, J. K. S., Lømo, T., Nicolaysen, K., and Westgaard, R. H. (1973). Science 181, 559-561.
Jones, R., and Tuffery, A. R. (1973). J. Physiol. 232, 13-15P.
Jones, R., and Vrbová, G. (1974). J. Physiol. 236, 517-538.
Katz, B., and Kuffler, S. W. (1941). J. Neurophysiol. 4, 209-223.
Katz, B., and Miledi, R. (1972). J. Physiol. 224, 665-699.
Korneliussen, H., and Jansen, J. K. S. (1976). J. Neurocytol. 5, 591-604.
Korneliussen, H., and Sommerschild, H. (1976). Cell Tissue Res. 167, 439-452.
Kuffler, D., Thompson, W., and Jansen, J. K. S. (1977). Brain Res. 138, 353-358.
Kuffler, D., Thompson, W., and Jansen, J. K. S. (1980). Proc. R. Soc. B 208, 189-222.
Kuno, M., Turkanis, S. A., and Weakly, J. N. (1971). J. Physiol. 213, 545-556.
Laing, N. G., and Prestige, M. C. (1978). J. Physiol. 282, 33-34P.
Landmesser, L. (1978). J. Physiol. 284, 391-414.
Landmesser, L., and Pilar, G. (1976). J. Cell Biol. 68, 357-374.
Lentz, T. L. (1974). Exp. Neurol. 45, 520-526.
Letinsky, M. D., Fischbeck, K. H., and McMahan, U. J. (1976). J. Neurocytol. 5, 691-718.
Lewis, J. (1978). Zoon 6, 175-179.
Lømo, T., and Rosenthal, J. (1972). J. Physiol. 221, 483-513.
Lømo, T., and Slater, C. R. (1978). J. Physiol. 275, 391-402.
Lømo, T., and Slater, C. R. (1980a). J. Physiol. 303, 173-189.
Lømo, T., and Slater, C. R. (1980b). J. Physiol. 303, 191-202.
Lømo, T., and Westgaard, R. H. (1975). J. Physiol. 292, 603-626.
Lømo, T., and Westgaard, R. H. (1976). Cold Spring Harbor Symp. 40, 263-274.
McMahan, U. J., Sanes, J. R., and Marshall, L. M. (1978). Nature (London) 271, 172-174.
Mark, R. F., and Marotte, L. R. (1972). Brain Res. 46, 131-148.
Miledi, R. (1963). In "The Effect of Use and Disuse on Neuromuscular Functions" (E. Gutmann and R. Hnik, eds.). Publishing House of the Czechoslovac Academy of Sciences, Prague.
Morris, D. G. (1978). J. Neurophysiol. 41, 1450-1465.
O'Brien, R. A. D., Østberg, A. J. C., and Vrbová, G. (1978). J. Physiol. 282, 571-582.
Olson, L., and Malmfors, T. (1970). Acta Physiol. Scand. Suppl. 348, 1-112.
Pestronk, A., and Drachman, D. B. (1978). Science 199, 1223-1225.
Pestronk, A., Drachman, D. B., and Griffin, J. W. (1976). Nature (London), 260, 352-353.
Pittman, R. H., and Oppenheim, R. W. (1978). Nature (London) 271, 364-365.
Prestige, M. C. (1970). In "The Neurosciences: Second Study Program" F. O. Schmitt, ed.), pp. 73-82. Rockefeller Univ. Press, New York.
Prestige, M. C. (1976). J. Comp. Neurol. 170, 123-134.
Purves, D. (1976a). J. Physiol. 259, 159-175.
Purves, D. (1976b). Int. Rev. Physiol. Neurophysiol. II 10, 125-177.
Purves, D., and Sakmann, B. (1974). J. Physiol. 237, 157-182.
Redfern, P. A. (1970). J. Physiol. 209, 701-709.
Riley, D. A. (1978). Soc. Neurosci. Abstr. 4, 1717.
Rosenthal, J. L., and Taraskevich, P. S. (1977). J. Physiol. 270, 299-310.
Rotshenker, S. (1978). Brain Res. 155, 354-356.
Rotshenker, S., and McMahan, U. J. (1976). J. Neurocytol. 5, 719-730.
Rubin, L. L., Schuetze, S. M., Weill, C. L., and Fischbach, G. D. (1978). Soc. Neurosci. Abstr. 4, 1193.

Sanes, J. R., Marshall, L. M., and McMahan, U. J. (1978). *J. Cell Biol.* **78**, 176–198.
Schuetze, S. M., and Fischbach, G. D. (1978). *Soc. Neurosci. Abstr.* **4**, 1195.
Schuetze, S. M., Frank, E. F., and Fischbach, G. D. (1978). *Proc. Natl. Acad. Sci. U.S.A.* **75**, 520–523.
Skau, K. A., and Brimijoin, S. (1978). *Nature (London)* **275**, 224–226.
Slack, J. R. (1978). *Brain Res.* **146**, 172–176.
Slater, C. R. (1978). *In* "Intercellular Junctions and Synapses" (J. Feldman, N. B. Gilula and J. D. Pitt, eds.), pp. 217–239. Chapman & Hall, London.
Thompson, W. (1978). *Acta Physiol. Scand.* **103**, 81–91.
Thompson, W., and Jansen, J. K. S. (1977). *Neuroscience* **2**, 523–536.
Thompson, W., Kuffler, D. P., and Jansen, J. K. S. (1979). *Neuroscience* **4**, 271–282.
Tonge, D. A. (1974). *J. Physiol.* **239**, 96–97P.
Van Essen, D., and Jansen, J. K. S. (1974). *Acta Physiol. Scand.* **91**, 571–573.
Van Essen, D. C., and Jansen, J. K. S. (1977). *J. Comp. Neurol.* **171**, 433–454.
Vigny, M., Koenig, J., and Rieger, F. (1976). *J. Neurochem.* **27**, 1347–1353.
Wake, K., (1976). *Cell Tissue Res.* **173**, 383–400.
Woolf, A. L., and Coërs, C. (1974). *In* "Disorders of Voluntary Muscle" (J. N. Walton, ed.), pp. 274–309. Churchill, London.
Younkin, S. G., Brett, R. S., Davey, B., and Younkin, L. H. (1978). *Science* **200**, 1292–1295.

CHAPTER 8

COLONY CULTURE OF NEURAL CELLS AS A METHOD FOR THE STUDY OF CELL LINEAGES IN THE DEVELOPING CNS: THE ASTROCYTE CELL LINEAGE

S. Fedoroff and L. C. Doering

DEPARTMENT OF ANATOMY
UNIVERSITY OF SASKATCHEWAN
SASKATOON, SASKATCHEWAN, CANADA

I. Introduction

The so-called Boulder Committee, in an attempt to reduce the confusion caused by the terms accumulated over the years as knowledge and concepts of the developing central nervous system (CNS) changed, undertook to sort out, simplify and, in short, bring some order out of the chaos. They introduced a conceptual model of the embryonic CNS in which four fundamental zones were identified and named ventricular, subventricular, intermediate, and marginal (Boulder Committee, 1970). Two zones, the ventricular and the subventricular, are composed mainly of proliferating cells. The other two consist of nonproliferating postmitotic, differentiating, or differentiated cells. When this model is extended a step further so that it relates temporally to the development of the adult CNS, it can be viewed in terms of three major compartments, namely, the ventricular cell compartment, the subventricular cell compartment, and the end cell compartment. Thus, the developing nervous system can be thought of in terms of proliferative and nonproliferative cell compart-

283

CURRENT TOPICS IN
DEVELOPMENTAL BIOLOGY, Vol. 16

ments with development characterized by a progression of cells from one compartment to another. The ventricular cell compartment is believed to be composed of proliferative pluripotential stem cells which give rise mainly to type I neurons or macroneurons and to subventricular cells. The subventricular cell compartment is composed of proliferative progenitor cells with somewhat limited capabilities but able to give rise to more than one cell lineage. In addition to progenitor cells, the compartment also contains committed precursor cells in various stages of differentiation but still able to divide and committed to only one cell lineage. The subventricular cells give rise mainly to type II neurons or microneurons, astrocytes, and oligodendrocytes. The end cell compartment or nonproliferative cell compartment consists of committed precursor cells after they have entered the postmitotic stage of differentiation, and fully differentiated end cells. (It also includes glial cells, even though they have not entirely lost their ability to divide and do so usually at a low rate.)

From the point of view of compartments and cellular progression, development is basically sequential; i.e., large type I neurons form first and then type II neurons, followed by the formation of astrocytes which overlaps the formation of type II neurons. The last cells formed are oligodendrocytes (Paterson et al., 1973; Imamoto et al., 1978). This implies that, if one takes slices of nervous tissue from the same anatomical location at different stages of development, they will be composed of cell populations differing from the ones preceding in frequency of occurrence of cells of the three compartents and in degree of differentiation of cells within the compartments. Of course, the ideal situation in studying cell lineages would be to have uninterrupted observation of representative cells as they progress through the lineages throughout neurogenesis. Since this is impossible, an acceptable compromise is to study cell progression in isolated nervous tissue in cultures and to relate the *in vitro* observations to those made *in vivo*.

II. Colony Cultures

Tissue culture comes close to providing ideal conditions for the study of cell lineages. Cells in primary cultures are direct descendants of cells *in vivo* and provide continuation of the *in situ* cell lineage. The behavior and function of cells in cultures is greatly influenced by the genetic program of the cells, their past history, and the conditions *in vitro*. It has been shown that neural tube–spinal cord primordium from 1.5- to 10-day old embryos in fragment cultures gives rise to large, multipolar type I neurons (Fisher and Fedoroff, 1977); dissociated brain cells of 5- to 13-day chick embryo brain give rise to a monolayer of glial precursor cells with small type II neurons on top of the monolayer (Booher and Sensenbrenner, 1972); and dissociated brain cells of 14-day or older chick embryos give rise only to a monolayer of glial precursor cells (Booher and Sensenbrenner, 1972). This indicates that cultures of glial precursor cells free of neurons can be obtained provided the embryos are selected at developmental stages at which most neurons are already in the postmitotic state of differentiation but at which glial cell precursors are still proliferating. Such cultures usually are initiated with an inoculum of (2 to 2.25) × 10^5 cells/ml (Booher and Sensenbrenner, 1972) or higher numbers (Hertz et al., 1978a), and within a relatively short time the cells form a monolayer covering the entire surface of a culture dish. Although such preparations are composed mainly of glial precursor cells, they may also contain individual cells at different stages of differentiation. It is very difficult to identify or follow individual cells within the monolayer. In addition, study is limited to glial precursor cells in embryos in which the production of neurons has ceased. Younger embryos would yield cultures that are even more heterogeneous, because cells of neuronal lineage would also be present.

However, if dissociated cell suspensions from the neopallium of chick or mouse embryos are plated in smaller number, e.g., 2 × 10^4 cells/ml or less, then cell colonies form rather than a monolayer. In our studies we used DBA/1J, C_3H/HeJ, and Swiss mice, as well as chick embryos. The usual procedure was to isolate the neopallium aseptically and carefully to remove the meninges. The hemispheres were freed of the basal ganglia, olfactory lobe, and hippocampus and then, by means of microscalpels, divided into small fragments which were dissociated by gently forcing the tissue through sterile Nitex mesh (pore size 75 μm). The cells were then suspended in a growth medium consisting of Eagle's minimum essential medium (EMEM) containing a fourfold concentration of vitamins, a double concentration of amino acids (except glutamine, which was kept at the 2 mM level), and 7.5 mM glucose with 5% horse serum (v/v). Cell viability was determined by using the nigrosine dye exclusion technique (Kaltenbach et al., 1958). The cells, in various dilutions in a total volume of 2.5 ml growth medium, were plated into 60-mm Falcon petri dishes. The cultures were incubated at 37°C with a humidified atmosphere of 5% CO_2 in air. After 3 days of incubation, round, refractile cells could be seen attached to the plastic dish surface, and many single cells, cell clumps, and cell debris floating in the medium. At this stage the cell debris and nonattached cells were washed out, and fresh medium was added.

The cultures were then reincubated until needed. At that time, the cultures were fixed in absolute methanol and examined under a dark-field or phase-contrast microscope, or stained with 0.25% Coomassie brilliant blue (acid blue 83; C14 2660, Bio-Rad Laboratories) and examined under the light microscope.

A colony culture method selects cells from a dissociated neopallium that can attach to the plastic substratum and proliferate to form a colony. Such cells, therefore, must belong to the proliferative cell compartment of the brain. Regardless of their lineage or degree of differentiation, we call cells that can form colonies in cultures "colony-forming cells" (CFCs).

In our work, usually 95% of each cell suspension consists of single cells; the remainder consists of cell aggregates of two or more cells. Cells that attach to the plastic substratum after 3–4 days in culture begin to proliferate and, depending on the duration of culturing, form colonies varying in size from 25 to several hundred cells per colony. In a sample of dissociated cells of neopallium the CFCs are distributed according to the Poisson distribution and there is direct proportionality between the size of the inoculum and the number of colonies formed (S. Fedoroff, R. White, and C. Hall, unpublished). The plating efficiency, defined as the percentage of individual cells that give rise to colonies when inoculated into culture vessels (Fedoroff, 1966), was determined by a method used by Brunette et al. (1976). It entailed adding a varying number of cells to the wells of a microtest culture plate (Falcon 3040) in a total volume of 0.20 ml growth medium. The principle underlying this method is that, since the number of wells and the number of cells inoculated into each well are known, it is possible to calculate the total number of cells. At a lower concentration of cells, not all the wells will have proliferating cells forming colonies, and it is probable that wells that do contain a colony were initiated by a single viable cell. Hence the number of wells in the microtest plate with growing cells, divided by the total number of cells, is an estimate of the plating efficiency. Using this method we established the plating efficiency of cells from newborn mouse neopallium to be approximately 1% (Table I; S. Fedoroff and R. White, unpublished). This is considerably higher than ordinarily observed with primarily dissociated cells (Paul, 1975). It is possible to dissociate a single colony by using, for example, a weak trypsin solution and subculturing the cells to other petri dishes (S. Fedoroff and C. Hall, unpublished). The number of times such cells can be subcultured has not yet been established.

For colony cultures, the first 24–48 hours of culturing are critical.

TABLE I

PLATING EFFICIENCY OF CFCs

Cells added per well	Proportion of wells having colonies of cells	Plating efficiency (%)[a]
390	40/40	—
195	39/40	—
98	30/40	—
49	22/40	—
24	10/40	1.04
12	5/40	1.04
560	320/326	—
256	403/413	—
128	333/514	—
64	210/480	—
32	128/480	—
16	103/480	—
8	48/576	1.04
4	26/480	1.28

[a] Plating efficiency determined only in series in which there was no more than one colony per well.

During this time small cell aggregates may form, especially if the inoculum is too large and the culture vessels are shaken. Aggregates are more likely to form in cultures from young embryos than in those from older ones. However, we have the impression that, when such small cell aggregates are formed, they appear to consist of cells of the same type. Because some cells may aggregate, it is impossible to be certain which colonies in culture were formed from single cells and which from small aggregates. To take this situation into account, the term "colony-forming unit" (CFU) was borrowed from hematology (Metcalf and Moore, 1971). The term is noncommittal; i.e., it does not suggest that every colony is definitely a clone.

III. Embryological Development of Colony Forming Cells

In order to determine when CFCs begin to appear in the CNS, we examined dissociated cells from the neural tube–spinal cord primordium of chick embryos at various developmental stages.

We found that the neural tubes of 2.5-day chick embryos at Hamilton–Hamburger (H and H) stage 17 were the youngest to yield a few cells that could form colonies in culture (Fig. 1). The plating effi-

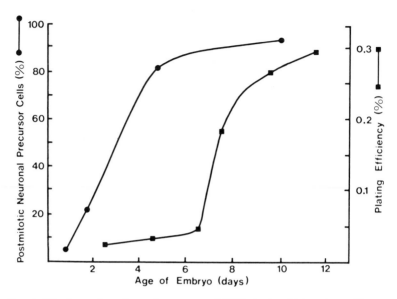

FIG. 1. The formation of type I neurons and CFCs in the chick embryo. The percentage of postmitotic neural precursor cells was determined by growing neural tubes or spinal cord primordia continuously in medium containing 1.0 μCi/ml of [³H]thymidine. The medium was changed twice weekly. After 21 days the cultures were washed, fixed in Cajal's formol NH₄Br, washed in alcohol, dried, covered with Kodak AR-10 stripping film, and exposed for 2-3 weeks. After developing, the cultures were stained with cresyl violet or thionine, buffered at pH 3.75. Cultures were examined microscopically, and the labeled and nonlabeled neurons were scored. It was assumed that, if a neuron was not labeled, it must have originated from a precursor cell that was already postmitotic at the time the explant was removed from the embryo (Fisher and Fedoroff, 1977, 1978). The plating efficiency of CFCs was determined as described in Section II.

ciency of such cell suspensions was approximately 0.025%. Between day 6.5 (H and H stage 30) and day 7.5 (H and H stage 32) there was a sudden exponential increase in the number of cells from the spinal cord primordium that could form colonies in cultures (Fig. 1). After day 9 (H and H stage 35) the number of CFUs produced began to reach a plateau. In mice, the number of CFUs formed seems to be constant during the later part of embryonic development but begins to decline postnatally (Fedoroff, 1977; S. Fedoroff and C. Hall, unpublished).

It is of interest to compare the time of origin of type I neurons and CFCs. Fisher and Fedoroff (1977, 1978) cultured neural tubes or spinal cord primordia from chick embryos at various developing stages continuously in medium containing [³H]thymidine. After 21 days in culture they found well-developed multipolar neurons (type I), some labeled and some not. This was interpreted to mean that unlabeled neurons originated from neuronal precursor cells that were already postmitotic at the time of explantation, i.e., since they did not divide in cultures, they were not labeled. The number of such unlabeled neurons in cultures increased with advancing age of donor chick embryos (Fig. 1). If the time of origin of type I neurons in the spinal cord of chick embryos is compared to that of CFCs, it becomes obvious that the origin of the two types of cells is sequential. It seems that only after ventricular cells have completed the formation of type I neurons does formation of CFCs begin (Fig. 1). This coincides also with the beginning of the decline of the ventricular zone in the spinal cord (Doering, 1980).

Previously, we mentioned that a cell must be able to proliferate a number of times before a colony of reasonable size forms. Therefore, CFCs must be proliferative cells and must come from the proliferative cell compartment of the chick CNS. A graphic analysis of the origin, level throughout development, and decline of CFCs, particularly in the CNS, probably reflects the state of the proliferative cell compartment in general, and by using the colony culture method it may become possible to monitor it in the future.

IV. Heterogeneity of Cell Colonies

Colonies formed in cultures by the dissociated cells of neopallium or spinal cord are not all identical. They vary in the morphology of their cells and in their size and compactness. Cells of dissociated neopallium from newborn (P0) mice form six distinct colony types in cultures (Fedoroff, 1977). For convenience, in our laboratory we have designated them types A, B, C, D, E, F, and P. The frequency of occurrence of the various colony types varies according to the age of the embryo or postnatal mouse, the anatomical region of the CNS, the duration of culturing, and the culture medium used (Fedoroff, 1977, 1978; Fedoroff and Hall, 1979; Juurlink et al., 1980). Generally, the frequency of occurrence of colonies after 4 and 7 days of culturing is about the same, indicating that during the first week of culturing there is no visible change or transformation in the cells, and at 7 days the size of the colonies is such that the various types can be easily identified. For these reasons, 7-day cultures are used in our laboratory in all assays for determination of the frequency of occurrence of various types of CFCs present in the original cell suspensions.

Using this assay method we determined the frequency of occurrence of various colony types in 7-day cultures initiated from chick neural tube or spinal cord primordia of various developmental stages. The spinal cord primordia of 3.5-day chick embryos (H and H stage 21) yielded a high frequency of type-P colonies (70.1 ± 6.8%) and a low frequency of type-A colonies (29.9 ± 6.8%). Type-P colonies consist of closely apposed epithelial cells with small amounts of cytoplasm and large round-to-oval nuclei with distinct nucleoli (Fig. 3). Type-P colonies are formed only in cell cultures initiated by CFCs from very young embryos, before the ventricular zone begins to decline. With increasing age of the embryo the frequency of occurrence of type-P colonies decreases, and they form a mere 2.2 ± 0.8% of the total number of colonies per culture plate at 11.5 days of incubation (Fig. 2). In contrast, the frequency of occurrence of type-A colonies increases from 29.9 ± 6.8 to 88.1 ± 2.3% at 7.5 days of incubation and then gradually declines to 70.1 ± 8.9% of the total colonies per plate at 11.5 days of incubation (Fig. 2).

Type-A colonies (Fig. 4) are also composed of closely apposed epithelial cells, with smaller cells toward the center. Nuclei tend to be oriented toward the center of the colony, rather than being centrally

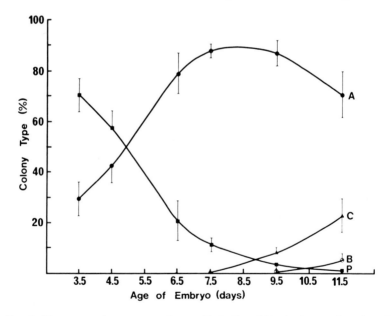

FIG. 2. Frequency of occurrence of types P, A, C, and B colonies as a function of the developmental age of the chick embryo in 7-day cultures (before colonies begin to transform) initiated with dissociated cells from neural tubes or spinal cord primordia. The experiment was repeated twice. Vertical lines indicate 1 SD from the mean. Letters indicate colony types. The detailed procedure is described in Section II. ■, Colony type P; ●, colony type A; ▲, colony type C; ○, colony type B.

placed in the cells. The peripheral cells contain thinly spread cytoplasm, outlining the colony boundaries (Fedoroff, 1977, 1978, 1980).

Type-C colonies (Fig. 5) were found only in cultures initiated from 9.5- and 11.5-day embryos, comprising 8.4 ± 2.3 and $22.7 \pm 6.7\%$ of the total colonies per plate, respectively (Fig. 2). In appearance, type-C colonies are compact and have many short, slender, interdigitating processes connecting the cells. The nuclei are centrally placed (Fig. 5).

Type-B colonies were observed only in cultures containing dissociated spinal cord primordia of 11.5-day embryos and occurred with a frequency of $5.1 \pm 2.4\%$ of the total number of colonies (Fig. 2). Type-B colonies are composed of pleomorphic, fibroblast-like cells separated from each other within the colony (Fig. 6).

After 7 days in culture the colonies begin to change and transform; consequently, we have been able to use shifts in the frequency of occurrence of colony types on further culturing to trace cell progression through the lineage *in vitro* (Fedoroff, 1978, 1980).

Cultures were initiated from 4.5-day chick embryos. After 7, 10, 14, and 21 days of culturing the cultures were fixed and analyzed for the frequency of occurrence of the four colony types. At 7 days in culture, $58.0 \pm 8.9\%$ of the colonies were type P and $41.9 \pm 8.9\%$ were type A (Fig. 9). Type-P colonies decreased continually throughout the 21 days of culturing, whereas the frequency of occurrence of type-A colonies per plate ($71.4 \pm 7.3\%$) increased at 10 days and then declined to reach $34.7 \pm 2.4\%$ at 21 days of culturing (Fig. 9). Type-C colonies were first observed in the cultures at 10 days of incubation, forming $9.2 \pm 3.1\%$ of the total number of colonies per plate. The number of type-C colonies increased gradually throughout the remaining incubation period, forming the majority of colonies ($53.4 \pm 4.6\%$) at 21 days (Fig. 9). Type-B colonies were not observed until 14 days of culturing, and at 14 and 21 days they formed $12.2 \pm 3.6\%$ and $11.6 \pm 4.6\%$ of the total number of colonies, respectively (Fig. 9).

These observations *in vitro* indicated that throughout the culturing period the morphology of colony types changes continuously. The sequence of change is type P to type A, and then to type C. Type B is last to appear in cultures (Fig. 9). It is significant that cultures initiated from embryos of different ages had this same sequence of predominant colony types in 7-day cultures (Fig. 2). These results are identical to those obtained with mouse embryos, in which cultures were initiated by dissociated neopallium of embryos of different ages. Here, also, the sequence observed was type A, type C, and finally type B (Fedoroff, 1978; Juurlink *et al.*, 1980). Type-P colonies were not observed in cultures from mouse embryos because, in the series

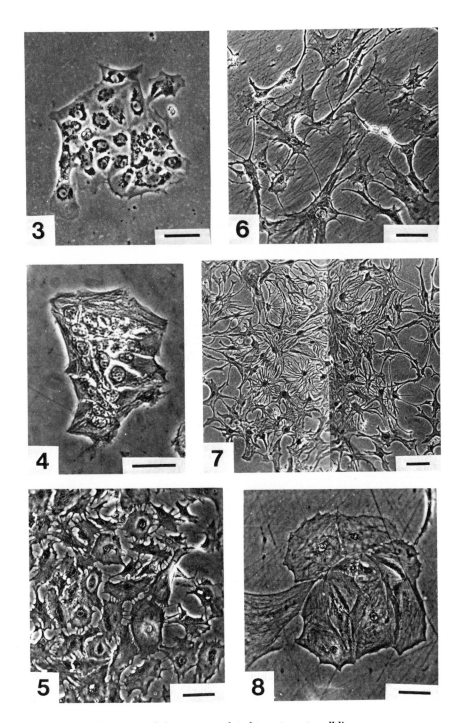

FIGS. 3-8. Colony types related to astrocyte cell lineage.

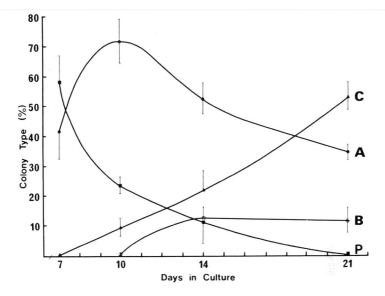

FIG. 9. Frequency of occurrence of types P, A, C, and B colonies as a function of time in culture. Cultures initiated with dissociated cells from 4.5-day (H and 6 H stage 25) neural tubes. The experiment was repeated three times. Vertical lines indicate 1 SD from the mean. Letters indicate colony types. The detailed procedure is discussed in Section II. ■, Colony type P; ●, colony type A; ▲, colony type C; ○, colony type B.

FIG. 3. Type-P colony. Seven-day culture of dissociated cells of chick embryo neural tubes. Epithelial cells with centrally placed nuclei. Phase-contrast microscopy. Bar = 100 μm.

FIG. 4. Type-A colony. Seven-day culture of dissociated cells of mouse neopallium. Epithelial cells with nuclei oriented toward the center of the colony. The peripheral cells contain thinly spread cytoplasm. Phase-contrast microscopy. Bar = 60 μm.

FIG. 5. Type-C colony. Seven-day culture of dissociated cells of mouse neopallium. Colony is compact, consisting of cells with many short, slender cytoplasmic processes and centrally placed nuclei. Phase-contrast microscopy. Bar = 100 μm.

FIG. 6. Type-B colony. Seven-day culture of dissociated cells of mouse neopallium. Colony consists of loosely arranged pleomorphic cells with irregular processes and centrally placed nuclei. Phase-contrast microscopy. Bar = 100 μm.

FIG. 7. Astrocyte colony. Four-week culture of dissociated cells of mouse neopallium. Culture was grown for the first 2 weeks in growth medium and for the last 2 weeks in growth medium to which dBcAMP was added. The colony originally began as a type-A colony (Section VI) and now is composed of star-shaped astrocytes rich in GFA protein (see Figs. 10 and 11). Phase-contrast microscopy. Bar = 100 μm. (Juurlink et al., 1980.)

FIG. 8. Senescent cells. Three-week culture of dissociated cells of mouse neopallium. Larger, nondividing cells with stress fibers are seen. Phase-contrast microscopy. Bar = 100 μm.

studied, very young embryos were not included. Transformation of colonies in cultures initiated by dissociated cells of embryonic mouse neopallium followed the same sequence as was observed with chick embryos (Fedoroff, 1978; Fedoroff and Hall, 1979). We also were able to observe with time-lapse cinemicrophotography that type-A colonies did indeed transform into type C and that type C transformed into type B as well as forming large, nondividing cells which we have referred to as senescent cells (Fedoroff et al., 1979a,b; Fedoroff, 1980) (Fig. 8).

The observation of colonies initiated from chick embryo spinal cords and mouse embryo neopallium in samples from embryos at different ages, in primary cultures during 21 days of culturing, and in time-lapse cinemicrophotographic records of the cultures, has led us to conclude that CFCs progress through a definite sequence of lineage *in vitro* and that the sequence is the same *in vivo*.

V. Origin of Type-A Colonies

Type-A colonies in cultures are composed of epithelial cells (Fig. 4) and occur in high frequency in cultures from younger embryos, as mentioned in Section VI. Type-A colonies occur in high frequency in cultures before colonies of type C or B do. We did not think it likely that type-A colonies originate from fibroblasts, endothelial cells, blood cells, or smooth muscle cells, all of which have, when cultured, morphological appearances and growth patterns quite different from those of cells of type-A colonies. The most likely source of the cells of type-A colonies seemed to be the neuroectoderm. The only other derivatives of neuroectoderm that have an epithelial-like appearance and which might have been present in our cultures are cells of the pia mater. However, when cells of known pia mater origin were cultured, type-A colonies did not form. We concluded, therefore, that cells comprising the type-A colonies were not descendants of pia mater but originated from neuroectoderm.

Ultrastructural analysis of the cells of type-A colonies has indicated that they are indeed very immature. They have large, oval or irregular nuclei with evenly distributed chromatin and a few patches of condensed chromatin. The moderate amount of cytoplasm, rich in free ribosomes arranged in rosettes, contains only a few mitochondria, a few profiles of short cisternae of rough endoplasmic reticulum, occasionally a few profiles of the Golgi apparatus, inclusion bodies, and very few microtubules (Juurlink et al., 1980).

On examining the fragments from mouse neopallium with electron microscopy before they were dissociated, we have found, as many others have observed previously, that the subventricular zone of newborn (P0) mice is composed of two major cellular subpopulations. One consists of "dark" cells with electron-dense nuclei and cytoplasm, and the other of "pale" cells with large electron-lucid nuclei and small amounts of moderately electron-dense cytoplasm (Juurlink et al., 1980). These cells correspond to the previously described dark and pale cells of the subventricular zone (Smart, 1961; Fisher, 1967; Lewis, 1968; Blakemore, 1969; Blakemore and Jolly, 1972; Skoff et al., 1976; Privat, 1978; Imamoto et al., 1978). The fine structure of the cells of type-A colonies closely resembles that of the pale cells of the subventricular zone of P0 mice (Juurlink et al., 1980).

To substantiate the origin of cells of type-A colonies from the subventricular zone, this zone was carefully dissected from the rest of the neopallium, dissociated, and plated into cultures. The subventricular zone of P0 mice yielded 73% of all CFUs and 70% of all type-A CFCs in the neopallium (Juurlink et al., 1980). These observations, together with the facts that the fine structure of the pale cells of the subventricular zone is very similar to that of the cells forming type-A colonies, that CFCs begin to form when the ventricular zone begins to decline, and that the frequency of occurrence of type-A colonies is high in cultures initiated from very young embryos, strongly suggest that cells that form type-A colonies come mainly from the subventricular zone of the neopallium (Fedoroff et al., 1979a,b; Juurlink et al., 1980).

VI. Astrocyte Cell Lineage in Tissue Cultures

As discussed in Section IV, type-A colonies form type-C colonies in culture and, in time, these form type-B colonies. If type-A colonies are marked and grown for 2 weeks with subsequent addition of dibutyryl cAMP (dBcAMP) to the medium for another 2 weeks of culturing, the colonies originally marked as type-A colonies will now be composed of star-shaped astrocytes (Juurlink et al., 1980). Such cells stain with Cajal's $AuCl_3$ sublimate (Fig. 10), specific for astrocytes (Ramón y Cajal, 1913; Vaughn and Pease, 1967; Mori and Leblond, 1969), and contain an abundance of glial fibrillary acidic (GFA) protein (Figs. 11–13). GFA protein is considered a specific marker for astrocytes (Bignami and Dahl, 1977; Schachner et al., 1977). Many other workers have shown that, if monolayer cultures initiated with dissociated cells from the brain of newborn mice or rats are grown for 2 weeks in cultures and then treated with dBcAMP, star-shaped astrocytes form which morphologically and biochemically very closely resemble astrocytes in

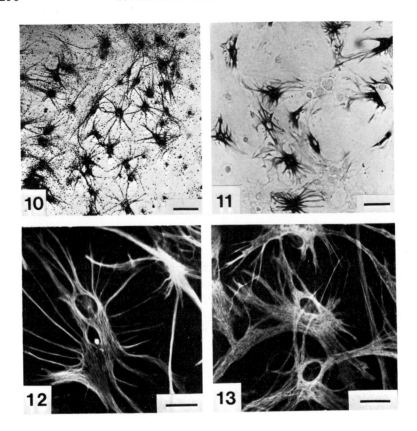

FIGS. 10–13. Astrocyte cultures.

FIG. 10. A culture of rat astrocytes stained with Cajal's gold chloride sublimate, a specific stain for astrocytes. The astrocyte processes are clearly visible. Light microscopy. Bar = 100 μm.

FIG. 11. A culture of mouse astrocytes stained with immunoperoxidase for GFA protein. After fixing in 95% ethanol, cultures were first treated with a 1:20 dilution of rabbit anti-GFAP serum (courtesy of Dr. E. Bock), stained with a 1:10 dilution of peroxidase-labeled swine anti-rabbit IgG (DAKO-Immunoglobulins, Copenhagen, Denmark), then treated with a solution of 3′,3′-diaminobenzidine and hydrogen peroxide. Finally, they were postfixed in 10% neutral buffered formalin and viewed with light microscopy. Note that the processes and the perikarya are stained, but not the nuclei. The unstained cells are probably type-B cells, which are not sensitive to dBcAMP and do not form astrocytes. Light microscopy. Bar = 100 μm.

FIG. 12. Immunofluorescent staining of mouse astrocytes in culture. Cells grown on glass coverslips were fixed at −20°C first in methanol and then in acetone, air-dried, washed in PBS, labeled with 1:50 dilution of rabbit anti-GFAP serum (courtesy of Dr. E. Bock), washed in PBS, then treated with a 1:4 dilution of goat anti-rabbit IgG serum

vivo (Hertz *et al.*, 1978a,b; Lim *et al.*, 1973; Moonen *et al.*, 1975; Moonen *et al.*, 1976; Schousboe *et al.*, 1975).

In order to determine cell type and the stage of the cell lineage at which the cells become sensitive to dBcAMP, colony cultures initiated from P0 mouse neopallium were cultured for 7–35 days. During this period, dBcAMP was added at various times for periods of 1 week. At the end of the culturing period the frequency of occurrence of type-A, -B, and -C colonies, as well as colonies with astrocytes, was determined (Table II). The data indicate that during the first week in culture dBcAMP inhibits the formation of type-C colonies from type-A colonies. However, during the second and third week in culture it is the cells of type-C colonies that are sensitive to dBcAMP, and under its influence they form star-shaped astrocytes (Fedoroff, 1980). This conclusion was based on the observation that whether or not dBcAMP is present the frequency of type-A colonies declines on culturing and the frequency of occurrence of type-B colonies in cultures remains about the same. In contrast, the frequency of occurrence of type-C colonies in cultures without dBcAMP increases as that of type-A colonies decreases, and in cultures with dBcAMP the frequency of occurrence of type-C colonies increases up to the second week and then decreases. At the same time the number of colonies with star-shaped astrocytes increases (Table II). Thus the only change in colony frequency that can be correlated with the increase in astrocytes is that of type-C colonies. From this it has been concluded that cells of type-C colonies acquire sensitivity to dBcAMP and transform (differentiate) into astrocytes (Fedoroff, 1980). Cells of type-A and type-B colonies do not possess such sensitivity to dBcAMP. It is possible that the cells of type-C colonies have to mature for a period of time before their sensitivity to dBcAMP becomes maximal. Colonies with the most

---

conjugated with fluorescin (Hyland), washed in PBS, and finally mounted in 50% glycerol in PBS at pH 7.9. Cells were observed with a Zeiss fluorescent microscope. A fine intracytoplasmic network of filaments containing GFAP can be seen. Bar = 25 μm. (V. I. Kalnins, S. Fedoroff, and R. White, unpublished.)

Fig. 13. Immunofluorescent staining of mouse astrocytes in culture. The cells grown on glass coverslips were fixed at −20°C first in methanol and then acetone, air-dried, washed in PBS, labeled with a 1:25 dilution of rabbit antiglial filament protein (55,000 MW) serum, washed in PBS, treated with a 1:4 dilution of goat anti-rabbit IgG serum conjugated with fluorescin (Hyland), washed in PBS, and finally mounted in 50% glycerol in PBS at pH 7.9. Cells are observed with a Zeiss fluorescent microscope. A fine intracytoplasmic network of glial filaments can be seen. Cultures treated with preimmune serum from the same animal were negative, i.e., did not give the staining pattern seen when immune serum was used. Bar = 25 μm. (V. I. Kalnins, S. Fedoroff, and R. White, unpublished.)

TABLE II

EFFECT OF TIME OF ADDITION OF dBcAMP ON FREQUENCY OF OCCURRENCE
OF A-, C-, AND B-TYPE COLONIES AND COLONIES WITH ASTROCYTES[a]

Period of culturing (days)	Period of culturing with dBcAMP (days)[b]	Frequency of colonies (%)[c]			
		Type A	Type C	Type B	With astrocytes
7	1–7	90.0 ± 5.3	9.3 ± 6.0	0.7 ± 1.2	0
	0	78.0 ± 2.0	18.0 ± 2.0	4.0 ± 2.0	0
14	7–14	42.7 ± 10.1	46.7 ± 1.2	2.0 ± 3.5	8.7 ± 7.0
	0	12.0 ± 2.0	82.7 ± 5.8	5.3 ± 4.2	0
21	14–21	4.7 ± 3.1	41.3 ± 7.6	4.0 ± 4.0	49.3 ± 9.5
	0	2.0 ± 2.0	90.0 ± 4.0	6.8 ± 6.4	0
28	21–28	0	23.3 ± 5.0	6.7 ± 3.1	68.7 ± 2.3
	0	0.7 ± 1.2	82.7 ± 3.1	14.7 ± 3.1	0
35	28–35	0	17.9 ± 9.3	15.2 ± 8.0	64.9 ± 13.0
	0	0	80.0 ± 3.5	14.0 ± 5.3	2.0 ± 2.0

[a] From Fedoroff (1980).
[b] 0.25 mmoles of dBcAMP was added to the medium; medium was changed every 2 days.
[c] Mean percentage ± 1 SD from the mean.

astrocytes were found in cultures in which dBcAMP was present during the fourth week of culturing. During the third week, even though sufficient numbers of type-C colonies were present, the number of colonies forming astrocytes was significantly less ($P < 0.05$) (Table II).

Cells of type-C colonies differ from cells of type-A colonies by having processes that form complex junctions at the sites of attachment to other cells of type-C colonies (Scott and Oteruelo, 1977). Many more mitochondria are found in the cytoplasm, as are large numbers of filaments which may be arranged in bundles either parallel and close to the cell membrane or in random directions in the cytoplasm. The nuclei of these cells are electron-lucid, as in cells of type-A colonies (Fedoroff, 1980). Because the cells of type-C colonies acquire sensitivity to dBcAMP and have characteristic filaments in their cytoplasm, they can be considered astrocyte-committed precursor cells, probably corresponding to the previously described astroblasts *in vivo* (Imamoto *et al.*, 1978; Skoff *et al.*, 1976; Vaughn, 1969). Cells of type-A colonies that form cells of type-C colonies probably can be considered progenitor cells, corresponding to the previously described glioblasts in the astrocyte cell lineage (Imamoto *et al.*, 1978; Sturrock, 1976; Vaughn, 1969; Fedoroff, 1980). Whether or not cells of type A colonies are pluripotential, i.e., whether under certain environmental conditions they can give rise to cells in lineages other than the astrocyte lineage, remains to be seen.

VII. Astrocyte Cell Lineage *in Vivo*

It has been reported previously (Fedoroff, 1978) and in this chapter (Section IV) that frequencies of occurrence of types of colonies formed vary according to the age of the cell donor animal. The significant point, however, is that the sequence of variation due to the age of the animal was the same as the sequence observed on culturing colonies for 3 or 4 weeks; i.e., type-A colonies gave rise to type-C and type-C to type-B colonies or, in the presence of dBcAMP in cultures, also to star-shaped astrocytes.

Juurlink *et al.* (1980) performed experiments in which parts of the neopallium were isolated from newborn (P0) and 1-week-old (P7) mice. The parts of the neopallium were carefully divided into fragments containing subventricular zone fragments (Sv fragments) and fragments containing the rest of the neopallium (C fragments), thus providing four kinds of tissue fragments. These were dissociated, and the cells were plated in cultures. It was assumed that P0 Sv fragments were composed of the least differentiated proliferative cell population, that P7 Sv fragments contained a proliferative cell population that had shifted a degree further along the cell lineage and that the P7 C fragments consisted of cells that had migrated from the subventricular zone. It was, therefore, expected that P7 C fragments would consist of the most differentiated cell populations as compared to other kinds of fragments. On assaying the cell populations of the four kinds of fragments by the colony culture method it was found that the frequency distributions of colonies of type A, C, and B in P0 Sv fragments and P0 C fragments were very similar; i.e., type-A colonies occurred in the highest frequency, then type C, and in the lowest frequency, type B (Fig. 14). This indicated that the two kinds of fragments had the same proliferative cell populations (Juurlink *et al.*, 1980). On the other hand, the P7 Sv fragments had a low frequency of occurrence of type-A colonies, a high frequency of type-C colonies, and approximately the same frequency of type-B colonies as was found in P0 Sv fragments and P0 C fragments (Fig. 14). The P7 C fragments had a much higher frequency of type-B colonies than any of the other three kinds of fragments and a low frequency of type-C colonies. As expected, if the sequence of type A, type C, and type B is an indication of the direction in the cell lineage, then the least differentiated

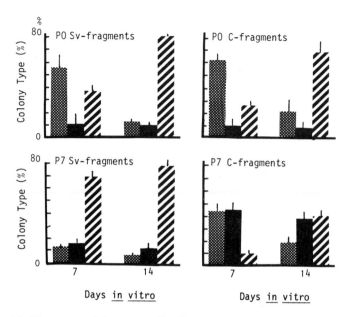

F𝚒ɢ. 14. Histogram of frequency distribution of types A, B, and C colonies in cultures from Sv fragments and C-fragments of P0 and P7 mouse neopallium. The fragments were dissociated and cultured for 7 or 14 days, at which time the frequency distributions of three types of colonies were determined. Bars represent 1 SD of the mean. Stippling, colony type A; solid, colony type B; stripe, colony type C. (From Juurlink *et al.*, 1980.)

cell population was in the P0 Sv fragments and P0 C fragments and then in the P7 Sv fragments; the most differentiated cell population in comparison to the other types of fragments was that of the P7 C fragments.

Juurlink *et al.* (1980), using the colony culture assay method, also estimated the number of CFUs present in two parts of the brain (the subventricular zone and the rest of the neopallium) at two stages of development (P0 and P7). There was a dramatic reduction in the number of CFUs present in P7 mice in both regions, i.e., the subventricular zone as well as the rest of the neopallium, when compared to P0 mice. There is a relationship between the age of the embryo (chick or mouse) and the shift in frequency of occurrence of colonies from type A to type C and eventually to type B and a relationship between shifts in frequency of colony types and predicted stage of differentiation in two regions of the brain during two stages of differentiation; furthermore, the shift in the frequency of colony types is the same as that observed on culturing colonies for several weeks. These facts strongly support the notion that astrocyte precursor cells progress through the lineage *in vivo* and *in vitro* according to the same sequence.

VIII. Conclusions

The colony culture method can assay only proliferative cells that can attach to the plastic surface of culture dishes and form colonies. Ventricular cells, although they are proliferative, because of their apparent interdependence, can grow only as a monolayer out of the fragment on top of the collagen and not as single dissociated cells on the plastic surfaces. Cells are not able to form colonies in cultures, it seems, until a certain degree of cell independence is reached, and in chick embryo neural tube this is at 2.5 days of incubation (H and H stage 17).

There is an exponential increase in the number of CFCs between 6.5 (H and H stage 30) and 7.5 (H and H stage 32) days of incubation. The precise time of origin of CFCs in mice is at present still not known. The absolute plating efficiency of CFCs remains at approximately the same level until birth. It begins to decline postnatally, and by 30 days the CFCs can hardly be detected (Fedoroff, 1977). CFCs are mainly located in the subventricular zone, although some are found in other parts of the neopallium. During postnatal development, the CFCs in all regions of the neopallium decrease, but the major decrease is in the subventricular zone.

Cells forming type-A colonies in culture are considered progenitor cells, and they probably correspond to glioblasts in the astrocyte cell lineage. They are found mainly in the subventricular zone and resemble the pale cells of this zone. Type-A colonies are composed of epithelial-like cells (Fig. 4), and the first steps in their differentiation along the astrocyte cell lineage are manifested by the cells' attempts to separate from each other. Any factors that prevent cell separation will also inhibit differentiation along the astrocyte cell lineage; conversely, factors that favor separation of the cells will speed up differentiation. We have observed that dissociation of type-A colonies and plating the cells on polylysine-coated culture dishes accelerate cell differentiation (S. Fedoroff and C. Hall, unpublished) but that increasing the concentration of horse serum in the culture medium inhibits differentiation (Fedoroff and Hall, 1979).

Type-A colonies give rise to cells that form type-C colonies (Figs. 4

and 5). Study of the fine structure of cells of type-C colonies has revealed that they contain a large number of filaments and some dense bodies. We consider cells of type-C colonies astrocyte-committed precursor cells which correspond to astroblasts previously described. These cells, in the presence of dBcAMP in culture, form star-shaped astrocytes (Fig. 7). They also can form cells of type-B colonies (Fig. 6) and large uni- or multinucleated senescent cells (Fig. 8). The highest frequency of occurrence of type-B colonies is found in cultures initiated from the neopallium of postnatal animals (Fig. 14). At present, we consider the cells of type-B colonies dedifferentiated cells. They are not sensitive to dBcAMP and do not form astrocytes; therefore, their capabilities and relation to the astrocyte cell lineage is not clear at present.

It is of interest to note that, regardless of the age of the donor animal and the initial frequency distribution of the colony types, on culturing for 2 or 3 weeks all cultures end up with approximately the same frequency of occurrence of colony types, i.e., a low frequency of type-A colonies, a high frequency of type-C colonies and, in cultures originating from postnatal animals, an increase in type-B colonies (Fig. 14). It seems that present culture conditions block the progression of astrocyte precursors through the lineage of differentiation at the type-C (astroblast) colony level (Fig. 14). In the presence of dBcAMP the cells overcome this block and complete their differentiation by forming star-shaped astrocytes.

The star-shaped astrocytes found in cultures morphologically and biochemically resemble closely the astrocytes found in vivo and represent, in our concept, end cells of the astrocyte cell lineage. It is significant that the sequence of differentiation of astrocyte progenitor cells (cells of type-A colonies) in cultures seems to be the same as in vivo and relates to the increasing postnatal age of the mice. Because of the parallelism of cell differentiation in vivo and in vitro, it is hoped that such studies will eventually lead to a more complete understanding of the astrocyte cell lineage and its regulatory mechanisms and that the colony culture assay method will be applied to the study of other neural cell lineages.

ACKNOWLEDGMENTS

We acknowledge with thanks the valuable help of Mr. C. Hall and Mr. R. White, also that of Mr. O. Kademoglu for the preparation of the photographic prints and Ms. I. Karaloff for final preparation of the manuscript. This investigation was supported by grant MT 4235 from the Medical Research Council of Canada.

REFERENCES

Bignami, A., and Dahl, D. (1977). *J. Histochem. Cytochem.* **25**, 466–469.
Blakemore, W. F. (1969). *J. Anat.* **104**, 423–433.
Blakemore, W. F., and Jolly, R. D. (1972). *J. Neurocytol.* **1**, 69–84.
Booher, J., and Sensenbrenner, M. (1972). *Neurobiology* **2**, 97–105.
Boulder Committee. (1970). *Anat. Rec.* **166**, 257–262.
Brunette, D. M., Melcher, A. H., and Moe, H. K. (1976). *Archiv. Oral Biol.* **21**, 393–400.
Doering, L. C. (1980). M.Sc. Thesis, University of Saskatchewan, Saskatoon, Canada.
Fedoroff, S. (1966). *Exp. Cell Res.* **46**, 648–672.
Fedoroff, S. (1977). *In* "Cell, Tissue, and Organ Cultures in Neurobiology" (S. Fedoroff and L. Hertz, eds.), pp. 215–221. Academic Press, New York.
Fedoroff, S. (1978). *In* "Dynamic Properties of Glial Cells" (E. Schoffeniels, G. Franck, D. B. Tower, and L. Hertz, eds.), pp. 83–92. Pergamon, New York.
Fedoroff, S. (1980). *In* "Tissue Culture in Neurobiology" (A. Vernadakis and E. Giacobini, eds.), Raven, New York (in press).
Fedoroff, S., and Hall, C. (1979). *In Vitro* **15**, 641–648.
Fedoroff, S., Doering, L., Juurlink, B. H. J., and Hall, C. (1979a). *Soc. Neurosci. Abstr.* **5**, 159.
Fedoroff, S., Hall, C., Oteruelo, F. T., and Juurlink, B. H. J. (1979b). *In Vitro* **15**, 212.
Fisher, K. (1967). *Acta Neuropathol.* **8**, 242–253.
Fisher, K. R. S., and Fedoroff, S. (1977). *In Vitro* **13**, 569–579.
Fisher, K. R. S., and Fedoroff, S. (1978). *In Vitro* **14**, 878–886.
Hertz, L., Bock, E., and Schousboe, A. (1978a). *Dev. Neurosci.* **1**, 226–238.
Hertz, L., Schousboe, A., Boechler, N., Mukerji, S., and Fedoroff, S. (1978b). *Neurochem. Res.* **3**, 1–14.
Imamoto, K., Paterson, J. A., and Leblond, C. P. (1978). *J. Comp. Neurol.* **180**, 115–138.
Juurlink, B. H. J., Fedoroff, S., Hall, C., and Nathaniel, E. J. H. (1980). *J. Comp. Neurol.* (in press).
Kaltenbach, J. P., Kaltenbach, M. H., and Lyons, W. B. (1958). *Exp. Cell Res.* **15**, 112–117.
Lim, R., Mitsunobu, K., and Li, W. K. P. (1973). *Exp. Cell Res.* **79**, 243–246.
Lewis, P. D. (1968). *Brain* **91**, 721–738.
Metcalf, D., and Moore, M. A. S. (1971). "Haemopoietic Cells." North-Holland Publ., Amsterdam.
Moonen, G., Cam, Y., Sensenbrenner, M., and Mandel, P. (1975). *Cell Tissue Res.* **163**, 365–372.
Moonen, G., Heinen, E., and Goessens, G. (1976). *Cell Tissue Res.* **167**, 221–227.
Mori, S., and Leblond, C. P. (1969). *J. Comp. Neurol.* **137**, 197–226.
Paterson, J. A., Privat, A., Ling, E. A., and Leblond, C. P. (1973). *J. Comp. Neurol.* **149**, 83–102.
Paul, J. (1975). "Cell and Tissue Culture" (5th ed.), p. 228. Churchill, London.
Privat, A. (1978). *In* "Dynamic Properties of Glial Cells" (E. Schoffeniels, G. Franck, L. Hertz, and D. B. Tower, (eds.), pp. 55–64. Pergamon, New York.
Ramón y Cajal, S. (1913). *Trab Lab. Invest. Biol. Univ. Madrid* **11**, 255–315.
Schousboe, A., Fosmark, H., and Hertz, L. (1975). *J. Comp. Neurol.* **180**, 115–138.
Schachner, M., Hedley-White, E. T., Hsu, D. W., Schoonmaker, G., and Bignami, A. (1977). *J. Cell Biol.* **75**, 67–73.
Scott, B., and Oteruelo, F. T. (1977). *In* "Cell, Tissue and Organ Cultures in Neurobiology" (S. Fedoroff and L. Hertz, eds.). Academic Press, New York.

Skoff, R. P., Price, D. L., and Stocks, A. (1976). *J. Comp. Neurol.* **169**, 291–311.
Smart, I. (1961). *J. Comp. Neurol.* **16**, 325–347.
Sturrock, R. R., (1976). *J. Anat.* **122**, 521–537.
Vaughn, J. E. (1969). *Z. Zellforsh Mikrosk. Anat. Abt. Histochem.* **94**, 293–324.
Vaughn, J. E., and Pease, D. C. (1967). *J. Comp. Neurol.* **131**, 143–154.

CHAPTER 9

GLIA MATURATION FACTOR

Ramon Lim

BRAIN RESEARCH INSTITUTE
THE UNIVERSITY OF CHICAGO
CHICAGO, ILLINOIS

I. Introduction

One of the central issues in neurobiology lies in the problem of how the nervous system, with its complexity of cellular organization, is developed from simple embryonic cells. Although descriptive neurogenesis is well established, our knowledge on dynamic neurogenesis is sketchy. It is generally agreed that the problem can best be attacked by reducing the developing brain into simple systems in which variables can easily be controlled. Thus, tissue culture appears to be particularly suitable for this approach and promises to provide us with chemical clues on cellular interactions in the developing brain.

The use of cloned cell lines of neuronal or glial origin has gained popularity over the last decade. This approach carries the obvious advantage that a homogeneous population of cells can be obtained in relatively large quantities for chemical studies. However, some important drawbacks exist. First, such cell lines are usually of tumor origin, and it is not easy to distinguish the neoplastic from the normal properties of the cells. Second, with the number of passages, the prob-

CURRENT TOPICS IN
DEVELOPMENTAL BIOLOGY, Vol. 16

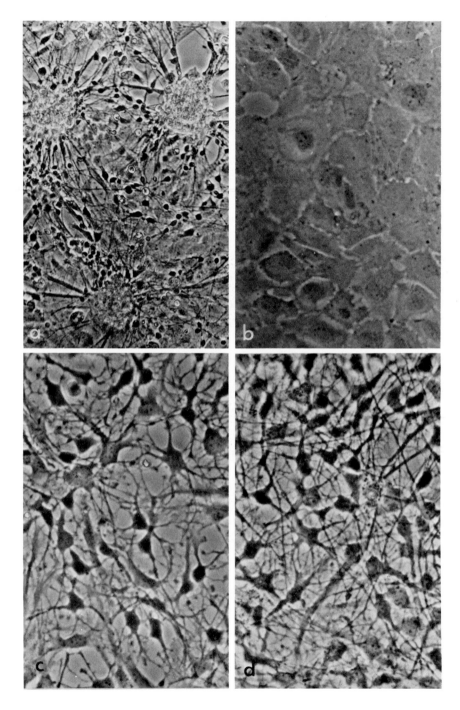

ability of mutation increases, which leads to the heterogeneity of the cell population and results in the necessity of recloning. The problem of spontaneous change makes the cells susceptible to selective influences inadvertently introduced by the investigators, creating diversities in cells carrying the same name. Third, cell lines are susceptible to chronic infections such as PPLO which may not be readily detectable and may alter the properties of the cells.

An alternative to the cell line approach is the use of primary cell culture. Since in this procedure one deals with normal cells freshly obtained from the brain, the problems encountered with tumor cell lines are absent. The obvious disadvantage here is the heterogeneity of the cell population.

Over the last 6 years, our laboratory has developed a cell culture system modified from the primary culture. With this method a primary culture from dissociated fetal rat brain cells is carried to the second passage. By means of differential attachment during the transfer, which eliminates the neuroblasts, a homogeneous population of glioblasts, predominantly astroblasts, is obtained (Fig. 1). With this procedure the problem of a mixed cell population is greatly minimized, while the cells isolated retain many of their original characteristics. Furthermore, this short-term culture system does not permit contaminants such as fibroblasts to take over the culture.

The use of this glial system led us to the identification of a protein factor, endogenous to the adult brain, that is capable of promoting reversibly the morphological and chemical differentiation of glioblasts. The factor is operationally designated the glia maturation factor (GMF). The use of this term does not preclude the possibility of other functions the factor may have, nor does it imply that it is the only factor needed for the differentiation of glial cells. It could well be one of several variables influencing gliogenesis. The purpose of this chapter is to update our knowledge of this factor, with emphasis on its cellular effects and chemical nature. Speculations on the biological implications are extrapolated from currently available data.

FIG. 1. Live phase-contrast photographs demonstrating morphological changes in rat glioblasts. (a) Primary culture of dissociated fetal rat brain cells showing a mixture of glioblasts (epithelial cells in the background) and neuroblast-like cells (those growing above the glioblasts). × 150. (b) Secondary culture of fetal brain cells after elimination of the neuroblasts showing a homogeneous population of glioblasts forming a cell carpet. × 300. (c) Secondary culture of glioblasts 1 day after exposure to GMF. × 300. (d) Secondary culture of glioblasts 1 week after exposure to GMF, the factor as well as the medium being renewed every other day. × 300. (From Lim et al., 1977a.)

II. How Glia Maturation Factor Was Detected

When dissociated rat embryonic brain cells are selectively cultured under conditions that eliminate neuroblasts, the major cell type that survives is epithelial in appearance (Fig. 1). These cells form a monolayer over the flask surface and assume a flat appearance as observed by both phase and scanning electron microscopy (Fig 2). If given a chance to reaggregate and cultured in a three-dimensional matrix, mature glial cells will eventually emerge. However, if left in the monolayer with renewal of the culture medium at regular intervals, with or without subculture, the cells will survive but remain epithelial in appearance for an indefinite time. In January 1972, as part of our search for growth or differentiation factors in the brain, we added a dialyzed brain extract (a high-speed supernatant fraction) to the monolayer and were excited to observe that, after an initial lag period of about 12 hours, the histotypic pattern changed from that of a cell carpet to that of an interconnecting cell net resembling mature astrocytes, best observed at 24 hours or later (Figs. 1 and 2) (Lim *et al.*, 1972, 1973; Lim and Mitsunobu, 1974). Because of this striking effect on cell organization, we first designated this factor the "morphological transforming factor." The term was subsequently changed to "glia maturation factor" as we learned more about the cells and the chemical changes.

That the morphological effect of GMF is not an artifact of tissue culture is evidenced by the following observations: (1) The effect is not related to variability in the pH of culture media (Lim and Mitsunobu, 1975). (2) The effect is demonstrable at various serum concentrations (Lim *et al.*, 1977a). (3) The effect is inhibited by cycloheximide at 0.1 μg/ml medium, indicating a dependence on protein synthesis (Lim and Mitsunobu, 1974). (4) The effect is abolished by colchicine (0.1 μg/ml medium) and vinblastine (0.5 μg/ml medium), suggesting a dependence on the integrity of the microtubules (Lim and Mitsunobu, 1974). (5) The effect is demonstrable at 37°C but not at 20°C, indicating the involvement of cellular metabolism in the histotypic rearrangement. (6) Preincubation of culture flasks with GMF does not lead to morphological change, indicating that the effect of GMF is not mediated through alterations in the surface texture of the flask (Lim *et al.*, 1977a).

Withdrawal of GMF from the medium leads to eventual reversion of the differentiated glial cells to the flat appearance within a few days (Lim *et al.*, 1977a). On the other hand, renewing the medium containing fresh factor every other day not only sustains but also augments

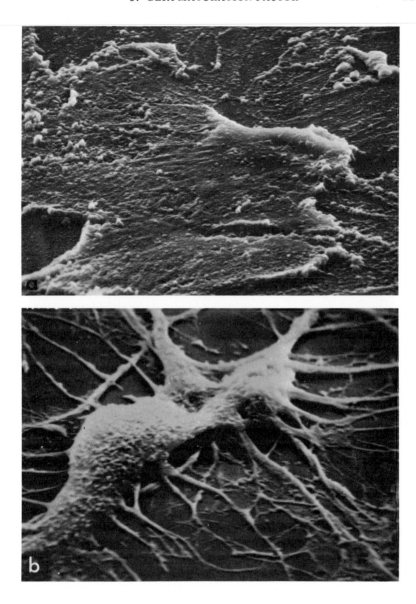

FIG. 2. Scanning electron microscopy of monolayer glioblasts before (a) and 1 day after (b) exposure to GMF. × 2500. (From Lim *et al.*, 1977a.)

the outgrowth of cell processes. A temporal summation of the GMF effect appears to be present in that a second stimulation by the factor elicits a greater morphological response than the first (Lim et al., 1977a).

III. Effect of GMF on Cellular Dynamics and the Ultrastructure of Glioblasts

Studies with cinematography (Lim et al., 1976) have revealed that cells round up without detaching from the flask and extrude numerous processes which are branched, multipolar, and interconnected. There is a sudden surge in cell division, which takes place concomitantly with the outgrowth of processes. The processes are highly motile, exhibiting strong tugging activities and creating an impression of "tug-of-war," as originally described by Lumsden and Pomerat (1951) for cultured glial cells. In contrast to the processes, the cell bodies remain stationary but pulsate at a periodicity of about 5–10 minutes (Fig. 3). Since its initial observation (Lumsden and Pomerat, 1951) cell pulsation has been considered a hallmark of glia. (Myoblasts fom the heart contract, but they show a completely different rhythm.) The general appearance of an undulating cell net in our system differs greatly from the dynamic pattern of fibroblast cultures, where the cells are usually engaged in random walk, without extruding processes or making intercellular connections. The physical activities of the stimulated glioblasts slow down after the third day, while the processes continue to elongate. At the end of a week, the appearance is that of a field of crisscrossing tree branches (Fig. 1).

Electron microscopy (Lim et al., 1977c) shows great contrasts between the flat glioblasts and the morphologically differentiated cells. Whereas the former are dominated by microfilaments common to many cell types, the latter are packed with 100-Å filaments consistent with gliofilaments. Although the number of mitrotubules remains constant, their orientation has changed from a random pattern to a parallel formation along the axes of the processes. The flat cells contain numerous desmosome junctions typical of epithelial cells. These junctions almost completely disappear form the differentiated cells, which now contain puncta adhaerentia types of junctions at points of intercellular contact. Puncta adhaerentia are frequently observed among brain cells in vivo, both between neurons and between glia (Peters et al., 1970). Features characteristic of neurons, such as synaptic vesicles and synaptic junctions, have never been found in differen-

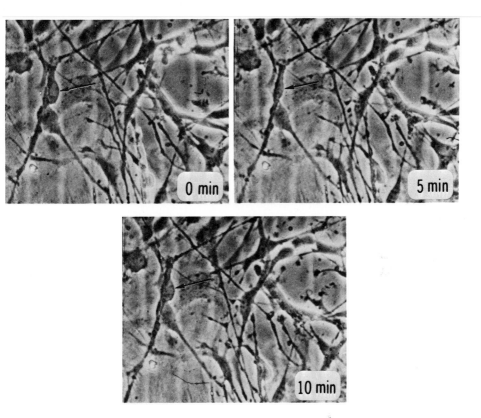

FIG. 3. Time-lapse photography of glioblasts 3 days after exposure to GMF, showing pulsation of a cell body (arrow) with a periodicity of about 10 minutes. The times indicate intervals with respect to the first picture. Live phase-contrast. × 280. (From Lim et al., 1976.)

tiated glioblasts. Electrical stimulation and recording have also failed to detect an action potential.

IV. Effect of GMF on Chemistry of Glioblasts

The first chemical change in response to GMF consists of an increase in intracellular cyclic GMP, which reaches as high as five times the control level and occurs as early as 1 hour after stimulation (Turriff and Lim, 1979). The next detectable chemical event consists of an increase in thymidine incorporation into DNA, which is best observed between hours 12 and the 24 (Table I). The increase in DNA synthesis,

TABLE I

EFFECT OF GMF ON MACROMOLECULAR SYNTHESIS IN GLIOBLASTS[a]

Period of exposure	DNA	RNA	Protein
12 hours			
Control	1049.5 ± 63.7	259.0 ± 7.0	134.6 ± 3.0
Experimental	1632.5 ± 167.8	212.1 ± 19.2	114.0 ± 15.8
	(+56%)	(−18%)	(−15%)
1 day			
Control	809.9 ± 9.0	212.2 ± 30.3	87.1 ± 2.9
Experimental	1718.6 ± 5.6	324.5 ± 6.0	139.9 ± 7.2
	(+112%)	(+53%)	(+61%)
4 days			
Control	508.9 ± 12.5	415.7 ± 37.5	83.6 ± 7.3
Experimental	337.1 ± 31.4	168.9 ± 8.0	38.6 ± 4.6
	(−42%)	(−59%)	(−54%)
7 days			
Control	324.8 ± 76.1	433.3 ± 2.7	60.3 ± 3.2
Experimental	164.4 ± 39.5	92.2 ± 4.7	34.4 ± 0.6
	(−49%)	(−78%)	(−43%)

[a] Control and experimental cells were paired cultures grown under identical conditions except for the absence or presence, respectively, of the factor. The times indicate the periods of exposure of the experimental cells to the factor. Tritiated thymidine, uridine, and proline were used for labeling DNA, RNA, and protein, respectively, for a 1-hour pulse. Values presented are cpm × 10^2/mg cell protein (mean ± SD) from four pools of two culture flasks. Numbers in parentheses are the differences between experimental and control with reference to the control. (Lim *et al.*, 1976.)

which subsides after 3 days, agrees well with the surge in cell division revealed by cinematography. After DNA synthesis, there is a secondary increase in RNA and protein synthesis. The brain-specific protein S–100, which is predominantly a glial protein (although its presence in neurons has been reported), increases 10-fold after 4 days of stimulation (Fig. 4) (Lim *et al.*, 1977b). We have detected a 3-fold increase in cyclic AMP after the fourth day (Fig. 5), which coincides with a decrease in cellular motility and cell division (Lim *et al.*, 1977b). GMF also induces the cytoplasmic enzyme glycerophosphate dehydrogenase (EC 1.1.1.8) by as much as 3-fold, mainly on days 4 and 7 (Fig. 6). That these changes are not secondary to differences in cell density is evident from the fact that both the experimental and control cultures arrive at the same cell density (Fig. 7) when chemical maturation is detected in the former. Apparently, while cell division in the experimental cells is sequentially turned on and off by GMF, that in the control cells goes on at a slow but steady rate. On the other hand, GMF does not induce the predominantly neuronal 14–3–2 protein (recently identified as an isozyme of enolase).

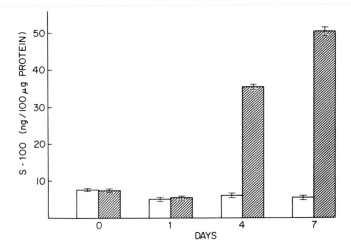

FIG. 4. Effect of GMF on S-100 protein level in glioblasts. Open bars, Control cells; hatched bars, cells exposed to GMF. Values are means ± SD from four pools of two culture flasks each. Days indicated are the lengths of time after initial exposure to the factor. (From Lim *et al.*, 1977d.)

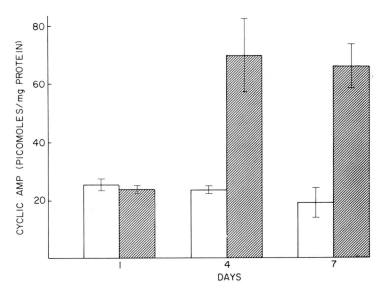

FIG. 5. Effect of GMF on cyclic AMP level in glioblasts. For explanations see legend for Fig. 4. (From Lim *et al.*, 1977d.)

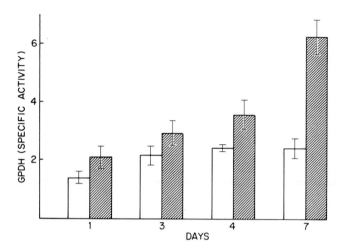

FIG. 6. Effect of GMF on glycerophosphate dehydrogenase (GPDH) in glioblasts. Specific activity is defined as the amount of enzyme catalyzing 1 nmole of substrate per minute per milligram of total cell protein. Values are means of four flasks ± SD. For other explanations see legend for Fig. 4. (From Lim *et al.*, 1977d.)

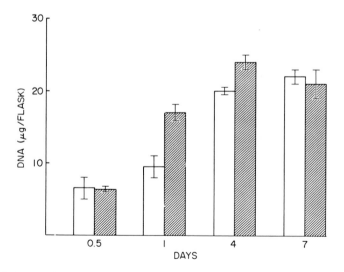

FIG. 7. DNA content of glioblasts grown in the absence and presence of GMF. The surface area of the flask was 75 cm². For other explanations see legend for Fig. 4. (From Lim *et al.*, 1977d.)

The response of glioblasts to GMF can best be characterized as a biphasic change, where the early events (first 3 days) consist of (1) an increase in cyclic GMP, (2) an increase in DNA synthesis and cell division, (3) an increase in RNA and protein synthesis, (4) a rounding up of cell bodies and an outgrowth of processes, and (5) high motility of the processes. The late events (after 3 days) consist of (1) inhibition of DNA synthesis and cell division, (2) a slowing down of process motility, (3) continuing elongation of the processes, (4) an increase in cyclic AMP, (5) an increase in S–100 protein, and (6) an increase in glycerophosphate dehydrogenase. The overall response of the cells fits into the picture of initial cell proliferation followed by chemical maturation.

V. Evidence that the Cells Are Glia

That the cells in our system are not neuroblasts or fibroblasts is supported by the following distinctive features of glial cells: (1) the general appearance of astrocytes such as multipolar, branched processes, (2) the presence of numerous gliofilaments inside the processes, (3) the presence of glycogen granules inside the differentiated cells, (4) the presence of cell junctions normally found in glia, (5) the presence of movements characteristic of glia such as tugging of the processes and pulsation of the cell bodies, and (6) the presence of glial chemical markers such as S–100 and glial fibrillary acidic (GFA) protein. That the cells indeed undergo differentiation is evidenced by the appearance of or increase in most of these glial features after exposure to the factor. It is interesting to point out that, even before stimulation by GMF, the flat monolayer glioblasts already contain large amounts of GFA protein, indicating that the appearance of this protein precedes the emergence of other astrocytic properties, including the visible gliofilaments. Table II compares the level of GFA protein in rat glioblasts with that in other cell types.

VI. Chemical Nature of GMF

GMF is an acidic protein readily extractable from the adult brain at physiological pH (Lim and Mitsunobu, 1975). The initial evidence in favor of a protein nature of the factor comes from its susceptibility to several proteases. Other pieces of evidence include its thermolability and inactivation by the extremes of pH and denaturing agents. The activity is not affected by DNase, RNase, or periodate oxidation, nor

TABLE II

GFA Protein in Various Cell Cultures [a]

Source	GFA protein (ng/μg DNA)
Rat glioblasts (monolayer culture)	60.40
C6 cells (monolayer culture)	0.32
C6 cells (organ culture)	2.29
L cells (monolayer culture)	0.02
L cells (organ culture)	0.03
KB cells (suspension culture)	0.01

[a] All values were obtained by radioimmunoassay. Rat glioblasts were from secondary cultures not exposed to GMF (L. F. Eng and R. Lim, unpublished). Data for other cells are from Bissell et al. (1975). In an independent study using rat glioblast monolayer cultures similar to ours, Bock et al. (1975) reported that the specific concentration of GFA protein as determined by rocket immunoelectrophoresis was 46.6 times that of the whole adult rat brain.

is the activity eliminated by lipid extractants such as ethanol and butanol. We have demonstrated that GMF from pig brains exists in two forms (Kato et al., 1979), the high-molecular-weight form (HMW–GMF) and the low-molecular-weight form (LMW–GMF). When estimated with a calibrated Sephadex column, HMW–GMF is 200,000 in molecular weight and LMW–GMF 40,000. The former comprises 85% of the total biological activity of GMF in the crude brain extract. The proportion of LMW–GMF increases following freezing and thawing and on further purification, probably as a result of dissociation of the high-molecular-weight form. In addition to the morphological effects, both forms possess mitogenic activity but no esteropeptidase activity. Both forms show similar enzyme susceptibility, being inactivated by papain, ficin, and pronase but resistant to subtilisin, thermolysin, and trypsin. On the other hand, the high-molecular-weight form is more resistant to low pH, heating, and urea denaturation than the low-molecular-weight form. The stability of HMW–GMF and its preponderance in the crude brain extract suggests that it is the native form of GMF in the brain. The differences in the properties of the two forms of GMF are summarized in Table III.

VII. Distribution of GMF

GMF is not species-specific. Brain extracts from rats, pigs, cattle, sheep, monkeys, and humans are equally active on rat glioblasts. GMF-like activity is demonstrable in other rat organs, such as kidney

TABLE III

COMPARISON OF PROPERTIES OF TWO FORMS OF GMF[a]

Property	LMW–GMF	HMW–GMF
Molecular weight	40,000	200,000
Isoelectric point	5.04	4.27
pH stability	pH 5–11	pH 3–11
Heat stability	Less stable	More stable
Stability in urea	Less stable	More stable
Susceptibility to proteases	More susceptible	Less susceptible
Esteropeptidase activity	None	None

[a] Data from Kato et al. (1979).

and heart. Fetal rat brain (17-day gestation) contains only $\frac{1}{5}$ of the activity found in adult brain. The factor is also low in rat astrocytomas and mouse neuroblastomas when obtained as solid tumors, where less than $\frac{1}{10}$ of the activity in adult brain can be extracted. Human red blood cells contain only $\frac{1}{25}$ of the rat brain activity, whereas the activity in microorganisms such as *Escherichia coli*, *Bacillus subtilis*, and yeast is practically nil (less than $\frac{1}{200}$ of that of adult brain). All these comparisons are made on the basis of extractable protein. The following body fluids, when used at concentrations up to 15% (v/v) in the medium, do not possess any GMF activity: fetal calf serum, bovine serum, horse serum, rat serum, chicken serum, and human cerebrospinal fluid. Thus, the general distribution indicates that the factor is limited to solid organs of normal adult animals (Lim *et al.*, 1977a).

We have demonstrated (Lim *et al.*, 1977c) that a homogeneous population of monolayer glioblasts, if dislodged from the flask surface and permitted to reaggregate, will eventually undergo morphological differentiation even in the absence of exogenous GMF. This indicates that at least under certain conditions (aggregate culture) the glioblasts can produce GMF and effect their own maturation, and that in a monolayer culture GMF either is absent or fails to function properly.

Recently, we have detected large amounts of endogenous GMF-like activity in monolayer rat glioblasts (Turriff and Lim, 1980). Even though these cells are epithelial in appearance, extract prepared from them can stimulate another monolayer of identical glioblasts to assume the histotypic pattern of mature astrocytes, in a manner and time course indistinguishable from the effect of adult brain extract. Conditioned media, on the other hand, are without any activity even after a 15-fold concentration. This observation indicates that GMF is

present in the flat glioblast cells but is somehow prevented from exerting its effect; once GMF is extracted into solution, it is free to act on
the receptors of another monolayer culture to bring about differentiation. The specific activity of the endogenous GMF in the monolayer
glioblasts is about three times higher than that in the adult brain extract. The high level of GMF may be explained by a compensatory increase in GMF synthesis when it is in a nonfunctioning state. Even
though the GMF level in fetal brain is low, the 2-week duration of *in
vitro* existence of our cells may be sufficient time for the GMF level to
rise. The properties of the endogenous GMF from the glioblasts are in
many respects similar to those of HMW–GMF from pig brains, including molecular size, enzyme sensitivity, and susceptibility to
chemical and physical denaturing agents (Turriff and Lim, 1980). The
question, then, is, What prevents the endogenous GMF from acting
on the cells themselves? One can conceive of three possible explanations. First, GMF may be a soluble cytoplasmic protein and not released by the cells. The second explanation is that the factor may be
entrapped in one of the cytoplasmic organelles, as in the case of
lysosomal enzymes. The third possibility is that it is a membrane-
bound protein. The second possibility is ruled out by our observation
that purified lysosomes do not possess GMF activity even after repeated freezing and thawing, despite the presence of high specific activities of lysosomal enzymes such as β-glucuronidase and acid
phosphatase (Turriff and Lim, 1980). The other two possibilities are
discussed in the next section.

VIII. Possible Biological Role of GMF

The exact biological significance of GMF is presently unknown.
However, two lines of evidence seem to emerge: first, the need for the
continuous presence of GMF for cytodifferentiation of glioblasts and
the fact that the GMF level is high in the adult brain but low in the
embryonic brain indicate a sustaining rather than an inducing role;
second, GMF appears to be a mediator of short-range cellular interaction. The second point is elaborated below.

Mammalian cells exhibit high degrees of interdependency. When
grown individually *in vitro,* they either fail to survive or frequently
lose their differentiated characteristics. In a living organism, there are
three possible mechanisms by which cell integrity can be maintained
through mutual influence: (1) by humoral factors mediating long-
range communications between cells in different organs, as ex-
emplified by the well-known hormones, (2) by diffusible factors acting

locally between cells in the same organ (either by secretion into intercellular space or by transfer across a cell junction), (3) by nondiffusible or membrane-bound molecules on cell surfaces acting on adjacent cells. The absence of GMF activity in sera and cerebrospinal fluid indicates that it is not a humoral factor. The absence of GMF in conditioned media suggests that it is not secreted by the cells. Its large molecular size precludes its transfer across cell junctions.

The existing data are consistent with a working hypothesis I previously proposed (Lim, 1977): (1) that GMF is a membrane-bound protein existing on the outside cell surface of the glioblast; (2) that GMF exerts its effect on another glioblast by stimulating the receptor sites of the target cell when the two are in physical contact; (3) that every glioblast contains both GMF and the receptors; (4) that the geometric arrangement of GMF and the receptors on the surface is such that self-stimulation in the same cell is prevented, but mutual influence in a group of adjacent cells is possible; (5) that the degree of cell maturation depends on, and is proportional to, the number of receptors being stimulated. A schematic representation of this hypothesis is shown in Fig. 8.

This hypothesis explains our observation that glioblasts in a monolayer culture do not spontaneously differentiate, whereas they do in an aggregate culture. I assume that both the monolayer and the aggregated glioblasts produce endogenous GMF, the only difference between the two being the possibility that in the former instance GMF fails to function properly (geometric constraint). The addition of exogenous GMF to cells in a monolayer could promote their maturation by saturating the unoccupied receptor sites, reconstituting the

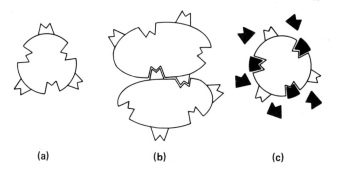

(a)   (b)   (c)

FIG. 8. Hypothesis on how GMF works. (a) A cell containing GMF (protruding structures) and the receptors (notched areas) on the cell surface. (b) Two cells in physical contact, with GMF from one cell acting on the receptor of the other cell. (c) A cell stimulated by exogenous GMF (solid areas) added to the culture medium. (From Lim, 1977.)

chemical environment the cells normally enjoy in a solid organ. That the monolayer glioblasts contain endogenous GMF is supported by our observation that GMF activity can be extracted directly from these cells.

A question that arises from this model is, If GMF is a membrane-bound protein, why is it readily extractable into the soluble fraction? The difficulty can be resolved by proposing one or both of the following possibilities: (1) GMF is loosely bound to the surface membrane and thus can be easily dislodged by the shearing effect of homogenization. (2) GMF exists in two pools: a soluble pool in the cytoplasm and a membrane-bound pool on the cell surface. An example of this type of bimodal distribution exists in the case of discoidin and pallidin, the cell-aggregating proteins from slime molds (Barondes and Rosen, 1976); for these proteins it has been found that the amount in the cytoplasmic pool (nonfunctioning) exceeds the amount in the surface membrane pool (functioning).

IX. General Implications in Cell Biology

Is the function of GMF limited to glial cells, or is it a general biological phenomenon found in many cell types? Since GMF is not a humoral factor, and since GMF-like activity can be detected in several other solid organs, I tend to think that GMF belongs to a class of protein that exerts a common function. The GMF-like protein in various organs need not be identical but may exhibit a certain degree of cross-reactivity. If so, the study of GMF in glioblasts can serve as a model for the understanding of short-range cellular interaction.

Mammalian cells differ from unicellular organisms in that they exist in communities and exert mutual influence. I wish to propose the term "neighboring effect" to denote the complex relationship among adjacent cells. Neighboring effect differs from the phenomenon of cell recognition in that the former explains the cooperative action of cells after the formation of a community, while the latter explains how the community is formed (Moscona, 1976). Neighboring effect is best studied in cultured cells, since under tissue culture conditions the interorgan relationship is abolished and only the intercellular relationship remains, and can be manipulated with relative ease.

Various manifestations of neighboring effect have been described before. One is the fact that mammalian cells need to adhere to the flask surface (or to a feeder layer) for survival. In this instance the flask surface serves as a surrogate plasma membrane of an imaginary giant cell of infinite size. Second is the well-known contact inhibition of growth and movement. Third is the intercellular communication through the gap junction, which permits ions and molecules less than 2000 daltons to pass (Pitts, 1977; Loewenstein *et al.*, 1978). To these examples I wish to add the effects of GMF and GMF-like molecules, which, in my opinion, convey one type of mutual influence when cells are in close apposition.

The effect of GMF in rearranging the histotypic pattern of a confluent culture of glioblasts should be interpreted as a partial restoration of the neighboring effect the cells normally enjoy *in vivo*. Thus, the apparent release from contact inhibition elicited by GMF should be viewed as a step-up change toward the more differentiated state, as opposed to the release in neoplastic transformation, which represents a step-down change in the differentiation scale. In other words, the change in GMF-stimulated cells is due to the satisfaction of certain requirements the cell have, whereas in neoplastic transformation the morphological change is due to the loss of such requirements, i.e., cellular interdependency.

One of the perplexing problems in developmental biology is the role of the cell environment in determining the phenotypic expression of the gene. I hope the study of GMF and related compounds will help us understand one aspect of this problem.

X. Summary

Glioblasts in a monolayer culture are stimulated by a protein factor from the adult brain to differentiate morphologically and chemically. The morphological effect is reversible, requiring the continuous presence of the factor in the culture medium. The response of glioblasts to GMF consists of a biphasic change. The early events (first 3 days) include an increase in cyclic GMP, an increase in DNA synthesis and cell division, an increase in RNA and protein synthesis, and a dramatic rearrangement of the histotypic pattern as a result of cell retraction and process outgrowth. The processes are highly motile, and the cell bodies show pulsatile movements. The late events (after 3 days) include the cessation of cell division, the slowing down of tugging activity with continuous elongation of the processes, and the onset of chemical maturation as indicated by increases in cyclic AMP, the glial protein S-100, and the enzyme glycerophosphate dehydrogenase. The factor, which is extractable from the brain at neutral pH, is an acidic protein of 200,000 molecular weight. GMF has no species specificity, and GMF-like activity can be detected in organs other than the brain. The exact biological role of GMF is unclear. Currently available evidence suggests that it may be a mediator of short-range cellular interaction.

ACKNOWLEDGMENT

This work was supported by NIH grants NS-09228, NS-14316, NS-07376, and CA-19266. The data presented here were made possible by contributions from the following research associates who have worked in my laboratory at one time or another: Drs. Katsusuke Mitsunobu, David E. Turriff, Shuang S. Troy, Taiji Kato, Tien-Cheng Chiu, Satoe Nakagawa, Yng-Jiin Wang.

REFERENCES

Barondes, S. H., and Rosen, S. (1976). In "Neuronal Recognition" (S. H. Barondes, ed.), pp. 331–356. Plenum, New York.
Bissell, M. G., Eng, L. F., Herman, M. M., Bensch, K. G., and Miles, L. E. M. (1975). *Nature (London)* 255, 633–634.
Bock, E., Jörgensen, O. S., Dittmann, L., and Eng, L. F. (1975). *J. Neurochem.* 25, 867–870.
Kato, T., Chiu, T.-C., Lim, R., Turriff, D. E., and Troy, S. S. (1979). *Biochim. Biophys. Acta* 579, 216–227.
Lim, R. (1977). In "Mechanisms, Regulation and Special Functions of Protein Synthesis in the Brain" (S. Roberts, A. Lajtha and W. H. Gispen, eds.), pp. 299–306. Elsevier/North-Holland, Amsterdam.
Lim, R., and Mitsunobu, K. (1974). *Science* 185, 63–66.
Lim, R., and Mitsunobu, K. (1975). *Biochim. Biophys. Acta* 400, 200–207.
Lim, R., Li, W. K. P., and Mitsunobu, K. (1972). *Abstr. Annu. Meet. Soc. Neurosci., 2nd, Houston,* p. 181.
Lim, R., Mitsunobu, K., and Li, W. K. P. (1973). *Exp. Cell Res.* 79, 243–246.
Lim, R., Turriff, D. E., and Troy, S. S. (1976). *Brain Res.* 113, 165–170.
Lim, R., Turriff, D. E., Troy, S. S., and Mitsunobu, K. (1977a). In "Cell Culture and its Application" (R. T. Acton, ed.), pp. 461–480. Academic Press, New York.
Lim, R., Turriff, D. E., Troy, S. S., Moore, B. W., and Eng, L. F. (1977b). *Science* 195, 195–196.
Lim, R., Troy, S. S., and Turriff, D. E. (1977c). *Exp. Cell Res.* 106, 357–372.
Lim, R., Turriff, D. E., Troy, S. S., and Kato, T. (1977d). In "Cell, Tissue and Organ Cultures in Neurobiology" (S. Fedoroff and L. Hertz, eds.), pp. 223–235. Academic Press, New York.
Loewenstein, W. R., Kanno, Y., and Socolar, S. J. (1978). *Fed. Proc.* 37, 2645–2650.
Lumsden, C. E., and Pomerat, C. M. (1951). *Exp. Cell Res.* 2, 103–114.
Moscona, A. A. (1976). In "Neuronal Recognition" (S. H. Barondes, ed.), pp. 205–226. Plenum, New York.
Peters, A., Palay, S. L., and Webster, H. de F. (1970). "The Fine Structure of the Nervous System." Harper, New York.
Pitts, J. D. (1977). In "International Cell Biology: 1976–1977" (B. R. Brinkley and K. R. Porter, eds.), pp. 43–49. Rockefeller Univ. Press, New York.
Turriff, D. E., and Lim, R. (1979). *Brain Res.* 166, 436–441.
Turriff, D. E., and Lim, R. (1980). *Dev. Neurosci.* (in press).

CHAPTER 10

MOLECULAR AND LECTIN PROBE ANALYSES OF NEURONAL DIFFERENTIATION

S. Denis-Donini

LABORATORY OF MOLECULAR EMBRYOLOGY
NAPLES, ITALY

and

G. Augusti-Tocco

LABORATORY OF QUANTUM ELECTRONICS
FLORENCE, ITALY

How cells migrate in the course of embryogenesis and how they change shape, adhere selectively to certain types of cells, and finally come to rest in an organized tissue are not yet understood. This problem is of particular relevance in neurogenesis where young neurons migrate from their site of origin to their final locations, send out axons in search of their specific targets, and establish functional connections with them by means of very specialized and elaborate structures: the synapses. The experimental facts strongly suggest that such aspects of cell interaction as the ability to distinguish through contact one kind of cell from another are related to the organization of the membrane components which seem to evolve with the progressive differentiation of the cells. One system that seems particularly suitable to investigation of the membrane properties correlated with adhesion and neurite outgrowth is the neuroblastoma (NB) where the expression of differentiated neuronal functions can be modulated according to the culture conditions. This chapter will deal mainly with the surface changes occurring during neuronal maturation in NB and dorsal root ganglia (DRG) in culture. The general properties and advantages of the NB system will be reviewed first.

CURRENT TOPICS IN
DEVELOPMENTAL BIOLOGY, Vol. 16

I. General Properties of Neuroblastoma Clones

The C1300 NB is a spontaneous tumor of the neural crest which has been maintained for many years as a transplantable tumor in A/J mice. In spite of the fact that the tumor cells are morphologically undifferentiated, they express several neuronal markers such as neurotransmitters and their biosynthetic enzymes [tyrosine hydroxylase (TH) and choline acetyltransferase (CAT)] and acetylcholine esterase (AChE)(Augusti-Tocco and Sato, 1969; Schubert et al., 1969). Clonal lines of neurons have been isolated from the tumor, and the cultured cells present many neuronal characteristics among which are electrical excitability (Nelson et al., 1969, 1970a,b; Harris and Dennis, 1970; Augusti-Tocco et al., 1970), enzymes involved in the synthesis or breakdown of neurotransmitters (Augusti-Tocco and Sato, 1969; Amano et al., 1972; Anagnoste et al., 1972; Prasad et al., 1973a; Hamprecht et al., 1974; Breakefield, 1976), transmitter receptors (Vogel et al., 1972; Nelson and Peacock, 1973), and the neuron-specific protein 14-3-2 (Augusti-Tocco et al., 1973a; McMorris et al., 1974). These cells have been adapted to grow under culture conditions where they either maintain the undifferentiated morphology described for the cells present in the tumor or, on the contrary, send out processes, thus assuming the morphology of mature neurons (Schubert et al., 1969; Augusti-Tocco et al., 1970; Schubert and Jacob, 1970; Seeds et al., 1970; Furmanski et al., 1971; Prasad and Hsie, 1971). This morphological differentiation is accompanied by a number of other events characteristic of neuronal maturation. In fact, the cells forming fibers show an increase in size (Prasad, 1972; Prasad and Sheppard, 1972; Kimhi et al., 1976), in RNA and protein content (Augusti-Tocco et al., 1973b; Blume et al., 1970; Miller and Levine, 1972; Prasad et al., 1973b) and, in many cases, in the level of neurospecific enzymes (Blume et al., 1970; Furmanski et al., 1971; Waymire et al., 1972; Richelson, 1973). The acetylcholine (ACh) receptors become specifically distributed on the cell body and on one fiber as in mature neurons (Harris and Dennis, 1970), and new neuron-specific surface antigens appear (Akeson and Hershman, 1974a; Martin, 1974). In most of the clonal lines isolated, this morphological and biochemical differentiation can be induced in various ways; these are summarized in Table I and will be discussed in greater detail in the next section. However, NB cells are not able to reach complete maturation; in fact, there is no evidence of synapses, and the cells appear to be chemically isolated from their neighbors (Harris and Dennis, 1970; Augusti-Tocco et al., 1970). This failure to accomplish the final step of neuronal maturation can be overcome by somatic hybridization with glioma cells which produce cell lines able to form functional synapses with myotubes (Nelson *et al.*, 1976; Puro and Nirenberg, 1976).

An interesting and technically very helpful feature of NB cell cultures is that it is possible to derive clones with different capacities for expressing differentiated neuronal properties from the same tumor. In some clones all the neuronal differentiated properties investigated can be induced concomitantly, whereas in others only some of these properties are expressed. For instance, fiber formation can be observed in the absence of transmitter synthesis (i.e., as in the case of the "inactive" clone N18 described by Amano *et al.*, 1972), and vice versa (i.e., the axon-minus clones of Amano *et al.*, 1972). Similarly, the action potential mechanism can be found in the absence of fibers [suspension culture (SP) cells] (Augusti-Tocco *et al.*, 1970; Harris and Dennis, 1970). The ability to produce action potentials, AChE, and prostaglandin E_1 (PGE) receptors seems to be retained in many clones, while the ability to form fibers and to synthesize and store neurotransmitters varies greatly. Since NB originates from the neural crest and, although serotoninergic clones have been described (Knapp and Mandel, 1974), attention has been mainly focused on the synthesis of ACh and norepinephrine (NE), and the enzymes TH and CAT have been measured in a large number of clones. On the basis of their TH and CAT content, the clones can be divided into categories that may be considered the developmental analogs of neural crest derivatives:

1. Cells that possess enzymes involved in the synthesis of both neurotransmitters. When this phenomenon was observed, the existence of dually functioning neurons was believed to be exceptional; the presence of both enzymes in a clonal population was often considered as being due to malignancy or to the presence of stem cells in the tumor giving rise to a heterogeneous clonal population. More recently, it has been established that a single sympathetic neuron at an immature stage can express both transmitter systems simultaneously (Furshpan *et al.*, 1976; Landis, 1976) and that such neurons are at least transiently plastic with respect to their choice of transmitter. This choice can be influenced by other cells including neurons (Patterson and Chun, 1977; Walicke *et al.*, 1977) and nonneuronal cells (Patterson and Chun, 1974). In light of these findings, these types of NB clones could be thought of as neurons that are still at an immature stage and which are "multipotential" with respect to their choice of transmitter.

2. Clones synthesizing only one neurotransmitter (ACh or catecholamines). These clones would correspond to more advanced neu-

TABLE I

INDUCTION OF DIFFERENTIATION

Culture condition	Clone	Reference	Neurite formation	Inhibition of growth	Increase in			TH	Induction of			Change in protein pattern
					Size	Protein	RNA		CAT	AChE	cAMP	
Serumless medium	N18	Seeds et al. (1970)	+	+	+							+
		Blum et al. (1970)								+		
	C1	Schubert et al. (1971a)	+	+								
	C1A	Schubert et al. (1971b)	+	+								
	NBP$_2$	Waymire et al. (1972)	+	+		+		−		+		
	N1E-115	Richelson (1973)	+			+		−				
	N1E-113		+					−				
	NB42B	Kates et al. (1971)	+									
	NBE-NBA$_2$	Prasad et al. (1973)	+			+	+		−	+		
BrdU	C1	Schubert and Jacob (1972)	+	−								+
	7	Simantov and Sachs (1972)	+									
	NBE-NBA$_2$	Prasad et al. (1973a)	+			+	+			+	+	
dBcAMP	NB60	Furmanski et al. (1971)	+	+							+	
		Furmanski and Lubin (1973)								+		
	NBP$_2$	Waymire et al. (1972)	+	+				+				
	N1E-115	Richelson (1973)	+	+				+				

N1E-113									+ + +
NS-20									
N1E-115									
N18	Prashad et al. (1977)	+							
NBE		+							
PGE NBA₂(1)	Prasad et al. (1973b)	+	+	+	+			+	
NBA₅	Prasad (1972)	+	+	+	+			+	
N18	Prasad (1975)								
	Gilman and Nirenberg (1971)								
cAMP NBA₂(1)	Prasad and Sheppard (1972)	+	+	+				+	+ +
phospho- diesterase inhibitors									
FdU Ara-C NB42B	Kates et al. (1971)	+					−	+	
C1	Schubert et al. (1971a)	−							
ACh NB42B	Harkins et al. (1972)	+						+	
DMSO N1E-115	Kimhi et al. (1976)	+	+						−
N18		+							
NS-20		+							
N1A-103		−							
Glass 2A									
surface or N4 collagen N18	Miller and Levine (1972)	+	−	+	+		−		
ML or SP	Augusti-Tocco et al. (1973b)	−	−	+	+				
41A₃	Augusti-Tocco and Chiarugi (1976)	+		+	+				+

rons that have already become restricted or committed to a specific pathway in development.

3. Clones in which CAT or TH activity cannot be detected. This group may represent very immature neurons that have not yet acquired the ability to synthesize neurotransmitters, or they might produce another type of neurotransmitter and represent the other neural crest derivative, the sensory neuron. The report that the activity of CAT in hybrid lines (NB–glioma) is considerably higher than in the parental lines (Hamprecht et al., 1977) favors the first hypothesis. This specific induction of CAT upon hybridization with glial cells resembles the CAT induction occurring when young sympathetic neurons are cocultured with glial cells (Patterson and Chun, 1974).

Thus, with respect to transmitter synthesis, NB cultures present a whole spectrum of cells that can be compared not only to various neural crest derivatives but also to sympathetic neurons at successive developmental stages. Since an immature neuroblast must acquire several specialized properties in order to reach its ultimate and complex function, the advantage of the various NB clones is that they express only certain differentiated functions, thus making it possible to dissect out single steps of neuronal specialization.

II. Differentiation

A. FACTORS CONTROLLING FIBER FORMATION

It has already been mentioned that NB cells can be grown under culture conditions that do or do not allow fiber formation. Analysis of the experimental procedures used to induce fiber formation may lead to an understanding of the nature of the factors controlling this process in the embryo. NB cultures are a particularly suitable experimental system with which to approach this problem, since they allow us to dissociate cell survival from fiber formation, while in primary cultures neuron survival is strictly associated with fiber formation.

The ability of NB cells to send out processes varies greatly among clones. It ranges from cases where the attachment of the cells to the culture dish is sufficient to induce them to form fibers (Augusti-Tocco and Sato, 1969). [In this instance cells are cultured in suspension (SP) or monolayer (ML) to obtain cells, respectively, without or with fibers. Under such conditions fiber outgrowth is maximal when the cells reach the stationary growth phase.] In other cases, it is necessary to subject the cells to additional treatment after they have attached to the culture dish (Schubert and Jacob, 1970; Seeds et al., 1970; Prasad and Hsie, 1971; Furmanski *et al.*, 1971; Kimhi *et al.*, 1976). Finally, clones have been described that are unable to form fibers, at least under the experimental conditions tested so far (Amano *et al.*, 1972).

Various procedures have been devised in order to promote fiber formation. They are differentially effective in the various clones and can be grouped into three main categories: (1) conditions that act primarily on cell adhesiveness, for example, serumless medium (Schubert *et al.*, 1971a), SP or ML culture (Augusti-Tocco *et al.*, 1970; Schubert *et al.*, 1969), and the addition of bromodeoxyuridine (BrdU) (Schubert and Jacob, 1970; Holtzer, 1970); (2) conditions that primarily block cell division by inhibiting DNA synthesis, among which are the addition of cytosine arabinoside (AraC), fluorodeoxyuridine (FdU) (Klebe and Ruddle, 1969), or 6-thioguanine (Prasad, 1973); (3) conditions directly increasing the cellular level of cAMP, for example, the addition of dibutyryl cAMP (dBcAMP) (Prasad and Hsie, 1971; Furmanski *et al.*, 1971), phosphodiesterase inhibitors (Prasad and Sheppard, 1972), or PGE (Prasad, 1972).

In addition, other compounds have been used and do not clearly belong to the categories illustrated above. Among these, dimethyl sulfoxide (DMSO) (Kimhi *et al.*, 1976) and the ionophore valinomycin (Koike, 1978) modify surface membrane properties.

It is relevant to note that cell adhesiveness, cell division, and cAMP level are interrelated phenomena. In fact, serumless medium increases cell adhesiveness, but it also blocks cell division and elevates the cAMP level (Sheppard and Prasad, 1973). The block of cell division also affects adhesion and the cAMP level, which are lower in mitotic than in resting cells (Sheppard and Prescott, 1972), while elevation of the cAMP level blocks cell division and increases adhesiveness (Rebhun, 1977). Therefore, these three kinds of procedures, in spite of their diversity, produce the same effects. However, a block of cell division per se is not sufficient to promote fiber formation in all clones (Schubert *et al.*, 1971a; Furmanski and Lubin, 1973), and fibers can be formed without elevating the cAMP level (Kimhi *et al.*, 1976). The only absolute requirement for fiber outgrowth is a strong adhesive interaction with a substratum. This is in agreement with studies on primary cultures of DRG neurons where the formation of fibers parallels the strength of adhesion (Letourneau, 1975). The fact that different clones respond preferentially to agents acting primarily on adhesion, cAMP level, or DNA synthesis might be explained on the basis of different surface properties. Thus, a clone that is highly adhesive will spontaneously form fibers upon attachment to the substratum. On the other hand, a clone that is moderately adhesive will respond better to conditions that increase cell adhesion such as a blocking of cell division or an elevation of the cAMP level, and a clone characterized by a low degree of adhesivity will need conditions directly modifying the type of interaction with the substratum. Furthermore, Miller and Levine (1972) have reported evidence that some specificity of interaction is necessary for fiber formation. They have observed that collagen-coated dishes do not support fiber formation in NB cells, even though adhesion on collagen is as strong as on glass. Moreover, agents such as BrdU, FdU, and cAMP, which induce fibers in cultures on glass, are ineffective on collagen. This finding, although interesting, is surprising, since collagen has been widely used as a substratum for neuron culture; the reported lack of fibers on collagen could be a peculiarity of NB or at least of the clones used.

B. RNA SYNTHESIS

The requirement for RNA synthesis in differentiating NB cells was studied by adding actinomycin to the culture under conditions promoting axon formation (Table II). In all cases, irrespective of the inducing agent used (BrdU, PGE, or serumless medium), actinomycin did not prevent neurite outgrowth; this led to the conclusion that RNA synthesis was not necessary and that the undifferentiated cells already had all the necessary "information" to form fibers (Schubert and Jacob, 1970; Schubert et al., 1971a; Prasad, 1972). However, differentiated NB cells have a higher RNA content under at least three different inducing conditions, i.e., SP or ML culture (Augusti-Tocco et al., 1973b), collagen or glass culture (Miller and Levine, 1972), and the addition of PGE (Prasad et al., 1973a). Since such an increase in RNA content is also found in the nervous system during development (Judes et al., 1973), the level and metabolism of the various RNA species have been investigated in NB.

Casola et al. (1974) studied the mechanism of rRNA increase in NB cells in ML and SP cultures and examined the rate of its synthesis, posttranscriptional processing, and turnover. They showed that the rate of synthesis of 45S remained unchanged but that the processing into 28 and 18S rRNA increased drastically when the cells were switched from SP to ML culture.

No changes were observed when the tRNA level and the amino acid-accepting ability were studied under both culture conditions (Augusti-Tocco, Metafora, and Felsani, unpublished results).

In differentiated cells, heterogeneous RNA (HnRNA) has been found in higher amounts than in SP cells. However, its rate of synthesis and the flow of mRNA molecules to the cytoplasm are un-

TABLE II

RNA Synthesis in Differentiating NB Cells

Inducing agent	Time of addition (hours after plating)	Inhibitor added	Concentration	Length of treatment (hours)	Neurite outgrowth[a]	Reference
Serumless medium	24	Cycloheximide	$1.8 \times 10^{-4} M$	2	+	Seeds et al. (1970)
	0	Cycloheximide	40 μg/ml	18	–	Schubert et al. (1971a)
	0	Puromycin	10 μg/ml	18	–	Schubert et al. (1971a)
	0	Actinomycin	5 μg/ml	18	+	Schubert et al. (1971a)
	24	Cycloheximide	10 μg/ml	48	+ –	Furmanski et al. (1971)
dBcAMP	24	Cycloheximide	10 μg/ml	48	–	Furmanski et al. (1971)
	0	Cycloheximide	5 μg/ml	24	–	Prasad et al. (1972)
	0	Actinomycin	5 μg/ml	24	+	Prasad et al. (1972)
PGE	24	Cycloheximide	5 μg/ml	24	–	Prasad (1972)
	24	Actinomycin	5 μg/ml	1	+	Prasad (1972)
BrdU	0	Cycloheximide	40 μg/ml	15	–	Schubert and Jacob (1970)
	24	Puromycin	10 μg/ml	15	–	Schubert and Jacob (1970)

[a] –, Absence of neurites; +, neurite outgrowth as in the control.

changed under the two growth conditions (Augusti-Tocco et al., 1974). This finding has been interpreted as an indication that HnRNA molecules become more stable in differentiated cells. Similarly, the analysis of polyadenylated mRNA species in the two types of cells has suggested an increased stability of some messengers in differentiated cells (Croizat et al., 1977). On the other hand, when the mRNA content was estimated on the basis of the poly(A) level, it was higher in PGE- and dBcAMP- induced cells (Prasad et al., 1975; Simantov and Sachs, 1975). However, in these studies, poly(A) was measured after [³H]adenosine labeling, and no attention was paid to the nucleotide pools or to uptake of the exogenous precursor, which is probably affected by dBcAMP. In addition, the discrepancy between these results could be explained by the different experimental conditions. While the metabolism of RNA was measured in SP and ML cultures both in the stationary phase of growth (Casola et al., 1974; Augusti-Tocco et al., 1974), the poly(A) level was measured in cells blocked in the G_1 phase of the cell cycle, while the control population was asynchronous (Prasad et al., 1973a). Therefore, the high poly(A) level could be due to the different positions in the cell cycle of the compared cell populations, rather than being related to differentiation.

C. Protein Synthesis

The first experiments on the relevance of protein synthesis in axon formation were performed with cycloheximide (Table II) (Seeds et al., 1970; Schubert and Jacob, 1970; Prasad, 1972). The results were not always consistent, and this may have been due to the diversity of the experimental conditions adopted. While fiber formation induced by cAMP, PGE, and BrdU was completely abolished by cycloheximide (Furmanski et al., 1971; Prasad, 1972; Schubert and Jacob, 1970), it was only partially or not at all affected in serumless medium (Furmanski et al., 1971; Seeds et al., 1970). However, in the first instance the inhibitor was added at the beginning of induction; in the second it was added 24 hours later. It seems therefore reasonable to think that protein synthesis is required at least for the initiation of axon formation. The increase in size and protein content (Prasad et al., 1972; Kimhi et al., 1976; Miller and Levine, 1972), as well as the induction of neurospecific enzymes in differentiated cells, suggests that protein synthesis is activated during neurite outgrowth. In fact, in cell-free systems, monolayer cells are more active in protein synthesis than SP cells. The various factors involved in protein synthesis were analyzed, and it was suggested that the initiation factors were most probably responsible for the observed activation (Zuco et al., 1975). The activation of protein synthesis in differentiated cells could be expected to be accompanied by the appearance of new proteins. However, one should bear in mind that the undifferentiated cells already express many neuronal markers and that the new functions appearing in the differentiated cells probably involve only a few new proteins. The spectrum of proteins synthesized in the two types of cells would then be largely overlapping and therefore difficult to resolve. On polyacrylamide gel electrophoresis (PAGE) a different distribution of radioactivity has been found in the pattern of newly synthesized proteins from SP, ML (Augusti-Tocco et al., 1973a), and BrdU-induced cells (Schubert and Jacob, 1970). More recently, the problem has been reexamined using bidimensional electrophoresis to analyze proteins from clones with different properties: cholinergic, adrenergic, and inactive (Prashad et al., 1977). This analysis has shown that a few proteins are charge-modified in the differentiated cells. Interestingly, different proteins are modified in adrenergic, cholinergic, and inactive clones.

Another approach that proved to be more rewarding in identifying new differentiated markers was the study of such specific proteins as tubulin. The differentiated cells are characterized by a large number of neurotubules (Schubert at al., 1969; Augusti-Tocco et al., 1970; Ross et al., 1975), and fiber formation is prevented in the presence of microtubule-disrupting agents such as colchicine and vinblastine (Seeds et al., 1970; Schubert and Jacob, 1970). The amount of tubulin in differentiated and undifferentiated cells is the same (Morgan and Seeds, 1975; Schmitt, 1976), showing that the appearance of neurotubular structures does not involve an increased synthesis of tubulin but is regulated only by factors acting on the assembly of the preexisting subunits. Indeed, only extracts of differentiated NB cells are able to induce the polymerization of lamb brain tubulin in vitro. The active factor has been partially purified and is a large protein of 170,000 molecular weight, which is not present in the undifferentiated cells (Seeds and Maccioni, 1978). This is a clear example of a major structural change (formation of the cytoskeleton sustaining the fibers) dependent on the appearance of a minor protein component.

D. The Cell Surface

1. Biochemical Analysis

Cell interactions of primary importance in neuron maturation are mediated through the cell surface either in the case of cell–cell contact in a tissue or cell attachment to a substratum in culture. Therefore, considerable effort has been devoted to study of the membrane changes occurring during neuronal maturation, both in NB and in developing neurons *in vivo* and *in vitro*. In NB these studies were expected to lead to an understanding of the biochemical events underlying the morphological changes correlated with cellular adhesion and neurite growth. Many membrane components were investigated: glycoproteins, proteins, and glycosaminoglycans (GAGs). The task of comparing and evaluating the results of such investigations is difficult, because most experiments were performed on clones whose characteristics and properties were often different and where differentiation was stimulated with a variety of agents. Brown (1971) studied the membrane glycopeptide pattern obtained by gel filtration of the trypsin-removable material and in BrdU-induced cells found a surface glycopeptide which was absent in the undifferentiated cells. Glick *et al.* (1973), using the same experimental approach, compared axon-forming and axon-minus clones. They found that the pattern of glycopeptides differed in the two types of clones as a result of the presence of more complex oligosaccharides in the latter. This type of difference is also found when normal and virus-transformed cells are compared, indicating that the axon-forming cells are more "normal" than axon-minus cells. Truding *et al.* (1974) studied on PAGE the polypeptide composition of isolated membranes from SP and ML cultures treated with BrdU or dBcAMP and found a very similar pattern. However, labeling experiments with glucosamine indicate that in differentiated cells the glycosylation of protein(s) (MW 104,000) increases. Using another clone Garvican and Brown (1971) isolated membranes from differentiated cells induced with BrdU or dBcAMP and showed an increased incorporation of fucose into low-molecular-weight proteins and into two additional components with molecular weights of 60,000 and 70,000. They found that dBcAMP stimulated both glycosylation and amino acid incorporation, while BrdU mainly stimulated glycosylation of the membrane proteins. Both these studies tend to indicate that the main difference between differentiated and undifferentiated cells lies in the glycosylation of surface proteins. This conclusion is also reached when the glycoproteins released into the culture medium are studied (Truding *et al.*, 1975). Also in this case, dBcAMP and BrdU mainly increase the degree of glycosylation. However, in all these experiments, since SP cells were not treated with inducing agents, one would think that the observed difference may be a nonspecific metabolic effect of BrdU or dBcAMP and not related to fiber formation. Most likely, this is not the case for at least BrdU since, as reported by Schubert and Jacob (1970), the overall changes in protein pattern caused by BrdU in ML cultures are not observed in BrdU-treated SP cultures.

In a very ingenious study, Charalampous (1977) showed that differentiation of NB cells was accompanied by changes in the amount or topography of plasma membrane proteins. He isolated "primary membranes" as well as those areas of plasma membrane that are internalized during phagocytosis of polystyrene beads. He found that the polypeptide composition, analyzed on PAGE, of primary membranes from both SP and ML cells was very similar; however, plasma membranes from ML cultures were richer in $Na^+ - K^+$-ATPase, AChE, and CAT. In contrast, the internalized membranes from ML cells have significantly lower specific activities of these enzymes when compared to those of SP cells, and the polypeptide pattern is very different. These differences between ML and SP cells in enzyme activities and in the PAGE pattern of the internalized plasma membrane clearly result from differences in the topographic distribution of the membrane components rather than from differences in the overall membrane composition.

Since GAGs also play a very important role in the mechanism of cell adhesion, Augusti-Tocco and Chiarugi (1976) have studied GAGs on the surface of ML and SP cells. They have shown that the former exhibit a higher level of heparan sulfate; however, the level of GAGs synthesized is similar in both types of culture. While SP cells release a higher amount of sulfated GAGs into the medium, ML cells retain these molecules on the surface.

Akeson and Herschman (1974a) chose a different approach for studying membrane changes; they investigated surface antigens related to the formation of processes. Such antigens were found only in the differentiated cells and were specifically adsorbed by a membrane-enriched fraction from brain. Schachner (1973), on the other hand, did not detect any difference in the reactivity of differentiated and undifferentiated cells to antisera against several murine cell surface antigens. All the foregoing results point to the fact that not only do the glycoproteins, GAGs, and antigenic properties of the cells change during differentiation but also that the membrane components are rearranged topographically (Charalampous, 1977).

2. Lectin Probe Analysis

Fluorescent lectins have been used as probes for the detection of specific carbohydrate residues on the cell surface and for direct visualization of possible topographical variations of the glycocomponents during neuronal maturation. The surface properties of NB clones grown in SP or ML (Denis-Donini *et al.,* 1978) culture of one NB–glioma hybrid line (Denis-Donini and Augusti-Tocco, unpublished results) and of primary cultures of embryonic DRG (Denis-Donini *et al.,* 1978) were compared in the hope of throwing some light on a possible correlation between modifications of the cell surface and various maturation events. The membrane changes were visualized with a battery of fluorescein isothiocyanate (FITC) lectins that differ in specificity. The lectins used were concanavalin A (Con A) which binds mainly to α-gluco- or -mannopyranoside residues, soybean agglutinin (SBA) which binds to *N*-acetylgalactosamine residues and to galactose, wheat germ agglutinin (WGA) which binds specifically to *N*-acetylglucosamine residues and unspecifically to sialic acid by electrostatic interactions, ricin agglutinin (RGA) which binds to galactose residues, and a fucose-binding protein (FBP) from *Ulex.* For a review on lectins see Lis and Sharon (1973).

 a. Neuroblastoma. The two following NB clones were studied.

 1. 41A3 which possesses both TH, a marker enzyme for adrenergic neurons, and CAT, a marker enzyme for cholinergic neurons. The cells form fibers after they attach to a substratum without any additional treatment.
 2. N18 which has low or no TH or CAT activity and forms neurites in the absence of serum in the culture medium. It was studied because it is one of the parents of the NB–glioma hybrid line.

 The undifferentiated NB cells grown in SP bind four of the lectins tested, Con A, SBA, WGA, and RGA. Single cells show the usual ring pattern of staining. The mobility of the lectin receptors is rather restricted. In fact, when the cells are incubated in the presence of lectins at 37°C, patches are formed, but very few cases of capping are observed.

 When differentiated cells grown in ML are incubated with Con A (Fig. 1A and B), the cell body is stained and not the fibers, thus showing that the NB surface is not homogeneous with respect to the exposure of glyco- or mannopyranosyl residues on the surface; the fibers apparently do not possess such receptors, or they are not available for binding. Augusti-Tocco (unpublished results) has found that cells grown in ML bind less Con A than SP cells (Fig. 2A). This difference is even more marked when the amount of bound lectin is calculated per unit area (Fig. 2B). A similar result has been reported by Rosenberg and Charalampous (1977) who also showed that the rate of association with Con A was lower in ML than in SP cells.

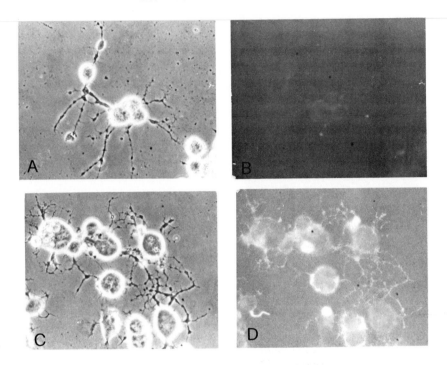

FIG. 1. Neuroblastoma cells grown in monolayer cultures. (A) Phase-contrast. (B) After staining with Con A. Only the cell bodies are stained. (C) Phase-contrast. (D) After staining with SBA. Both fibers and cell bodies are stained.

The fact that Con A receptors (at least when visualized with FITC lectins) are present only on the cell body and are absent or scarce on the fibers is reminiscent of the distribution of ACh receptors on the cell surface (Harris and Dennis, 1970). Therefore, competition experiments were carried out either with an excess of ACh ($5 \times 10^{-3}\,M$) or with bungarotoxin (10 μg/ml). Pretreatment of the cells with this drug, which binds to the ACh receptors, or with ACh did not cause any change in the lectin-binding pattern.

In the presence of SBA (Fig. 1C and D), WGA, and RCA, both fibers and cell body stain intensely. Staining is never observed with FBP, although fucosylated glycoproteins have been described on the cell surface of some NB clones (Glick et al., 1973; Truding et al., 1974; Hudson and Johnson, 1977).

When NB cells attach to a substratum, they form many microtubules and, since the membrane-associated cytoskeletal system seems to be involved in the mechanism of receptor mobility, the abil-

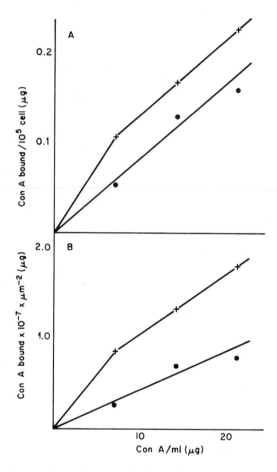

FIG. 2. (A) Abscissa: Amount of Con A bound per cell; ordinate: Con A concentration in the medium. ●, ML cells; +, SP cells. (B) Abscissa: amount of Con A bound per unit area; ordinate: Con A concentration in the medium. ●, ML cells; +, SP cells.

ity of the cells to reorganize their surface receptors has been investigated. When cells grown in ML are incubated with lectins at 37°C, caps are never observed, only patches are formed. Pretreatment of SP or ML cells with colchicine, cytochalasin, or trypsin does not favor cap formation. Although we know that microfilaments and microtubules are involved, either for anchorage of glycoproteins (Nicolson, 1976) or for directing the flow or membrane vesicles (Bretscher, 1976), it is likely that other factors are also involved in controlling the mobility of the glycoproteins in the plane of the membrane. It is also possible that the rigidity or the restricted mobility of glycoproteins in the membrane is imposed by glycans, or that the lectin receptors are themselves glycans (Buonassisi and Colburn, 1977). There is still much controversy over the concept and measurement of membrane fluidity or microviscosity, and the mobility of receptors in the plane of the membrane; the factors controlling or affecting them are still very obscure. In NB, a more rigid configuration of the cells in ML has been suggested by studies on the binding of fluorescent probes (ANS) (Erkell, 1977; Kawasaki *et al.*, 1978). Different results were obtained by spin-label studies (Struve *et al.*, 1977). However, in this case, the cells were induced to differentiate in serumless medium which by itself could cause an increase in fluidity. The lateral diffusion coefficient of Con A has been reported to be smaller in round than in differentiated cells. However, this measure may not at all be related to the movement of receptors resulting in capping (Zagyansky *et al.*, 1977). In fact, Con A receptors of malignant fibroblasts that collected into caps did not show any detectable lateral diffusion (Shinitzky and Inbar, 1976).

Although all the data point to a greater membrane rigidity in differentiated cells, the physiological significance of the reported changes in membrane microviscosity and fluidity accompanying the formation of neurite remains unclear.

b. Dorsal Root Ganglia. A parallel study was undertaken on one of the neural crest derivatives (DRG) at various developmental stages in order to establish whether the specific distribution of lectin receptors observed in NB could also be found at a defined developmental stage. Progressive and time-specific changes in the neuron surface components were observed. In fact, DRG neurons from 5- and 6-day chick embryos did not stain or stained very weakly with Con A, SBA, and WGA. In neurons from 7-day embryos, the Con A and SBA binding patterns were similar to those observed in ML NB cells. The cell bodies were stained with Con A, whereas the fibers were not (Fig. 3A and B). With SBA and WGA, an intense fluorescence was observed both on the cell body and on the fibers (Fig. 3C and D). In neurons from 8-, 9-, and 12-day embryos both cell bodies and fibers were labeled with Con A (Fig. 4A and B). With SBA and WGA the binding pattern was unchanged (Fig. 4C and D).

The absence of capping in this case also shows that the receptor mobility is very restricted, as in the case of NB cells. In mature cerebellar neurons, the restricted mobility of Con A receptors on the postjunctional membrane has been related to the very specific function assigned to this area (Kelly *et al.*, 1976). The restricted mobility and the specific display of receptors in the nervous system seem to be requisites for its specific function. Although one would expect that

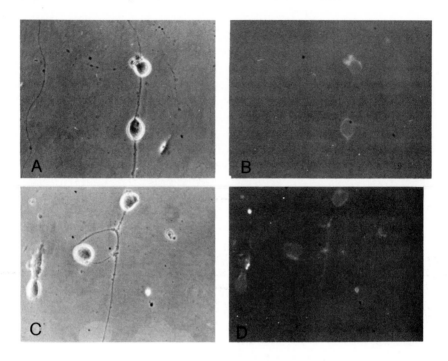

Fig. 3. DRG neurons from 7-day chick embryos. (A) Phase-contrast. (B) After staining with Con A. Only the cell bodies are stained. (C) Phase-contrast. (D) After staining with SBA. Both fibers and cell bodies are stained.

NB cells, like most malignant cells, present a high mobility of receptors, the restricted mobility observed both in NB and DRG indicates that the membrane of cells forming fibers are more "normal" (as also suggested by Glick *et al.*, 1973) and have the required rigidity of functioning neurons. The fact that the type of binding found in NB is also found in DRG neurons at a specific stage of development and the fact that this pattern evolves during maturation suggest that NB represents a definite step in neuronal maturation and that the exposure of certain carbohydrate residues on the cell surface is correlated with neuronal maturation. These observations could give some indication concerning the specific neuronal function acquired at this stage. In fact, at this stage (between days 7 and 8), when the fibers become labeled with Con A, at least two developmental events are known to occur and might involve changes in the neuron surface architecture (Pannese, 1975). The nerve fibers establish functional connections with their target cells, and the number of satellite cells in the ganglia increases rapidly, resulting in considerable change in the environment

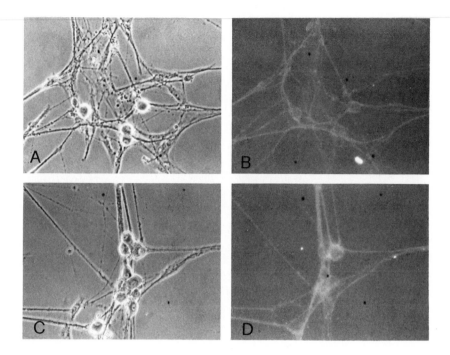

FIG. 4. DRG neurons from 8-day chick embryos. (A) Phase-contrast. (B) After staining with Con A. Both fibers and cell bodies are stained. (C) Phase-contrast. (D) After staining with SBA.

and in cellular interactions. A striking similarity is observed between the cellular environment of the 6- to 7-day DRG neurons and NB cultures where the neurons are not surrounded by glial cells and are functionally isolated. These facts suggest a correlation between the appearance of specific lectin receptors on the fibers and the formation of functional connections.

c. *Hybrid Line NB–Glioma.* Somatic cell hybridization has been used to investigate the expression of genes in various cell types. In the case of NB, cell fusion is an exceptionally valuable method for producing cell lines with interesting new combinations of neuronal properties. Of particular interest in this respect is a hybrid line obtained by fusing NB and glioma cells (Hamprecht, 1974, 1977; Minna *et al.*, 1974). The characteristics of the parental lines and of the hybrid are set out in Table III.

The lectin-binding properties of the hybrid line have shown new interesting features when compared to those of the parental line. The N18 line presents the same pattern as described previously for 41A3.

TABLE III

CHARACTERISTICS OF PARENTAL LINES AND HYBRIDS

	$N_{18}Tg_2$ NB	C_CBU_1 glioma	Hybrid
CAT	Low	−	+
DBH	Low	−	+
Respond to electrical stimulation and ACh	+	−	+
Synapse formation	−	−	+

The glioma cells bind Con A and WGA but not SBA, and their processes show very little binding. The Con A binding pattern of both parental lines is not modified by dBcAMP or when the cells are grown in mixed cultures. The Con A binding pattern of the hybrid is the same as that of the parental lines (Fig. 5A and B), the only difference being that in the presence of dBcAMP the fibers become lightly labeled (Fig. 5C and D). The Con A binding is further modified when the cells form synapses with myotubes: The fibers become definitely labeled with Con A, as in older DRG cells (Fig. 5E and F). This modification of Con A binding on the hybrid cells cultured under such conditions as to allow and promote synapse formation supports the idea that the Con A receptors on the fibers are related to the formation of functional contacts. It also emphasizes the importance of neuron–glia interactions for the proper functioning of neurons.

III. Concluding Remarks

Over the last 10 years, NB cultures have been widely used as a model system in approaching a number of problems related to developmental neurobiology. The possibility of culturing undifferentiated and differentiated cells has allowed the study of various molecular events occurring during differentiation.

Differentiated NB cells are often characterized by neurites, a higher RNA and protein content, and a higher activity of neurospecific enzymes. All these molecular events are regulated at a posttranscriptional level, but the mechanisms have not been clarified. However, all the evidence suggests that cAMP plays a key role in such regulation.

The importance of the cell surface in neuronal development and in particular in the process of fiber outgrowth, as originally suggested by Weiss (1958), emerges from the data reviewed in Section II. One

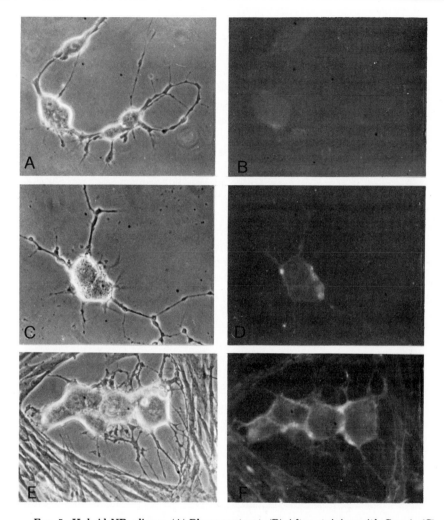

FIG. 5. Hybrid NB–glioma. (A) Phase-contrast. (B) After staining with Con A. (C) Phase-contrast of cells grown in the presence of dBcAMP. (D) After staining with Con A staining of the fibers is already more intense than in Fig. 1B. (E) Phase-contrast of cells cocultured with myoblasts in the presence of dBcAMP. (F) After staining with Con A. Both fibers and cell bodies are stained.

may envisage that the sequence of events in the course of neuronal maturation would be the following: an external change in the environment produces a change in the cell surface architecture, resulting in the activation of adenylate cyclase and therefore in an increase in the intracellular cAMP level. This increase in cAMP induces global physiological changes including the induction of neurospecific enzymes. In this context, study of the cell surface appears to be of major interest. During development, examples of changes in the surface carbohydrate-containing molecules are numerous, and glycocomponents have been implicated in many processes including specific cell interaction and recognition. In the developing nervous system, cases of transient lectin binding have been reported. Unfortunately, lectins clearly illustrate surface changes but do not provide information concerning the chemical nature or the biological roles of the lectin receptors, and one finds oneself in the realm of speculation. In the developing cerebellum, only growing or newly formed parallel fibers bind Con A. As they mature and establish synaptic connections, the binding disappears (Zanetta et al., 1978). In this case lectin receptors seem to be associated with membrane growth, as was also shown by Pfenninger (1977). Undifferentiated neural crest cells bind both Con A and WGA but, as they differentiate, only adrenergic cells and not melanocytes bind SBA at the time catecholamines are detected (Sieber-Blum and Cohen, 1978). This could be an example of surface changes closely correlated with the commitment to a developmental pathway. On the other hand, Con A receptors have been reported in synaptic clefts (Bittiger and Schnebli, 1971; Cotman and Taylor, 1974; Matus et al., 1973) where they could act as "stabilizers" of the synapses.

The comparative study of stable clones of NB, of DRG primary cultures at various developmental stages, and of the hybrid line NB-glioma has led to the assumption that Con A receptors on the fibers are related to synaptogenesis, and their role could be ascribed to any of the following processes: recognition, stabilization of an established connection, or signals that the neuron is already functionally coupled.

In view of the origin of the lectin receptors, it is still not clear whether they are inserted de novo in the membrane or whether they result from conformational changes or from the addition of carbohydrate residues. If it is assumed that they arise by glycosylation, this process could either occur in the intramembrane system (Golgi or endoplasmic reticulum) or directly on the surface through the ectoglycosyl transferase system (Shur and Roth, 1975). Since the lectin receptors appear under conditions where the neurons interact with other cells (glial and muscle cells in the case of DRG, or a mixed interaction with glia and myoblasts in the case of the hybrid), this suggests that ectoglycosyl transferases are in fact involved.

As far as the chemical nature of the receptors is concerned, among the cell coat components, glycans cannot be disregarded as possible candidates for binding Con A, since Con A binds not only to glycoproteins but also to glycans (Buonassisi and Colburn, 1977).

In conclusion, the use of lectins by themselves certainly does not allow these questions to be answered, and a combination of different experimental approaches is required.

REFERENCES

Akeson, R., and Herschman, H. (1974a). *Nature (London)* 249, 620–623.
Akeson, R., and Herschman, H. R. (1974b). *Proc. Natl. Acad. Sci. U.S.A.* 71, 187–191.
Amano, T., Richelson, E., and Nirenberg, M. (1972). *Proc. Natl. Acad. Sci. U.S.A.* 69, 258–263.
Anagnoste, B., Freedman, L. S., Goldstein, M., Broome, J., and Fuxe, K. (1972). *Proc. Natl. Acad. Sci. U.S.A.* 69, 1882–1886.
Augusti-Tocco, G., and Chiarugi, V. (1976). *Cell Differ.* 5, 161–170.
Augusti-Tocco, G., and Sato, G. (1969). *Proc. Natl. Acad. Sci. U.S.A.* 64, 311–315.
Augusti-Tocco, G., Sato, G., Claude, P., and Potter, D. D. (1970). *Symp. Int. Soc. Cell Biol.* 9, 109–120.
Augusti-Tocco, G., Casola, L., and Grasso, A. (1973a). *Cell Differ.* 2, 157–161.
Augusti-Tocco, G., Parisi, E., Zucco, F., Casola, L., and Romano, M. (1973b). *In* "Tissue Culture of the Nervous System" (G. Sato, ed.), pp. 87–106. Plenum, New York.
Augusti-Tocco, G., Casola, L., and Romano, M. (1974). *Cell Differ.* 3, 313–320.
Bittiger, H., and Schnebli, H. P. (1971). *Nature (London)* 249, 370–371.
Blume, A., Gilbert, F., Wilson, S., Farber, J., Rosenberg, R., and Nirenberg, M. W. (1970). *Proc. Natl. Acad. Sci. U.S.A.* 67, 786–792.
Breakefield, X. O. (1976). *Life Sci.* 18, 267–278.
Bretscher, M. S. (1976). *Nature (London)* 260, 21–22.
Brown, J. C. (1971). *Exp. Cell Res.* 69, 440–442.
Buonassisi, V., and Colburn, P. (1977). *Arch. Biochem. Biophys.* 183, 399–407.
Casola, L., Romano, M., Di Matteo, G., Augusti-Tocco, G., and Estenoz, M. (1974). *Dev. Biol.* 41, 371–379.
Charalampous, F. S. (1977). *Arch. Biochem. Biophys.* 181, 103–116.
Cotman, C. W., and Taylor, D. (1974). *J. Cell Biol.* 62, 231–242.
Croizat, B., Berthelot, F., Felsani, A., and Gros, F. (1977). *Eur. J. Biochem.* 74, 405–412.
Denis-Donini, S., Estenoz, M., and Augusti-Tocco, G. (1978). *Cell Differ.* 7, 193–201.
Erkel, L. J. (1977). *FEBS Lett.* 77, 187–190.
Furmanski, P., and Lubin, M. (1973). *In* "The Role of Cyclic Nucleotides in Carcinogenesis." Miami Winter Symposia, Vol. 6, pp. 239–261. Academic Press, New York.
Furmanski, P., Silverman, O. J., and Lubin, M. (1971). *Nature (London)* 233, 413–415.
Furshpan, E. J., MacLeish, P. R., O'Lague, P. H., and Potter, D. D. (1976). *Proc. Natl. Acad. Sci. U.S.A.* 73, 4225–4229.
Garvican, J. H., and Brown, G. L. (1977). *Eur. J. Biochem.* 76, 251–261.
Gilman, A. G., and Nirenberg, M. (1971). *Nature (London)* 234, 356–357.
Glick, M. C., Kimhi, Y., and Littauer, U. Z. (1973). *Proc. Natl. Acad. Sci. U.S.A.* 70, 1682–1687.
Haffke, S. C., and Seeds, N. W. (1975). *Life Sci.* 16, 1649–1658.
Hamprecht, B. (1974). *In* "Biochemistry of Sensory Functions" (L. Jacnicke, ed.), pp. 391–422. Springer-Verlag, Berlin and New York.

Hampbrecht, B. (1977). *Int. Rev. Cytol.* 49, 99-170.

Hamprecht, B., Traber, J., and Lamprecht, F. (1974). *FEBS Lett.* 42, 221-226.

Harkins, J., Arsenault, M., Schlesinger, K., and Kates, J. (1972). *Proc. Natl. Acad. Sci. U.S.A.* 69, 3161-3164.

Harris, J., and Dennis, M. J. (1970). *Science* 167, 1253-1255.

Holtzer, H. (1970). *In* "Control Mechanisms in the Expression of Cellular Phenotypes" (H. Padykula, ed.), pp. 68-88. Academic Press, New York.

Hudson, J. E., and Johnson, T. C. (1977). *Biochim. Biophys. Acta* 497, 567-577.

Judes, C., Sensenbrenner, M., Jacob, M., and Mandel, P. (1973). *Brain Res.* 51, 241-251.

Kates, J. R., Winterton, R., and Schlessinger, K. (1971). *Nature (London)* 229, 345-347.

Kawasaki, Y., Wakayama, N., Koike, T., Kawai, M., and Amano, T. (1978). *Biochim. Biophys. Acta* 509, 440-449.

Kelly, P., Cotman, C. W., Gentry, C., and Nicolson, G. L. (1976). *J. Cell Biol.* 71, 487-496.

Kimhi, Y., Palfrey, C., Spector, I., Barak, Y., and Littauer, U. Z. (1976). *Proc. Natl. Acad. Sci. U.S.A.* 73, 462-466.

Klebe, R. J., and Ruddle, F. H. (1969). *J. Cell Biol.* 43, 69a.

Knapp, S., and Mandell, A. J. (1974). *Brain Res.* 66, 547-551.

Koike, T. (1978). *Biochim. Biophys. Acta* 509, 429-439.

Landis, S. C. (1976). *Proc. Natl. Acad. Sci. U.S.A.* 73, 4220-4224.

Letourneau, P. C. (1975). *Dev. Biol.* 44, 77-91.

Lis, H., and Sharon, N. (1973). *Annu. Rev. Biochem.* 42, 541-574.

Lyman, G. H., Preisler, H. D., and Papahadjopoulos, D. (1976). *Nature (London)* 262, 360-363.

McMorris, F. A., Koller, A. R., Moore, B. M., and Perumal, A. (1974). *J. Cell. Physiol.* 84, 473-480.

Martin, S. E. (1974). *Nature (London)* 249, 71-73.

Matus, A., De Petris, S., and Raff, M. C. (1973). *Nature (London) New Biol.* 244, 278-279.

Miller, C. A., and Levine, E. M. (1972). *Science* 177, 799-802.

Minna, J., Glazer, D., and Nirenberg, M. (1974). *Nature (London) New Biol.* 235, 225-231.

Morgan, J. L., and Seeds, N. (1975). *J. Cell Biol.* 67, 136-145.

Nelson, P. G., and Peacock, J. H. (1973). *J. Gen. Physiol.* 62, 25-36.

Nelson, P., Ruffner, W., and Nirenberg, M. (1969). *Proc. Natl. Acad. Sci. U.S.A.* 64, 1004-1010.

Nelson, P. G., Peacock, J. H., and Amano, T. (1970a). *J. Cell Physiol.* 77, 353-362.

Nelson, P. G., Peacock, J. H., Amano, T., and Minna, J. (1970b). *J. Cell. Physiol.* 77, 337-352.

Nelson, P. G., Peacock, J. H., and Amano, T. (1971). *J. Cell. Physiol.* 77, 353-362.

Nelson, P., Christian, C., and Nirenberg, M. (1976). *Proc. Natl. Acad. Sci. U.S.A.* 73, 123-127.

Nicolson, G. L. (1976). *Biochim. Biophys. Acta* 457, 57-108.

Pannese, E. (1975). *Adv. Embryol. Cell Biol.* 47, 6-79.

Patterson, P. H., and Chun, L. L. Y. (1974). *Proc. Natl. Acad. Sci. U.S.A.* 71, 3607-3610.

Patterson, P. H., and Chun, L. L. Y. (1977). *Dev. Biol.* 56, 263-280.

Pfenninger, K. H. (1977). *In* "Neuronal Information Transfer," P and S Biomedical Sciences Symposium Series (H. Vogel, A. Karlin and V. Tennyson, eds.). Academic Press, New York.

Prasad, K. N. (1972). *Nature (London) New Biol.* 236, 49-52.

Prasad, K. N. (1973). *Int. J. Cancer* 12, 631-635.

Prasad, K. N. (1975). *Biol. Rev.* **50**, 129-265.
Prasad, K. N., and Hsie, A. W. (1971). *Nature (London) New Biol.* **233**, 141-142.
Prasad, K. N., and Sheppard, J. R. (1972). *Exp. Cell Res.* **73**, 436-440.
Prasad, K. N., Waymire, J. C., and Weiner, N. (1972). *Exp. Cell Res.* **74**, 110-116.
Prasad, K. N., Gilmer, K., and Kumar, S. (1973a). *Proc. Soc. Exp. Biol. Med.* **143**, 1168-1171.
Prasad, K. N., Kumar, S., Gilmer, K., and Vernadakis, A. (1973b). *Biochem. Biophys. Res. Commun.* **50**, 973-977.
Prasad, K. N., Mandal, B., Waymire, J. C., Lees, G. J., Vernadakis, A., and Weiner, N. (1973c). *Nature (London) New Biol.* **241**, 117-119.
Prasad, K. N., Bondy, S. C., and Purdy, I. L. (1975). *Exp. Cell Res.* **94**, 88-94.
Prashad, N., Wischmeyer, B., Evetts, C., Basin, F., and Rosenberg, R. (1977). *Cell Differ.* **6**, 147-157.
Puro, D. G., and Nirenberg, M. (1976). *Proc. Natl. Acad. Sci. U.S.A.* **73**, 3544-3548.
Rebhun, L. I., (1977). *Int. Rev. Cytol.* **49**, 1-47.
Richelson, E. (1973). *Nature (London) New Biol.* **242**, 175-177.
Rosenberg, S. B., and Charalampous, F. C. (1977). *Arch. Biochem. Biophys.* **181**, 117-127.
Ross, J., Olmsted, J. B., and Rosenbaum, J. (175). *Tissue Cell* **7**, 107-136.
Schachner, M. (1973). *Nature (London)* New Biol. **243**, 117-119.
Schmitt, H. (1976). *Brain Res.* **115**, 165-173.
Schubert, D., and Jacob, F. (1970). *Proc. Natl. Acad. Sci. U.S.A.* **67**, 247-254.
Schubert, D., Humphreys, S., Baroni, C., and Cohn, M. (1969). *Proc. Natl. Acad. Sci. U.S.A.* **64**, 316-323.
Schubert, D., Humphreys, S., De Vitry, F., and Jacob, F. (1971a). *Dev. Biol.* **25**, 514-546.
Schubert, D., Tarikas, H., Harris, A. J., and Heinemann, S. (1971b). *Nature (London) New Biol.* **233**, 79-80.
Seeds, N. W., and Maccioni, R.B. (1978). *J. Cell Biol.* **76**, 547-555.
Seeds, N., Gilman, A. G., Amano, T., and Nirenberg, M. W. (1970). *Proc. Natl. Acad. Sci. U.S.A.* **66**, 160-167.
Sheppard, J. R., and Prasad, K. N. (1973). *Life Sci.* **12**, 431-439.
Sheppard, J. R., and Prescott, D. M. (1972). *Exp. Cell Res.* **75**, 293-296.
Shinitzky, M., and Inbar, M. (1976). *Biochim. Biophys. Acta* **433**, 133-149.
Shur, B. D., and Roth, S. (1975). *Biochim. Biophys. Acta* **415**, 473-512.
Sieber-Blum, M., and Cohen, A. (1978). *J. Cell Biol.* **76**, 628-638.
Simantov, R., and Sachs, L. (1972). *Eur. J. Biochem.* **30**, 123-129.
Simantov, R., and Sachs, L. (1973). *Proc. Natl. Acad. Sci. U.S.A.* **70**, 2902-2905.
Simantov, R., and Sachs, L. (1975). *Eur. J. Biochem.* **55**, 9-14.
Struve, W. G., and Arneson, R. M., Cheveney, J. E., and Cartwright, C. K. (1977). *Exp. Cell Res.* **109**, 381-387.
Toole, B. P. (1976). *In* "Neuronal Recognition" (S. H. Barondes, ed.), pp. 275-329. Chapman & Hall, London.
Truding, R., Schelanski, M. L., Daniels, M. P., and Morell, P. (1974). *J. Biol. Chem.* **249**, 3973-3982.
Truding, R., Shelanski, M. L., and Morell, P. (1975). *J. Biol. Chem.* **250**, 9348-9354.
Vogel, Z., Shytkowski, A. J., and Nirenberg, M. W. (1972). *Proc. Natl. Acad. Sci. U.S.A.* **69**, 3180-3184.
Walicke, P. A., Campenot, R. B., and Patterson, P. H. (1977). *Proc. Natl. Acad. Sci. U.S.A.* **74**, 5767-5771.
Waymire, J. C., Weiner, N., and Prasad, K. N. (1972). *Proc. Natl. Acad. Sci. U.S.A.* **69**, 2241-2246.

Weiss, P. (1958). *Int. Rev. Cytol.* **7**, 391.

Zagyansky, Y., Benda, P., and Bisconte, J. C. (1977). *FEBS Lett.* **77**, 206–208.

Zanetta, J. P., Roussell, G., Ghandour, M. S., Vicendon, G., and Gombos, G. (1978). *Brain Res.* **142**, 301–319.

Zucco, F., Persico, M., Felsani, A., Metafora, S., and Augusti-Tocco, G. (1975). *Proc. Natl. Acad. Sci. U.S.A.* **72**, 2289–2293.

CHAPTER 11

CELLULAR METAPLASIA OR TRANSDIFFERENTIATON AS A MODEL FOR RETINAL CELL DIFFERENTIATION

T. S. Okada

INSTITUTE OF BIOPHYSICS
FACULTY OF SCIENCE
UNIVERSITY OF KYOTO
KYOTO, JAPAN

I. Introduction

In modern theory of cell differentiation, we have presumed the maintenance of a complete set of genomes in specialized somatic cells (e.g., Ebert and Sussex, 1970; Gurdon, 1977). The development of a complete plant from *in vitro* cultured cells using differentiated tissues can be considered one of the classic examples for establishing the basis of this theory (e.g., Steward *et al.*, 1966). In vertebrate animals, however, a sudden change in the differentiation repertoire of once-specialized cells, or the reprogramming of differentiation into a new pathway, is extremely rare. In this respect, studies on vertebrate eye

349

CURRENT TOPICS IN
DEVELOPMENTAL BIOLOGY, Vol. 16

tissues have provided a unique approach to the problem of programming of differentiation and stability in cell differentiation of animal cells, since specialized eye tissues are often reprogrammed into new pathways of differentiation.

The most well known example of reprogramming of differentiation is that of Wolffian lens regeneration in certain urodelen amphibians (Yamada, 1977), certain fishes (Sato, 1961) and, probably, in very early chick embryos (van Deth, 1940). There is little doubt that the reprogramming of progenitor cells of once-differentiated iris pigment cells into lens cells takes place in this process. Only recently has the reprogramming of cell types been unequivocally demonstrated under well-controlled *in vitro* culture conditions. Particularly, the successful introduction of an *in vitro* cell culture technique starting from singly dissociated cells of a given tissue, not tissue or organ fragments, has revealed that pigmented cells of the iris and retina as well as cells from the neural retina (NR) have a unique ability to alter extensively their state of differentiation. This ability is found not only in amphibian but also in avian and mammalian cells.

During early development of vertebrates, the retina develops from the optic cup which is formed by the invagination of the tip of the optic vesicles. The inner wall of the optic cup becomes the NR, while the outer wall develops into the pigmented retina (tapetum). In this chapter, emphasis will be placed on the unique properties of the NR and its instability during differentiation, while some attention will also be given to the pigmented retina. Although the aim of this chapter is to review the cellular aspects as observed in *in vitro* studies, reprogramming of differentiation in retinal regeneration *in situ* will also be briefly considered. For a discussion of classic Wolffian lens regeneration, readers should refer to recent extensive reviews (e.g., Scheib, 1965; Reyer, 1977; Yamada, 1976, 1977).

Alteration of the specificity of differentiated tissues or reprogramming of differentiation is referred to as metaplasia. Since metaplasia generally denotes changes occurring at the tissue level, it does not necessarily presume an alteration or a switch of differentiated cells into new pathways. Metaplasia may be equally induced by a sudden and selective increase in minor cell fractions with cellular specificity different from those that are predominant in a given tissue. In discussing cellular events, "cellular metaplasia" may be used to denote a switch of differentiated traits of once-specialized cells into new ones. Cell type conversion is also synonymous with cellular metaplasia (Yamada, 1977). In this article, "transdifferentiation" is often used to denote an alteration of the state of differentiation of cells that have already been, at least partially, specialized or programmed in a given direction under normal conditions *in situ*. This term was first applied to indicate the transition of cuticle-producing cells into salt-secreting cells during metamorphosis of the adult silk moth (Selman and Kafatos, 1974). In this original example, the alteration occurs *without* cell division. In the case of eye tissue cells of vertebrates to be discussed in the present chapter, this term will be used when the expression of new and altered specificities is preceded by progenitor cell division also.

Alteration of cell differentiation pathways that have been programmed but have not expressed a phenotype is called transdetermination (Hadorn, 1966). In principle, the events of transdetermination are a prerequisite for transdifferentiation. A distinction between transdetermination and transdifferentiation is not always clear in the systems discussed below, because it is hardly possible to define the stage of differentiation when cells have been determined in a given direction but do not express any differentiative traits. Thus, in the following description, "transdifferentiation" will be used to denote not only an alteration of differentiation of terminally differentiated cells but also a switch of cell differentiation pathways that may have been determined and are partially expressing specificities to distinguish them from other cell types.

II. Cellular and Tissue Metaplasia from Retinal Pigmented Epithelium

A. IN REGENERATION *IN SITU*

Differentiation of the NR from predifferentiated retinal pigmented epithelium (RPE) is, together with Wolffian lens regeneration, one of the most reliable cases of metaplasia so far reported in regard to regeneration. Since the discovery of this phenomenon by Colucci as early as 1891, extensive studies have been performed on this subject. This ability is retained in some urodeles throughout life (see Reyer, 1977, for recent review). In some adult anurans, this type of metaplasia can be observed both by cultivation of a piece of NR in a sandwich of RPE and by transplantation of a piece of RPE into the eye cavity of tadpoles (Lopaschov and Sologub, 1972, in *Rana;* Sologub, 1977, in *Xenopus*). The capacity of RPE to show metaplasia into NR after retinectomy has been reported in fishes (Dabaghian, 1959; Sologub, 1975) and in early embryos of birds (Orts-Lloca and Geniz-

Galvez, 1960; Coulombre and Coulombre, 1965) and mammals (Stroeva, 1962), but was not confirmed in adult mammals (Mandelcorn *et al.,* 1975; Muller-Jemses *et al.,* 1975).

In the process of this type of metaplasia, RPE cells gradually lose their melanin pigment, and they start DNA synthesis and mitosis before the expression of differentiative traits as NR. Thus, a loss (complete or partial) of the original differentiative traits (dedifferentiation) and several rounds of mitosis precede the metaplasic change, as is well known in Wolffian lens regeneration from the dorsal iris (e.g., Yamada, 1977). The process of differentiation of NR from depigmented cells essentially follows the same steps as differentiation of NR in ontogeny. However, regenerating NR differentiates very fast. Coulombre and Coulombre (1965) emphasize that the regeneration of NR by metaplasia in early chick embryos is so quick that they can attain a stage of maturation appropriate to the eye of the embryos in the same stage.

A great deal of evidence has been obtained indicating that the metaplasia of RPE into NR during regeneration requires a stimulus from NR. For instance, if a complete retinectomy is performed in chick embryos, it is necessary to reintroduce a piece of NR to elicit the metaplasia (Coulombre and Coulombre, 1965). RPE of *Rana* tadpoles transplanted into the orbit without NR does not result in metaplasia, while RPE transplanted with NR shows metaplasia into NR (Lopaschov and Sologub, 1972). However, NR is not an absolute necessity in some species (for instance, in the adult salamander; Stone and Steiniz, 1957) or under some experimental conditions (see Section II,B). The nature of the retinal factor(s) is not yet known.

In some urodeles, RPE retains the capacity to regenerate the iris also when both the iris and NR are extirpated together with the lens. Then, a new lens formation occurs from the regenerated iris to complete a restitution of the whole eye (Stone, 1955; Hasegawa, 1958, 1965). It is also highly probable that RPE can switch into a lens directly, not via the iris (Sato, 1951).

B. IN CULTURE *IN VITRO*

A successful demonstration of the reprogramming capacity of RPE cells in *in vitro* culture experiments has been made recently. When a piece of RPE of early chick embryo was cultured *in vitro* and spread on a plastic culture substratum, a number of depigmented foci appeared in the explanted pieces, and they differentiated into typical NR within 10 days under culture (Y. Tsunematsu and A. J. Coulombre, unpublished; also Tsunematsu, 1975). The structural character-istics of NR formed from such a foreign source are well documented by electron microscopic observations (Fig. 1). No stimulus from NR is needed in a switch of RPE to NR *in vitro*. The capacity for reprogram-ming in retained only in very early embryos, and it tends to be lost by the seventh day of incubation (Fig. 2).

RPE cells in these early embryos engage actively in DNA syn-thesis, and they are already pigmented, though very lightly. Repro-gramming of the differentiation of postmitotic RPE cells has been shown in cell cultures using adult *Triturus*. After successfully demonstrating *in vitro* transdifferentiation of dissociated pigmented cells of the iris into lens (Eguchi *et al.*, 1974), Eguchi (1979) recently extended his cell culture studies to RPE cells. Using L–15 medium supplemented with fetal calf serum, he showed that RPE cells, after depigmentation and repeated mitosis, could transdifferentiate into lens cells within about 40 days under cell culture.

For a direct demonstration of transdifferentiation, use of the *clonal* cell culture technique has unusual potential. [Clonal cell structure as discussed here refers to cell cultures at such low cell densities as less than 1500 cells per 5.5-cm culture dish. Statistically many colonies formed in such cultures can be considered clones, each originating from a single cell. A possibility that some colonies are of multiple cell origin cannot be strictly eliminated. For further discussion readers are referred to Okada *et al.* (1979).] Only by this method can we com-pletely eliminate the possibility that the appearance of cells with new specificities may be due to a selective outgrowth of a small cell popula-tion that may have been inadvertently included in the original in-oculum. Pigmented cells are particularly advantageous in this study, since their melanin content can serve as a reliable marker for identify-ing the specificity of single living cells to be used as starting material. Clonal studies were simultaneously conducted during transdifferentia-tion studies on RPE cells of adult *Triturus* (Eguchi, 1979). About 3% of the clonally inoculated cells gave rise to clones of appreciable size by 30 days. Some clones still retained the original specificities, i.e., consisted mostly of pigmented cells. In about 25% of all clones, the differentiation of lenslike structures, which were of authentic lens nature, was observed; these structures are designated lentoid bodies (LBs) here. A few clones, with or without LBs, contained neuronal cells which were characterized by long cytoplasmic processes and stained positive with silver by the Bodian technique. The results clearly show that at least some differentiated RPE cells can transdif-ferentiate to express new specificities in their progeny.

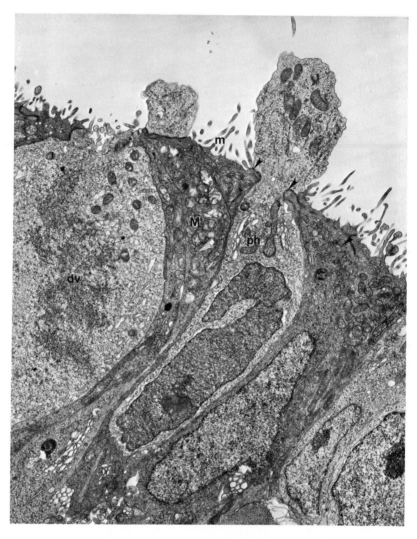

FIG. 1. Electron micrograph of cells of a NR-like focus formed in cultures of a piece of RPE from a 4-day chick embryo. M, Müller cells with microvilli (m); ph, photoreceptor cell; dv, dividing photoreceptor cell. Arrows indicate junctions of zonula adherens type. × 8640. (From Y. Tsunematsu and A. J. Coulombre, unpublished; by courtesy of the authors.)

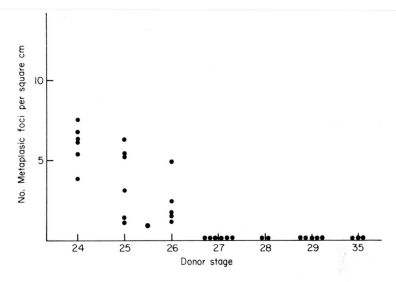

FIG. 2. Loss of the ability of chick embryonic RPE for metaplasia into NR with developmental stages (Hamburgur and Hamilton, 1951) as examined by tissue culture. The degree of the ability of metaplasia is determined by counting the number of NR-like foci formed in a unit area of the cultured piece. One point on the figure indicates the counting per single piece. (From Y. Tsunematsu and A. J. Coulombre, unpublished; by courtesy of the authors.)

Transdifferentiation of 8-day chick embryonic RPE cells into lens cells was convincingly demonstrated by means of the clonal culture technique (Eguchi and Okada, 1973). After a lag of about 100 days for depigmentation and repeated mitosis, a few clones derived from each pigment cell transdifferentiated into LBs. No indication of neuronal structure, however, was observed. Yasuda *et al.* (1978) showed that RPE cells of human fetuses could similarly transdifferentiate into LBs under cell culture.

III. Cellular and Tissue Metaplasia from Neural Retina

A. IN REGENERATION *IN SITU*

Metaplasia in NR into other eye tissues during regeneration is of very rare occurrence. In an early period of regeneration research on vertebrate eyes, Fischel (1900, 1903), using larval *Salamandra*, claimed that lenses were regenerated not only from the dorsal iris but also from NR after lentectomy. In his papers, there are several figures that indicate the formation of small lenslike structures seemingly derived from NR. To arouse the attention of modern readers, one of Fischel's original figures is reproduced in Fig. 3, in which LB-like structures are seen amid NR. A similar case suggesting lens formation from NR was also presented after a lentectomy in larval *Amblystoma* (Torö, 1932). Further studies examining the possibility of lens differentiation from NR were negative in adult *Bombina* and rats (Stroeva, 1956). In other comprehensive studies on the regeneration of NR, no indication of metaplasia of NR into other tissues has been indicated (e.g., Keefe, 1973a–d).

The earlier work by Fischel has never attracted much attention since the appearance of his papers at the beginning of this century. Interest has been focused on the studies of Wolffian lens regeneration

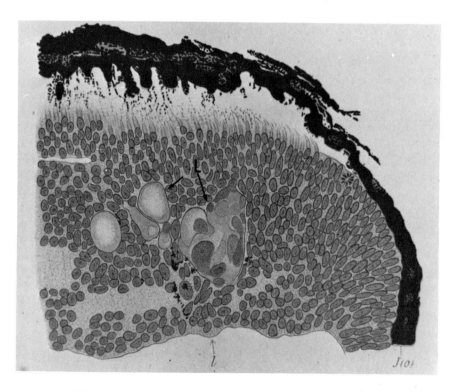

Fig. 3. Differentiation of lenslike structures (L) within NR after lentectomy of larval *Salamandra.* Photographic reproduction of the original color figure by Fischel (1903).

from the dorsal iris. It is of a great interest, however, to find the following sentence in his paper regarding the cellular origin of these lenslike structures formed in close association with NR: "So müsste man sich vorstellen, dass entweder die Nervenzelle in Stande bei, sich wieder zur indifferenzierten Epithelzelle zurückzubilden, oder dass in der Retina ausser den differenzierten Elemente noch undifferenzierte Zelle sich vorfinden." As will be discussed later, exactly the same question so clearly presented here still remains one of the most basic problems in interpreting the results of modern cell culture works concerning extensive transdifferentiation of NR cells.

B. EARLIER STUDIES *IN VITRO*

In organ culture of whole chick embryonic eye using large hanging drops in liquid nutrient medium or a plasma clot, several examples suggesting the differentiation of RPE from NR have been given (Dorris, 1938). The differentiation of pigment cells was also observed in cultures of dissociated cells of chick embryonic NR (Peck, 1964). Pigment granules appeared after short-term, 3-day culture. An important observation on the differentiation of lentoids from NR was first made in cultured reaggregates of dissociated NR cells of 3-day chick embryos (Moscona, 1957). The identification of lentoids in this study was based wholly on histological criteria. Further efforts to identify a possible lens nature in these lentoids by means of immunological techniques were not successful (Braverman *et al.*, 1969; Katoh *et al.*, 1971).

As already discussed, RPE is characterized by an unusual instability in differentiation for tissue metaplasia in the process of regeneration. Similar properties of NR, however, have only been sporadically reported as reviewed above, both *in situ* and *in vitro*. For many researchers, the possibilty of a switch of nervous tissue like NR to an altered course of differentiation appears to be very unlikely a priori. In 1975, extensive instability of the differentiation state of chick embryonic NR cells in cell culture was announced from the author's laboratory (Okada *et al.*, 1975). (The author is particularly indebted to Yoshiaki Itoh for his efforts in the first discovery of this phenomenon.) In the following part of this chapter further progress in the studies on transdifferentiation from NR cells will be reviewed as a model system for investigating the mechanisms in programming the differentiation of nervous tissue.

IV. Cellular Metaplasia or Transdifferentiation
of Neuroretinal Cells in Cell Culture

A. DISTRIBUTION OF THE ABILITY TO TRANSDIFFERENTIATE
   AMONG VERTEBRATE SPECIES

Table I enumerates all the cases of a switch of cell differentiation from RPE and NR to altered directions so far reported in culture experiments. It is worthwhile to note that in contrast to the number of reports on a switch of NR to lens and pigment cells in cell culture and a switch of RPE to lens, cases of a switch of RPE to NR, which have been widely known in metaplasia *in situ*, are still very few in *in vitro*. Another important point emerging from examination of this list is that the occurrence of transdifferentiation *in vitro* is not limited to materials that regenerate lost parts of eyes by metaplasia *in situ*. It can be speculated that retinal tissues of practically all vertebrate species may contain transdifferentiable cells. Their number depends on the species and also on the age of a given species. As mentioned below, almost all retinal cells become stable by the time of hatching in the chick (de Pomerai and Clayton, 1978; Nomura and Okada, 1979).

An example of differentiation of LBs and pigment cells from human retinoblastoma may be of particular interest (Okada, 1978). There is a special reason why we studied the ability of cells of this tumor to transdifferentiate. RPE and NR cells of human fetuses retain the ability to transdifferentiate (Okada *et al.*, 1977; Yasuda *et al.*, 1978). Since human retinoblastoma is a tumor found only in infant eyes, the initial step of alteration of the normal development of retina to the tumor is likely to occur during the fetal period. We determined whether human retinoblastoma contained any transdifferentiable cells, the presence of which characterizes the fetal retina. In cell cultures of four tumors so far investigated, typical epithelial foci of pigment cells were differentiated from elongated cells after about 1 month's culturing *in vitro*. By transferring these pigment cells into a clonal culture, the formation of LBs takes place (Fig. 4). However, the possibility of an inclusion of normal fetal cells in the tumors cannot be eliminated.

In most vertebrate embryos, both RPE and NR can be dissected out rather easily free of inadvertent inclusion of cells from other tissues (for RPE see Trinkhaus and Lentz, 1964; for NR, Okada *et al.*, 1975, and Araki and Okada, 1977). The specificities of the transdifferentiation products, i.e., pigment cells and LBs, can be convincingly identified morphologically, biochemically, and immunologically.

TABLE I

EXAMPLES OF THE TRANSDIFFERENTIATION OF RPE AND NR CELLS IN *in Vitro* CULTURE

Species[a]	Type of culture	Reference
From RPE		
To NR		
Triturus, adult	Cell culture and clonal cell culture	Eguchi (1979)
4- to 6-day CE	Tissue culture	Y. Tsunematsu and A. J. Coulombre (unpublished)
To LB		
8- to 9-day CE	Cell culture and clonal cell culture	Eguchi and Okada (1973)
12-week human fetus	Cell culture and clonal cell culture	Yasuda *et al.* (1978)
Triturus, adult	Cell culture and clonal cell culture	Eguchi (1979)
8- to 9-day CE	Cell culture	Yasuda (1979)
From NR		
To PC		
3-day C3	Organ culture	Dorris (1938)
6- to 14-day CE	Cell culture	Peck (1964)
8-day CE	Cell culture	Redfern *et al.* (1976)
To LB and PC		
3-day CE[b]	Culture of reaggregates	Moscona (1957)
Rana, neurula[b]	Organ culture	Hoperskaya (1972)
8- to 9-day CE	Cell culture	Okada *et al.* (1975)
8- to 9-day CE	Cell culture	Itoh *et al.* (1975)
8- to 9-day CE	Cell culture	Itoh (1976)
8- to 9-day CE and QE	Cell culture and clonal cell culture	Okada (1976)
8- to 9-day CE	Clonal cell culture	Okada (1977)
9- and 15-week human fetuses[b]	Cell culture	Okada *et al.* (1977)
3.5-day CE	Cell culture	Araki and Okada (1977)
8- to 9-day CE	Cell culture	Clayton *et al.* (1977)
8- to 9-day CE	Cell culture	de Pomerai *et al.* (1977)
3.5-day CE	Cell culture	Araki and Okada (1978)
8- to 9-day CE	Cell culture	Clayton (1978)
8- to 9-day CE	Cell culture	Thomson *et al.* (1978)
8- to 9-day CE	Cell culture	Jackson *et al.* (1978)
8- to 9-day CE	Cell culture	Pritchard *et al.* (1978)
6- to 17-day CE	Cell culture	de Pomerai and Clayton (1978)
Human retinoblastoma	Cell culture	Okada (1978)
3.5-day CE	Clonal cell culture	Okada *et al.* (1979)
3.5-day CE	Cell culture	Araki *et al.* (1979)
3.5- to 18-day CE	Cell culture	Nomura and Okada (1979)
16-day rat embryos	Cell culture	Y. Hamada and T. S. Okada (unpublished)

[a] CE, Chick embryos; QE, quail embryos.

[b] The differentiation of pigment cells has not been confirmed in these cases.

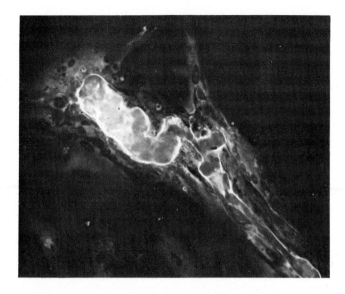

Fɪɢ. 4. Indirect immunofluorescent staining of LBs formed in cultures of cells from human retinoblastoma with anti-human crystallin serum. Approximately × 150.

These two points guarantee that the present system can serve as a very reliable model of cellular metaplasia or transdifferentiation.

B. An Outline of the Process of Transdifferentiation *In Vitro*

General features of transdifferentiating cultures of NR will be given with respect to NR from 8- to 9-day chick embryos. This material has so far been the most widely used (see Table I) because a clean separation of NR is particularly easy to make and substantial quantities are available for detailed biochemical studies.

Usually mass cultures of $1\text{--}10 \times 10^6$ dissociated NR cells are prepared in Falcon plastic dishes (5.5 cm in diameter) using Eagle's minimal essential medium (MEM) supplemented with selected batches of fetal calf serum. A number of small aggregates together with a few isolated cells adhered to the culture substratum and started to spread within 2 days after inoculation (Fig. 5). By about 10 days, the entire surface of the culture dishes was covered with a monolayer of flattened epithelial cells (E_1 cells, Okada *et al.*, 1975) upon which smaller cells (N cells) were superimposed (Fig. 6). According to the ultrastructural study, the former cell type probably

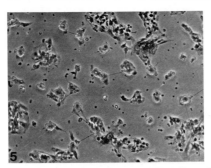

FIG. 5. Cell culture of 8-day chick embryonic NR cells 2 days after inoculation, when cells are plated on a culture substratum. Approximately × 100. (Okada *et al.,* 1975.)

represents Müller cells, while the latter may be neuroblast-like cells at different stages of differentiation (Crisanti-Combes *et al.,* 1977). Actually, a number of N cells are characterized by axon-like processes which interconnect the cells. Although many of the E_1 cells began mitosis soon after spreading, the total cell number per culture dish decreased because of the adhesion of only a limited fraction of the inoculated cells as well as a constant loss of N cells by about 10 days.

Then there was a rapid increase in cell number, perhaps due to a proliferation of E_1 cells. Many foci of small polygonal cells (E_2 cells, Okada *et al.,* 1975; or prepigment cells, Pritchard *et al.,* 1978) appeared within the sheet of E_1 cells. Many E_2 cells soon became pigmented, while a number of LB appeared all over dishes but most frequently on the peripheral edge of foci of E_2 cells (Figs. 7-9). Under

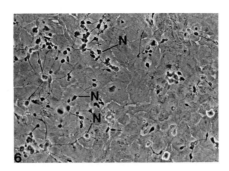

FIG. 6. Cell culture of 8-day chick embryonic NR cells 12 days after inoculation. Note the presence of a number of neuroblast-like cells with long cytoplasmic processes. N, N Cells. Approximately × 100. (Okada *et al.,* 1975.)

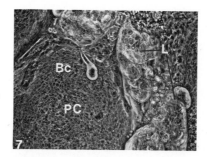

Fig. 7. Cell culture of 8-day chick embryonic NR cells 40 days after inoculation showing the differentiation of LBs (L) with bottle cells (Bc) (cf. Okada *et al.*, 1971) and the focus of pigment cells (PC). Approximately × 100. (From Okada *et al.*, 1975.)

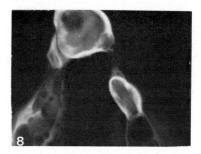

Fig. 8. Fluorescent micrograph of LBs formed in a cell culture of 8-day chick embryonic NR cells stained with FITC-conjugated anti-δ-crystallin serum. Approximately × 120.

standard culture conditions, the appearance of pigment cells and LBs can be first detected about 4 weeks after inoculation. In primary cultures at 40–50 days, more than two-thirds of the cells were lens cells and pigment cells. Transdifferentiation continued throughout the secondary and tertiary generation of cultures over a period of about 70 days.

Although the time required for transdifferentiation and the quantity of transdifferentiation products are affected by a number of factors as described below, the process as outlined above is essentially similar in most material. Somewhat basic differences are recognized in cultures started from very early embryonic NR, in which the expression of foreign phenotypes occurs very rapidly, while the number of N cells are maintained for a longer period (Araki and Okada, 1978).

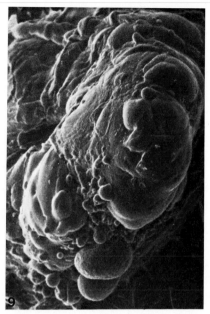

FIG. 9. Scanning electron micrograph of LBs formed in a cell culture of 8-day chick embryonic NR cells. Approximately × 600.

C. MOLECULAR EVENTS

Sequential changes in molecular constitution in the course of transdifferentiating cultures are indeed dramatic. In 52-day cultures from 6-day embryonic NR, about 60% of the total soluble proteins are crystallin (de Pomerai et al., 1977). In cultures of 3.5-day embryonic NR, the ratio of δ-crystallin synthesis to total soluble protein is as high as about 40% at 26 days (Araki et al., 1979; Fig. 10). Tyrosinase activity becomes detectable from about 4 weeks in cultures of 8-day embryonic NR, and further increase is very rapid (Itoh et al., 1975; Fig. 11).

Changes in the protein composition of cultures with time can be well visualized by means of sodium dodecyl sulfate (SDS) polyacrylamide gel electrophoresis (Fig. 12). In cultures of both 8- and 3.5-day embryonic NR, the appearance of new bands corresponding to crystallins is observed together with the disappearance of several other bands. In both SDS gel electrophoresis and immunoelectrophoresis the first appearance of detectable amounts of crystallin is

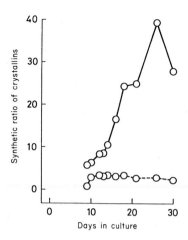

Fig. 10. Ratio of synthesis of α- and δ-crystallins to synthesis of total soluble protein at 10 different stages in the transdifferentiating cultures of 3.5-day embryonic NR cells. Cultured cells were labeled with [¹⁴C]leucine for 2 hours at a given stage. Dashed line, Radioactivity in the immunoprecipitate with anti-α-crystallin serum; solid line, radioactivity in the immunoprecipitate with anti-δ-crystallin serum. (From Araki *et al.*, 1979.)

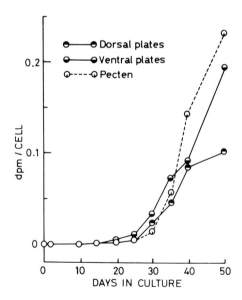

Fig. 11. The increase in tyrosinase activity in cultures of cells from three different parts (dorsal, ventral minus pectin, and pecten) of 8-day chick embryonic NR cells. The activity of tyrosinase is expressed by the radioactivity of ^3H_2O liberated from [³H]tyrosine incubated with the culture homogenates. (From Itoh *et al.*, 1975.)

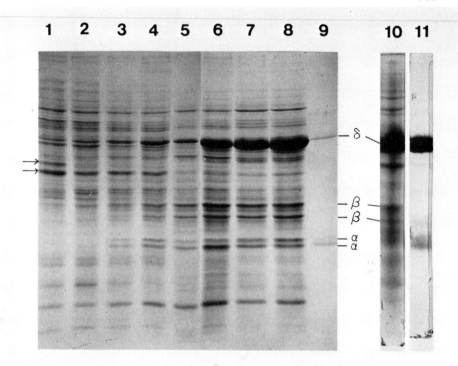

FIG. 12. Changes in patterns in SDS gel electrophoresis of the homogenates of transdifferentiating cultures of 3.5-day embryonic NR cells. (1) Six-day cultures; (2) 9-day cultures; (3) 12-day cultures; (4) 15-day cultures; (5) 18-day cultures; (6) 21-day cultures; (7) 26-day cultures; (8) 30-day cultures; (9) purified α- and δ-crystallins; (10) extract of the whole lenses of 3.5-day chick embryos; (11) homogenate of LBs from 30-day cultures. α-, β, and δ indicate the locations of bands formed by α-, β- and δ-crystallins, respectively. (From Araki et al., 1979.)

in about 12–15 days in cultures of 8-day embryonic NR and about 9 days in 3.5-day material (de Pomerai and Clayton, 1978; Araki and Okada, 1977). Both α- and δ-crystallins seem to appear simultaneously.

It has been repeatedly confirmed that LBs, terminal products of transdifferentiation, contain all three classes of crystallins, namely, α-, β-, and δ- (or γ- in mammalian cultures) crystallins. The complete molecular identity of α- and δ-crystallins in LB formed in transdifferentiating cultures with the proteins accumulated in the *in situ* lens was confirmed by affinity chromatography (Araki et al., 1979).

The appearance of crystallin molecules in transdifferentiating cultures is dependent on the massive synthesis of crystallin gene transcripts. Thomson et al. (1978) isolated total cytoplasmic mRNA from 42-day cultures of 8-day embryonic NR as well as from freshly isolated NR for cell-free translation experiments. Among the translation products, no crystallin was detected with the use of mRNA from fresh material, while substantial amounts of crystallins were present with the use of mRNA from 42-day transdifferentiating cultures (Fig. 13).

In immunofluorescence studies on transdifferentiating cultures, LBs fluoresce brightly in reaction with both anti-α- and anti-δ-crystallin sera (Fig. 8). Some nonpigmented epithelial cells localized in the periphery of LBs often react positively (Okada et al., 1975). In 26-day cultures of 3.5-day embryonic NR, there are some epithelial

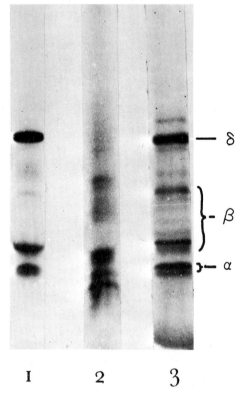

FIG. 13. Slab gel electrophoresis of *in vitro* translation products by polysomal mRNA from lenses of 1-day-old chicks (1), by total cytoplasmic mRNA from 8-day embryonic NR (2), and by total cytoplasmic mRNA from 42-day cultures of 8-day embryonic NR cells (3). Note the translation of substantial amounts of δ-, β-, and δ-crystallins by mRNA from transdifferentiating cultures. (From Thomson *et al.*, 1978.)

cells that react positively only to anti-α-crystallin (Araki *et al.*, 1979). De Pomerai *et al.* (1977) have reported that N cells in early cultures also react weakly with anticrystallin sera. Indeed, it is known that NR *in situ* contains antigens that cross-react with anti-lens sera (Clayton *et al.*, 1968; Bours and van Doorenmaalen, 1972), although it has not yet been decided whether the cross-reaction is due to the crystallin molecules or not.

Hybridization studies using cDNA prepared to the most abundant mRNA sequence of crystallins, however, indicate the presence of a similar sequence in mRNA from 8-day embryonic NR and RPE (Fig. 14). There is no detectable hybridization with mRNA from chick embryo muscle and from headless embryos (Jackson *et al.*, 1978; Clayton, 1978). How can we relate this finding to the fact that these two tissues are transdifferentiable to lens *in vitro*? The question will hopefully be solved in the near future.

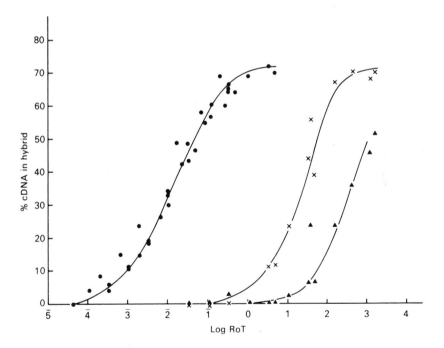

FIG. 14. Hybridization studies on cDNA to the most abundant class of 1-day-old chick lens mRNA with excess mRNA. ●, Hybridization with mRNA from 1-day-old chick lenses; ×, hybridization with mRNA from 8-day embryonic chick RPE; ●, hybridization with mRNA from 8-day embryonic NR. (From Clayton, 1978.)

Biochemical studies on neuronal cells, which are present particularly in earlier stages of transdifferentiating cultures of NR, are still few in number. Marked activity of choline acetyltransferase (CAT) has been detected in some stages of cultures of 6- to 9-day embryonic NR. The activity declines, however, probably concomitant with a loss of N cells from the cultures (Crisanti-Combes *et al.*, 1978). Monolayer cultures of 8-day chick embryonic NR can synthesize synapses and acetylcholine receptors (Vogel *et al.*, 1976). Unfortunately, in these studies, it was not shown whether transdifferentiation occurred under these culture conditions or not. Recently, Nomura *et al.* (1980) collaborated with N. Le Douarin and J. Smith in confirming the presence of some putative neurotransmitter molecules in cultures of 3.5-day embryonic NR containing a number of N cells. Sister plates of these cultures transdifferentiated into LBs and pigment cells after further culturing. It is highly probable that these neurotransmitter molecules are localized in N cells.

D. FACTORS AFFECTING TRANSDIFFERENTIATION OF NR *IN VITRO*

1. Different Parts of NR

It has long been known that Wolffian lens regeneration occurs only from the dorsal iris. Lens differentiation from pigment cells of the ventral iris has been reported only in cell culture experiments (Eguchi *et al.*, 1974; Abe and Eguchi, 1977) and after the administration of a carcinogen, N-methyl-N-nitro-N-nitrosoguanidine, to lentectomized eye (Eguchi and Watanabe, 1973). Then, it was important to determine if there was any difference in the ability to transdifferentiate *in vitro* among different parts of NR. No marked difference was noted in the dorsal quarters (D), the ventral quarters minus pecten (V), and pecten (P) of 8-day embryonic NR in this respect. At the terminal stages of primary cultures, however, some semiquantitative differences were detected in the tyrosinase activity as $P > V > D$ (Itoh *et al.*, 1975; see Fig. 11), while in LB formation, as $D > V > P$ (Okada *et al.*, 1975).

In the clonal culture of 8-day embryonic NR, colonies with LBs were formed only from the anterior part, whereas cells from the posterior part hardly gave rise to colonies of appreciable size (Table II; Okada, 1977). This result coincides well with the *in situ* observation that mitotically active cells in embryonic NR at this stage are exclusively distributed in its anterior portion (Coulombre, 1953; Fujita and Horii, 1963; Kahn, 1974).

TABLE II

FORMATION OF COLONIES WITH LBs IN CLONAL CULTURE
OF THE ANTERIOR (AR) AND POSTERIOR (PR) PARTS OF NR[a]

Material	Clonal culture in the medium	No. of cells inoculated	No. of plates examined	Plating efficiency[b]	Lentoid clones (%)[c]
AR	F	3000	3	0.62	0.10 (16.1)
		2000	2	0.70	0.13 (17.9)
	E	2000	5	0.43	0.10 (23.3)
PR	F	3000	4	0.17	0 (0)
		3000	6	0.10	0 (0)

[a] From Okada (1977).

[b] Percentage of inoculated cells that formed colonies.

[c] Colonies with LBs as a percentage of the number of total colonies; numbers in parentheses indicate colonies with LBs as a percentage of the number of inoculated cells

2. Developmental Stages

Systematic studies comparing the ability to transdifferentiate among NR isolated from different stages of donors were made by de Pomerai and Clayton (1978) with material from 6- to 17-day chick embryos, and by Nomura and Okada (1979) with material from 3-day embryos to 5-day-old chicks. NR from embryos at about 18 days of incubation is the oldest material able to transdifferentiate. Both studies clearly indicate that LBs and pigment cells develop more rapidly and in greater numbers in cultures of NR from earlier stages (Table III). The relative proportion of the three classes of crystallins differs according to donor stage. In the terminal stage of transdifferentiating cultures from 6-day NR, the δ/α crystallin ratio is 7:1, and this ratio further declines to 3:1, 2:1, and 1:1, respectively, in cultures from 8-,

TABLE III

DAYS IN CULTURE WHEN THE FIRST PIGMENT CELLS (PCs) OR LBs WERE OBSERVED[a]

Age (days)	Embryonic						3 days after hatching
	3	8.5	15	16	18	20.5	
PCs	9–13	17–25	20–25	N.T.	N.T.	N.T.	N.T.
LBs	6–12	19–30	21–27	21–27	23–27	N.T.	N.T.

[a] From Nomura and Okada (1979).

[b] About 60 culture dishes were examined for each age, and the range of days when the first PCs or LBs were observed tabulated; N.T. indicates that no transdifferentiation into PCs or LBs occurred.

12-, and 15-day material. LBs formed in cultures from earlier embryos are more similar to lens *in situ* with regard to a greater accumulation of δ-crystallin (de Pomerai and Clayton, 1978).

3. Genetic Difference

Hy-1 is a chick strain characterized by a high rate of body growth. Marked hyperplasia is recognized in the lens epithelium of this strain, while NR is morphologically normal (Clayton, 1975). NR from 8-day Hy-1 embryos transdifferentiate into LBs (de Pomerai *et al.*, 1977). The first indication of LBs is detected earlier here than in cultures of NR from normal embryos. The quantitative balance of crystallins in LBs is different between cultures of Hy-1 NR and normal NR. LB from Hy-1 NR is characterized by a high content of α-crystallin. Several differences in cell behavior in cultured cells between normal and Hy-1 NR are known (Pritchard and Clayton, 1978).

4. Culture Conditions

The differentiation of LBs and pigment cells from NR *in vitro* is much affected by the choice of culture medium and by varying medium constituents. More extensive transdifferentiation of LBs and pigment cells is always obtained with the use of Eagle's MEM than with the use of Ham's F-12 using the same batch of fetal calf serum, notwithstanding a similar rate of cellular outgrowth. The difference is particularly marked in cultures of NR from 3.5-day embryonic quail embryos (Araki and Okada, 1977). In the case of clonal cultures with low cell density, however, the expression of altered phenotypes of both LBs and pigment cells occurs in F-12 equally as well as in MEM (Okada, 1977; Okada *et al.*, 1979). Differentiation of pigment cells is promoted by the use of a medium based on Earle's salt formulation rather than Hanks' (Clayton *et al.*, 1977).

When cultures of 8-day embryonic NR are made with the use of dialyzed fetal calf serum, a monolayer of quiescent flattened cells is formed. By adding ascorbic acid to such cultures, cell growth and the expression of altered phenotypes are stimulated (Itoh, 1977). Cell density and folding of the cell sheet into multilayers are other factors that promote the development of LBs and pigment cells in transdifferentiating cultures (Clayton *et al.*, 1977). The total volume of medium and bicarbonate concentration also affect the end results of transdifferentiation (Pritchard *et al.*, 1978). Medium conditioned with cultures of NR with many N cells seems to contain factor(s) that inhibit transdifferentiation of pigment cells but not of LBs, from NR cells (Nomura

TABLE IV

CRYSTALLIN CONTENT IN SECONDARY CULTURES[a,b] of 3.5-Day Quail Embryonic NR

Medium of primary culture	Medium of secondary culture	Crystallin	Days in secondary culture	
			12-day	18-day
E	E	α	56	61
		δ	18	107
	F	α	42	39
		δ	0	15
F	E	α	54	52
		δ	50	88
	F	α	23	27
		δ	0	4

[a] From Araki and Okada (1978).

[b] Crystallin contents (μg/mg protein) are solely dependent on the medium used for secondary culture. E, Eagle's minimum essential medium; F, Ham's F-12 medium; both were supplemented with fetal calf serum.

and Okada, 1979). The nature of the effect of all the factors enumerated above is not understood at present.

It seems, however, that many of the factors listed above may not selectively stimulate an increase in a particular cell type in culture. This is clearly demonstrated by the results of the transfer of primary culture cells of quail embryonic NR. When primary cultures before the final stage of expression of transdifferentiation are transferred to the secondary generation by using a medium different from that used originally, the expression of terminal transdifferentiation is dependent only on the type of medium used for the secondary cultures (Table IV; Araki and Okada, 1977).

E. PROGENITOR CELLS FOR TRANSDIFFERENTIATION PRODUCTS

Before speculating on the mechanisms of transdifferentiation, it is critically important to know from which type of NR cells lens and pigment cells are derived. First, we have to examine the possibility that all LB and pigment cells may not be products of transdifferentiation but may be differentiated from totally unprogrammed (noncommitted) stem cells included in the NR. Naturally, the number of such stem cells, if present, must be so small that only through a highly selective outgrowth of these cells *in vitro* would extensive transdifferentiation be possible. Changes in cellular and molecular constituents in transdifferentiating cultures within a rather short culture term of about 20 days (Fig. 10) may be too dramatic to be ascribed only to the selection of a small subpopulation of stem cells. The result that the expression of "foreign" specificities is greatly influenced by a number of factors in a reversible manner is also difficult to explain only by this selection model. For the same reasons, it is difficult to assume that all the transdifferentiation products are derivatives of a small number of prelens and prepigment cells that have never been programmed toward NR characters (e.g., Clayton, 1977; Pritchard *et al.*, 1978). Then we can assume that most, though not all, of the foreign cell types formed in cultures of NR are due to transdifferentiation or cellular metaplasia from cells that have already been programmed and partially differentiated toward NR cells.

The next question to be answered is whether or not all the cells of NR are of equal potential for transdifferentiation into the directions of both lens and pigment cells. Clonal analysis is useful in order to solve this problem. When clonal cultures are made from 7- to 9-day mass cultures of 8-day chick or quail embryonic NR, only flattened and epithelial cells (E_1 cells), identified as immature Müller cells, can start clonal outgrowth. Of these colonies, *either* LB *or* pigment cells are differentiated. There is no single "mixed" clone that gives rise to both LBs and pigment cells (Okada, 1976, 1977).

The results are different when clonal cultures are started from singly dissociated cells of 3.5-day embryonic NR (Fig. 15; Okada *et al.*, 1979). The mixed clones appear together with the "monospecific" clones. It has also been noticed that, in some mixed clones and in some monospecific clones with only LBs, N cells have differentiated in their earlier stages of clonal outgrowth (Fig. 16), as schematically summarized in Fig. 17.

In 3.5-day chick embryos, NR is histologically well-identified neuroepithelium, although immature at the cellular level. There is little doubt that the cells of this epithelium have been programmed to develop into only NR *in situ* (e.g., Coulombre, 1965). The results of the development of various types of clones in clonal cultures from early embryonic NR (Fig. 17) can be hypothetically interpreted as indicating that cells of the neuroepithelium, although basically programmed, consist of a heterogeneous population with respect to the stages preceding the final step of programming. For instance, cell A in Fig. 17 must have a wide repertoire for further differentiation to take place.

In earlier stages of clonal outgrowth of 3.5-day embryonic NR, differentiation of N cells and E_1 cells occurs. The results of clonal culture of older embryonic NR suggest that the latter cell type can transdif-

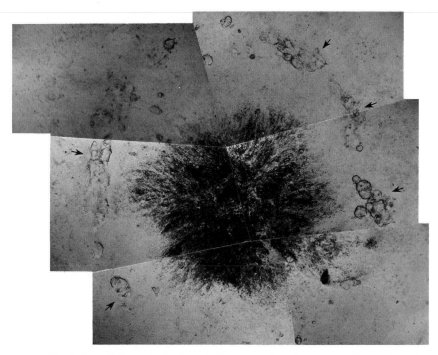

FIG. 15. A single living mixed colony 30 days after inoculation of 3.5-day embryonic NR in a clonal cell culture. Arrows indicate LB. × 16. (From Okada *et al.*, 1979.)

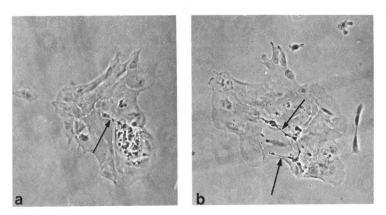

FIG. 16. Two living colonies with neuronal cells (indicated by arrows) 7 days after inoculation of 3.5-day embryonic NR into primary (a) and secondary (b) clonal cell culture. × 60. (a from Okada *et al.*, 1979.)

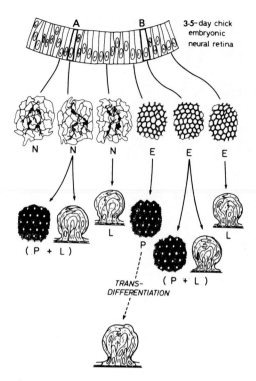

Fig. 17. Schematic representation of the multiple pathways in differentiation starting from early embryonic NR, which consists of a heterogeneous population with
respect to future potentiality in differentiation. E, "Unidentifiable" epithelial cells; L,
differentiation of LBs; N, differentiation of neuronal cells; P, differentiation of pigment
cells. Some of the pigmented cells differentiated in (P + L) can transdifferentiate
perhaps into LBs.

ferentiate into either lens or pigment cells. Are N cells also transdifferentiable? By means of time-lapse cinematographic observations of
transdifferentiating cultures of 3.5-day embryonic NR, we have
recently observed that N cells also can change directly into lens cells
(K. Yasuda and T. S. Okada, unpublished). Often, aggregates of N
cells transdifferentiate into LBs. On the other hand, there have been
no observations suggesting the transdifferentiation of pigment cells
from N cells. The whole series of cellular events leading to transdifferentiation, starting from multipotent progenitor cells of 3.5-day embryonic NR, can be hypothetically schematized as in Fig. 18.

As described earlier in this chapter, pigment cells in embryonic
RPE transdifferentiate into lens cells. Pigment cells formed in
cultures starting from NR also can achieve a secondary transdifferen-

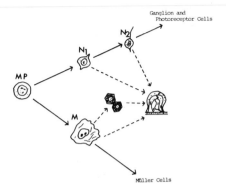

Fig. 18. Speculative scheme representing a sequence in the expression of repertoire in differentiation of multipotent progenitor cells (MP) contained in the neuroepithelium of early chick embryonic NR. Solid arrows, Differentiation to take place *in situ;* dashed arrows, transdifferentiation to take place *in vitro* cell culture. N_1 and N_2 denote two steps of differentiation toward a neuroblast; M denotes an immature Müller cell.

tiation into lens cells (Okaka *et al.,* 1979). However, conversion of cell types has never been observed in cultures of lens epithelial cells (for chick, Okada *et al.,* 1971, 1973: for mammals, van Venrooij *et al.,* 1974; Courtois *et al.,* 1978; Hamada and Okada, 1977). Therefore, we can assume the following sequence for transdifferentiation in *in vitro* propagating cell populations: NR → pigment cells → lens cells (the change from RPE to NR occurring only in very early embryos *in vitro*). In this respect, the events taking place *in vitro* reviewed here show a striking resemblance to well-known cases of transdetermination of imaginal disks of *Drosophila* (Hadorn, 1966; Kaufmann, 1974). In both cases, we are concerned with a sequential change in the expression of differentiative traits with a definite pattern, and new or altered expressions occur after a long lag during which time cells may be replicated.

F. Possible Mechanisms

The presence of a definite and stable pattern in a sequence of transdifferentiation indicates that progenitor cells have already been provided with certain reprogramming mechanisms. The latter mechanisms never operate *in situ*, at least not in most avians and mammals. Thus, we must first ask what is the "key" that opens the lock to (or derepresses) such mechanisms. During the metaplasic change of RPE to NR and of iris to lens *in situ*, the factor(s) emanating from NR is considered a rather specific inductor (e.g., Lopaschov, 1977). All cases observed in *in vitro* transdifferentiation, however, occur independently of such a NR factor(s). NR itself can alter the state of differentiation *in vitro*. As reviewed in Section IV,D, a number of factors have been known to affect the end results of transdifferentiation. Almost all these factors, however, influence transdifferentiation in a quantitative manner, but not qualitively enough to assume any one of them to be candidate for the specific inductor. Perhaps the only, but very reliable, way to initiate the sequential change of transdifferentiation is to culture the dissociated cells as a monolayer *in vitro*. Under such conditions, cell surfaces are, at least partially, directly exposed to both the culture medium and culture substratum. This may lead to an alteration of the cell's physiological properties, which may trigger the transdifferentiation. A possibility that nonspecific physiological changes, like those in the permeability of different ions, cause a change in the terminal quality of differentiation is shown in the ectodermal differentiation of amphibian gastrula (Barth and Barth, 1974).

Under conditions of cell culture, the original histological architecture is disrupted and, with further culturing, the spatial relationship between cells should be modified. Lopaschov (1977) has proposed a hypothesis that cell differentiation can be altered by removing a system of barriers existing in the tissue *in situ*. In the case of tissue metaplasia of RPE to NR *in situ*, he assumes that cell-free Bruch's membrane, which adheres to RPE, may be such a hypothetical barrier. Treatments for dissociating cells are likely to remove the barrier, and further culturing of the cells under monolayer conditions may not favor its regeneration. Thus, reprogramming mechanisms that have once been derepressed may autonomously continue to operate in order to complete the sequence of transdifferentiation.

Yasuda (1979) has presented clear evidence for the inhibition of differentiation of LBs from chick embryonic RPE cells by culturing them on collagen substratum which might serve as a barrier in *in situ* RPE. Collagen substrate does not inhibit the transdifferentiation from NR *in vitro*. Further efforts to reveal a possible inhibitor (barrier) for the alteration of differentiation of NR should be expected.

Finally, our attention should be turned to the fact that the expression of new and altered pathways of differentiation usually takes place after a long lag. What happens within this lag period? Are several rounds of mitosis in this period necessarily required for transdifferentiation, as in Wolffian lens regeneration (e.g., Yamada, 1977), in transdetermination of *Drosophila* imaginal disks (Hadorn, 1966), and in certain other systems during *normal* cell differentiation (e.g.,

Holtzer *et al.*, 1975)? Before differentiation of LBs from cultured cells of adult newt RPE is visible, it is preceded by at least eight cell cycles (Eguchi, 1979). As for other systems of transdifferentiation *in vitro*, similar information is not yet available. It is still difficult to distinguish clearly whether cell cycles are needed for the alteration of pathways in differentiation or for the mere expression of an already altered program. The relationship between cell replication and transdifferentiation still remains to be investigated.

V. Summary

Vertebrate eye tissues are characterized by a unique ability to perform extensive tissue metaplasia and to convert their specificities during regeneration *in situ*, after the removal of particular tissues. Studies on such instability in differentiation have now been extended to the cellular level by means of cell culture experiments. Besides the pigmented epithelial cells of the retina, cells from the NR of a number of vertebrate species can also extensively transdifferentiate into lens and pigment cells under conditions of cell culture. All the literature since 1975, when the phenomenon was first announced, has been reviewed. NR of early avian embryos contains multipotent cells with a potential for differentiation toward both lens and pigment cells, even after they have been partitially programmed to develop into NR cells *in situ*. Transdifferentiation *in vitro* is a sequential process starting from NR cells and terminating with lens cells, either through the differentiation of pigment cells or not. Several intrinsic and external factors affecting transdifferentiation have been described along with the molecular events observed in the process. There are a number of reasons, although they are not absolutely convincing, why transdifferentiation can be a regulatory change not mainly due to a selective outgrowth of mutated cells or a small subpopulation of cells programmed toward only lens or pigment cells. Possible mechanisms of transdifferentiation have been discussed. Cellular metaplasia or transdifferentiation of cells from NR *in vitro* certainly provides a promising experimental system as a model for studying the programming of differentiation in neural and other cells at the cellular and molecular levels.

ACKNOWLEDGMENTS

Participants in the work from my laboratory reviewed in this chapter were Drs. G. Eguchi, M. Takeichi, K. Watanabe, K. Yasuda, S. Abe, Y. Itoh, and Y. Hamada, and Messrs. M. Araki and K. Nomura (in chronological order). For their collaboration I am most grateful. I thank Dr. R. M. Clayton for many constructive discussions during the work and for giving me an opportunity to read a number of her unpublished manuscripts. Thanks are also due to Mrs. J. Kramer for correcting my English and to Miss Y. Katsurayama for her assistance in the preparation of the manuscript. The investigations in my laboratory were supported by a grant for basic cancer research from the Japan Ministry of Education, Science, and Culture.

REFERENCES

Abe, S., and Eguchi, G. (1977). *Dev. Growth Differ.* **19**, 309.
Araki, M., and Okada, T. S. (1977). *Dev. Biol.* **60**, 278.
Araki, M., and Okada, T. S. (1979). *Dev. Growth Differ.* **20**, 71.
Araki, M., Yanagida, M., and Okada, T. S. (1979). *Dev. Biol.* **68**, 170.
Barth, L. G., and Barth, L. J. (1974). *Dev. Biol.* **39**, 1.
Bours, J., and Van Doorenmaalen, W. J. (1972). *Exp. Eye Res.* **13**, 236.
Braverman, H., Cohen, C., and Katoh, A. (1969). *J. Embryol. Exp. Morphol.* **21**, 391.
Clayton, R. M. (1975). *Genet. Res.* **25**, 79.
Clayton, R. M. (1978). *In* "Stem Cells and Tissue Homeostasis" (B. I. Lord, C. S. Potten and R. J. Cole, eds.), p. 115. Cambridge Univ. Press, London and New York.
Clayton, R. M., Campbell, J. C., and Truman, D. E. S. (1968). *Exp. Eye Res.* **7**, 110.
Clayton, R. M., de Pomerai, D. I., and Pritchard, D. J. (1977). *Dev. Growth Differ.* **19**, 319.
Coulombre, A. J. (1953). *Am. J. Anat.* **96**, 153.
Coulombre, A. J. (1965). *In* "Organogenesis" (R. DeHaan and H. Ursprung, eds.). Holt, New York.
Coulombre, J. C., and Coulombre, A. J. (1965). *Dev. Biol.* **12**, 79.
Courtois, Y., Simonnean, L., Tassin, J., Laurent, M. V., and Malaise, E. (1978). *Differentiation* **10**, 23.
Crisanti-Combes, P., Privat, A., Pessac, B., and Calothy, G. (1977). *Cell Tissue Res.* **185**, 159.
Crisanti-Combes, P., Pessac, B., and Calothy, G. (1978). *Dev. Biol.* **65**, 228.
Dabaghian, N. V. (1959). *Dokl. Akad. Nauk. SSSR* **125**, 938.
Dorris, F. (1938). *J. Exp. Zool.* **78**, 385.
Ebert, J. D., and Sussex, I. (1970). "Interacting Systems in Development" (2nd ed.). Holt, New York.
Eguchi, G. (1979). *In* "Mechanisms in Cell Change" (J. D. Ebert and T. S. Okada, eds.). Wiley, New York.
Eguchi, G., and Okada, T. S. (1973). *Proc. Natl. Acad. Sci. U.S.A.* **70**, 1495.
Eguchi, G., and Watanabe, K. (1973). *J. Embryol. Exp. Morphol.* **30**, 63.
Eguchi, G., Abe, S., and Watanabe, K. (1974). *Proc. Natl. Acad. Sci. U.S.A.* **71**, 5052.
Fischel, A. (1900). *Anat. Hefte* **14**, 1.
Fischel, A. (1903). *Arch. Entwicklungsmech. Organ.* **15**, 1.
Fujita, S., and Horii, M. (1963). *Arch. Histol. Jpn.* **23**, 359.
Gurdon, J. B. (1977). *Proc. R. Soc. London B* **198**, 211.
Hadorn, E. (1966). *In* "Major Problems in Developmental Biology" (M. Locke, ed.), p. 85. Academic Press, New York.
Hamada, Y., and Okada, T. S. (1977). *Dev. Growth Differ.* **19**, 265.
Hamburger, V., and Hamilton, H. L. (1951). *J. Morphol.* **88**, 49.
Hasegawa, M. (1958). *Embryologia* **4**, 1.
Hasegawa, M. (1965). *Embryologia* **8**, 362.

Holtzer, H., Rubinstein, N., Fellini, S., Yeoh, G., Chi, J., Birnbaum, J., and Okayama, M. (1975). *Q. Rev. Biophys.* **8**, 523.
Hoperskaya, O. A. (1972). *Arch. Entwicklungsmech. Organ.* **171**, 1.
Itoh, Y. (1976). *Dev. Biol.* **54**, 157.
Itoh, Y., Okada, T. S., Ide, H., and Eguchi, G. (1975). *Dev. Growth Differ.* **17**, 39.
Jackson, S. F., Clayton, R. M., Williamson, R., Thomson, I., Truman, D. E. S., and de Pomerai, D. I. (1978). *Dev. Biol.* **65**, 383.
Kahn, A. J. (1974). *Dev. Biol.* **38**, 30.
Katoh, A., Braverman, M., and Yeh, C. (1971). *Exp. Cell Res.* **66**, 65.
Kauffman, S. (1974). *Science* **181**, 310.
Keefe, J. R. (1973a) *J. Exp. Zool.* **184**, 185.
Keefe, J. R. (1973b) *J. Exp. Zool.* **184**, 207.
Keefe, J. R. (1973c) *J. Exp. Zool.* **184**, 233.
Keefe. J. R. (1973d) *J. Exp. Zool.* **184**, 239.
Lopaschov, G. V. (1977). *Differentiation* **9**, 131.
Lopaschov, G. V., and Sologub, A. A. (1972). *J. Embryol. Exp. Morphol.* **28**, 251.
Mandelcorn, M. S., Machemer, R., Fineberg, E., and Hersh, S. B. (1975). *Am. J. Ophthalmol.* **80**, 227.
Moscona, A. A. (1957). *Science* **125**, 598.
Muller-Jensen, K., Machemer, R., and Azarina, R. (1975). *Am. J. Ophthalmol.* **80**, 530.
Nomura, K., and Okada, T. S. (1979). *Dev. Growth Differ.* **21**, 161.
Nomura, K., Takagi, S., and Okada, T. S. (1980). *Differentiation* **16**, 141.
Okada, T. S. (1976). *In* "Tests of Teratogenicity *in vitro*" (J. D. Ebert and M. Marois, eds.), p. 91. North-Holland, Amsterdam.
Okada, T. S. (1977). *Dev. Growth Differ.* **19**, 47.
Okada, T. S. (1978). *Proc. Int. Congr. Eye Res. VIth, Osaka* p. 2.
Okada, T. S., Eguchi, G., and Takeichi, M. (1971). *Dev. Growth Differ.* **13**, 323.
Okada, T. S., Eguchi, G., and Takeichi, M. (1973). *Dev. Biol.* **34**, 321.
Okada, T. S., Itoh, Y., Watanabe, K., and Eguchi, G. (1975). *Dev. Biol.* **45**, 318.
Okada, T. S., Yasuda, K., Hayashi, M., Hamada, Y., and Eguchi, G. (1977). *Dev. Biol.* **60**, 305.
Okada, T. S., Yasuda, K., Araki, M., and Eguchi, G. (1979). *Dev. Biol.* **68**, 600.
Orts-Lloca, F., and Geniz-Galvez, J. M. (1960). *Acta Anat.* **42**, 31.
Peck, D. (1964). *J. Embryol. Exp. Morphol.* **12**, 381.
de Pomerai, D. I., and Clayton, R. M. (1978). *J. Embryol. Exp. Morphol.* **48**, 179.
de Pomerai, D. I., Pritchard, D. J., and Clayton, R. M. (1977). *Dev. Biol.* **60**, 305.
Pritchard, D. J., and Clayton, R. M. (1978). *Exp. Eye Res.* **26**, 667.
Pritchard, D. J., Clayton, R. M., and de Pomerai, D. I. (1978). *J. Embryol. Exp. Morphol.* **48**, 1.
Redfern, N., Israel, P., Bergsma, D., Robison, W. G., Jr., Whikehart, D., and Chader, G. (1976). *Exp. Eye Res.* **22**, 559.
Reyer, R. W. (1977). *In* "Handbook of Sensory Physiology: The Visual System in Vertebrates" (F. Crescitelli, ed.), Vol. VII/5, p. 309.
Sato, T. (1951). *Embryologia* **1**, 21.
Sato, T. (1961). *Embryologia* **6**, 251.
Scheib, D. (1965). *Ergeb. Anat. Entwicklungsgesch.* **38**, 45.
Selman, K., and Kafatos, F. C. (1974). *Cell Differ.* **3**, 81.
Sologub, A. A. (1975). *Ontogenez* **6**, 39.
Sologub, A. A. (1977). *Arch. Entwicklungsmech. Organ.* **182**, 277.
Steward, F. C., Kent, A. E., and Mapes, M. O. (1966). *Curr. Top. in Dev. Biol.* **1**, 113.
Stone, L. S. (1955). *J. Exp. Zool.* **129**, 505.

Stone, L. S., and Steinitz, H. (1957). *J. Exp. Zool.* **135**, 301.

Stroeva, O. G. (1956). *Izv. Akad. Nauk, SSSR (Biol.).* **5**, 76.

Stroeva, O. G. (1962). *J. Embryol. Exp. Morphol.* **8**, 349.

Thomson, I., Wilkinson, C. E., Jackson, J. F., de Pomerai, D. I., Clayton, R. M., Truman, D. E. S., and Williamson, R. (1978). *Dev. Biol.* **65**, 372.

Torö, E. (1932). *Arch. Entwicklungsmech. Organ.* **126**, 185.

Trinkhaus, J. P., and Lentz, J. P. (1964). *Dev. Biol.* **9**, 115.

Tsunematsu, Y. (1975). Doctoral Thesis, University of Kyoto. Submitted.

van Deth, J. H. M. G. (1940). *Acta Neer. Morphol.* **3**, 151.

van Veroooij, W. J., Groeneveld, Ap A., Bloemendal, H., and Benedetti. (1974). *Exp. Eye Res.* **18**, 517.

Vogel, Z., Daniels, M. P., and Nirenberg, M. (1976). *Proc. Natl. Acad. Sci. U.S.A.* **73**, 2370.

Yamada, T. (1976). *In* "Progress in Differentiation Research" (M. Müller-Bérat, ed.), p. 355. North-Holland, Amsterdam.

Yamada, T. (1977). *In* "Monographs in Developmental Biology" (A. Wolsky, ed.), Vol. 13. Karger, Basel.

Yasuda, K. (1979). *Dev. Biol.* **68**, 618.

Yasuda, K., Okada, T. S., Eguchi, G., and Hayashi, M. (1978). *Exp. Eye Res.* **26**, 591.

CHAPTER 12

RNA SEQUENCE COMPLEXITY IN CENTRAL NERVOUS SYSTEM DEVELOPMENT AND PLASTICITY

Lawrence D. Grouse, Bruce K. Schrier,
Carol H. Letendre, and Phillip G. Nelson*

NEUROBIOLOGY AND INTERMEDIARY METABOLISM SECTIONS
LABORATORY OF DEVELOPMENTAL NEUROBIOLOGY
NATIONAL INSTITUTE OF CHILD HEALTH AND HUMAN DEVELOPMENT
NATIONAL INSTITUTES OF HEALTH
BETHESDA, MARYLAND

I. Introduction

Interest in gene expression of the mammalian brain arose when initial studies (Hahn and Laird, 1971; Brown and Church, 1972; Grouse *et al.*, 1972, 1973) showed an exceptionally large proportion of the nonrepetitive portion of the genome (the vast majority of all structural genes are copied from the nonrepetitive DNA) was transcribed in that organ. As much as 40% of all transcribable nonrepetitive DNA sequences were found represented as RNA copies in human (Grouse *et al.*, 1973) and in rodent brain (Bantle and Hahn, 1976; Grouse *et al.*, 1978a). Theoretically, such a great RNA complexity (the total number of nucleotides of individual RNA molecules is referred to as the complexity of the population of RNA) could code for more than 200,000 individual structural proteins. Recent data (Chikaraishi *et al.*, 1978;

* Present address: Journal of the American Medical Association, Chicago, Illinois.

CURRENT TOPICS IN
DEVELOPMENTAL BIOLOGY, Vol. 16

Kaplan *et al.*, 1978) confirm the impression of earlier work that the RNA complexity of mammalian brain is much greater than that of visceral organs. Although still not fully tested experimentally, it has been hypothesized that a significant fraction of brain RNA species is brain-specific. The discovery of the nature of these brain-specific gene products may be a key step in understanding the molecular basis of nervous function.

Several lines of research with brain RNA have developed since the initial RNA complexity studies began. It will be the purpose of this chapter to assess the present understanding of brain transcription, with reference to hybridization studies with nonrepetitive and complementary DNA. Recent advances in the understanding of RNA processing and its relation to brain RNA will be reviewed. Studies showing effects of development, aging, and sensory-environmental factors on gene expression will be discussed. The conclusions of studies on the RNAs of clonal cell lines of neural origin will be presented, and the implications of these data for characterization of the transcriptional pattern in brain will be discussed. Unlike studies focusing on the expression of individual gene products in brain or on the specific physiological properties of nerve cells, studies on the expression of nonrepetitive DNA are able to define the total transcriptional or translational response of the brain to such key neurobiological events as differentiation and synapse formation.

II. Nuclear RNA Complexity of Mammalian Brain

The primary RNA transcript is referred to as a heterogeneous nuclear RNA (HnRNA) molecule. The term arises naturally from the nuclear location and heterogeneous size observed for populations of these molecules. Their number-average size appears to be about 4000–5000 nucleotides in brain, depending upon the technique used to measure length (Bantle and Hahn, 1976). Early work (Grouse *et al.*, 1972) suggested that the entire RNA sequence complexity of brain was accounted for by HnRNA. Subsequent data (Bantle and Hahn, 1976) have documented this point. Metabolically, HnRNA molecules have unusual properties. They turn over with a half-life of only a few minutes, and 90% of all HnRNA synthesized is degraded within the nucleus (Harris, 1963). Many HnRNA molecules have polyadenylic acid [poly(A)] added posttranscriptionally to their $3'$-end (Edmonds and Caramela, 1969). The purpose of this addition is uncertain. The $5'$-end is often "capped" with methylated guanosine derivatives. This addition appears to facilitate translation of the mature mRNA species.

Evidence from a variety of experiments suggests that cytoplasmic mRNA is derived from the processing of precursor HnRNA forms. For example, globin mRNA sequences are present in HnRNA molecules several times longer than mature globin mRNA (Ross and Knecht, 1978). Ultraviolet sensitivity measurements indicate that the transcriptional units in the DNA for the sequences that become cytoplasmic poly(A)-containing RNA (RNA containing poly(A) on the 3'-end of the molecule) are two to five times larger than the average-sized mRNA (Sauerbier, 1976; Goldberg, et al., 1977), suggesting that HnRNA is the mRNA precursor (Goldberg et al., 1977). Restriction endonuclease techniques have made possible characterization of the base sequence of DNA coding for specific mRNAs (Tilghman et al., 1978; McGrogan and Raskas, 1978; Garapin et al., 1978; Konkel et al., 1978). Although such analyses have not yet been applied to brain-specific sequences, in all cases studied the primary HnRNA transcript (and the DNA gene from which it is copied) contains intervening sequences (introns) which do not contain the actual mRNA coding sequences (exons). These intervening sequences are excised from the HnRNA molecule, the coding sequences are ligated together, and the mature mRNA appears in the cytoplasm. This process appears to occur commonly, if not universally, during mRNA maturation. Maxwell et al. (1978) have found, using S_1 nuclease studies of hybrids of brain HnRNA and complementary DNA (cDNA) copied from poly(A)-mRNA by the avian myeloblastosis virus (A.M.V.) reverse transcriptase that a high percentage of brain genes may also contain these intervening sequences. Thus, HnRNA in brain appears to represent primary transcripts of brain genes, at least some of which are processed to yield mature mRNAs.

Various hybridization studies using total nuclear RNA to drive the reaction with nonrepetitive DNA have shown that adult rodent brain nuclear RNA is complementary to 15.6% (Chikaraishi et al., 1978), 16.9% (Grouse et al., 1978a), or 21.2% (Bantle and Hahn, 1976) of the nonrepetitive DNA. This amounts to 6.6–7.8×10^8 nucleotides of nuclear RNA sequence complexity (the nonrepetitive portion of rodent DNA has a complexity of 2×10^9 nucleotides). The large complexity value observed for brain nuclear RNA is particularly intriguing, because it represents a two- to four-fold greater overall genetic expression than is observed in other tissues (Chikaraishi et al., 1978). Brain poly(A)-HnRNA is a subpopulation of nuclear HnRNA. When brain total poly(A)-RNA [which contains poly(A)-HnRNA and poly(A)-mRNA] or brain nuclear poly(A)-HnRNA is used to drive the reaction with nonrepetitive DNA, 12.7% (Bantle and Hahn, 1976) or 12.3% (Kaplan et al., 1978) of the DNA probe reacts. Data to be reviewed in

Section IV suggest that the high complexity of brain RNA is a result both of the diverse cell types present in brain and of the fact that individual neurally derived cell types contain more complex RNA populations than nonneural cells. Aside from the great complexity of RNA, another property of brain HnRNA should be mentioned. The existence of separate components of brain nuclear RNA present at concentrations differing by more than 1000-fold has been demonstrated (Grouse *et al.*, 1978a). The implications of this finding will be discussed in Section IV.

Understanding the structure and role of HnRNA will be of great importance in molecular biology generally and in future neurobiological research. At the most basic level, the extent to which specific transcriptional changes are related to, and responsible for, the diverse phenomena associated with brain development and plasticity must be determined. Mechanisms by which intercellular events could lead to transcriptional changes must be explored. Recent data concerning HnRNA structure may have great relevance to the brain's information-processing capacity, as well as to the observed RNA sequence complexity. Sakano *et al.* (1979) have shown that protein functional units are coded in physically separated regions of DNA (exons), and they believe that unique proteins have been evolutionarily assembled from individual protein region "modules." Also, cellular mechanisms for somatic rearrangement of DNA sequences are known to occur (Tonegawa *et al.*, 1978). It is not known whether somatic rearrangements, as well as "split gene" RNA transcription, occur in brain cells. However, we believe that it is likely that both processes take place. The feasibility of constructing modular RNA molecules incorporating gene coding sequences widely separated on the genome could enable the central nervous system to multiply the number and variety of separate RNA products it could express, and could help to explain the great nuclear RNA complexity observed in brain.

Since a large fraction of HnRNA does not appear in the cytoplasm but is degraded in the nucleus, a nuclear regulatory function for HnRNA has been postulated (Davidson *et al.*, 1977). In this view, certain HnRNA molecules interact with the DNA to regulate gene expression. There are, at present, no data implicating this process as a regulatory step in brain. Detailed consideration of this interesting hypothesis is not possible within the scope of the present chapter.

III. mRNA Complexity of Mammalian Brain

mRNA is more easily defined and is physiologically less complex than nuclear mRNA. mRNAs are translated in the cytoplasm during the process of protein synthesis. They are physically bound to characteristic proteins (Irwin *et al.*, 1975) and may be present either actively engaged in the process of protein synthesis on polyribosomes or unassociated with polyribosomes. The current concept of non-polysomal mRNAs is that they may be blocked from translation by their association with other molecules, including certain RNA species termed "translation control RNAs" (Bester *et al.*, 1975). In the complexed state, translation is inhibited.

A fundamental characteristic of mRNA structure, important at least in terms of preparative concerns, is the presence of poly(A) on the 3'-end of the majority of mRNA molecules. This poly(A) moiety has facilitated the isolation of poly(A)-containing mRNAs by affinity chromatography on oligo-(dT) cellulose or poly(U)-Sepharose columns and, as a result, the vast majority of studies on brain mRNA have been performed on poly(A)-containing mRNA. Polyadenylation may have a stabilizing effect on mRNA species, increasing their mean survival; however, poly(A) addition does not appear to have a significant effect on the rate of messenger translation.

The sequence complexity of polyadenylated polysomal mRNA in adult rodent brain has been measured to be 3.5–3.8% of the nonrepetitive DNA (Bantle and Hahn, 1976; Grouse *et al.*, 1978a), which corresponds to 1.4×10^8 nucleotides or about 90,000 unique, average-sized mRNAs. By comparison, liver poly(A)-mRNA has been shown to contain about 24,000 different mRNAs (Savage *et al.*, 1978). It is apparent that gene expression at the mRNA level in brain is much more extensive than that observed in liver. Brain polyadenylated mRNA, like brain nuclear RNA, has been found to be composed of separate classes which differ in concentration by more than 1000-fold (Grouse *et al.*, 1978a). The details of this finding will be presented in Section IV.

Another component of brain mRNA that has only recently become accessible to accurate study is nonpolyadenylated mRNA. Nemer (1975) and Greenberg (1976) showed that the amount of non-polyadenylated mRNA in sea urchin embryos varied during development and varied in a pattern quite different from that observed for poly(A)-mRNA. This suggested that nonpolyadenylated mRNA might be a functionally separate class of mRNA species under separate cellular control from poly(A)-mRNA. This view was challenged by Kaufmann *et al.* (1977) in studies that demonstrated extensive overlapping of the *in vitro* translation of products of poly(A)- and nonpolyadenylated mRNAs in HeLa cells. However, recent work by Van Ness *et al.* (1978) on rodent brain nonpolyadenylated mRNA shows that the bulk of the nonpolyadenylated mRNAs are sequences not present in poly(A)-mRNA. In fact, nonpolyadenylated mRNA (purified from total nonpolyadenylated polysomal RNA on benzoylated cellulose) had a complexity of 1.5×10^8 nucleotides, approximately equal to that of poly(A)-mRNA and with very little overlapping. Thus, the total complexity of brain mRNA amounts to about 3.1×10^8 nucleotides. The total complexity of rodent brain nuclear RNA probably does not exceed 8×10^8 nucleotides, so almost 40% of the nuclear sequence complexity is found expressed in cytoplasmic polyribosomes. This percentage is higher than that found in other tissues (Hough *et al.*, 1975; Levy and McCarthy, 1975) and again suggests a uniqueness of gene expression in mammalian brain.

IV. RNA Sequence Compartmentation in Brain

As mentioned in earlier discussions, both nuclear RNA and poly(A)-mRNA of rodent brain have been found to be separable into several components on the basis of the kinetics of their reaction with nonrepetitive DNA (Grouse *et al.*, 1978a). RNA sequences present at high concentration react rapidly, while sequences in low abundance react more slowly. The ratio of the average concentration of the sequences in the highest-abundance class to that of the lowest-abundance class ranges from 3×10^3 [poly(A)-mRNA] to 10^4 (nuclear RNA). By considering the amount of RNA present in the brain and using the reaction parameters of the RNA components, it is possible to calculate the average number of copies of a typical sequence from each component that are present per brain and, by extension, the average number per cell (Grouse *et al.*, 1978a). These values (Table I) show that, in the rat brain, the high-abundance components of both nuclear and mRNA are comprised of sequences present at physiological levels (1–20 copies per cell for HnRNA; 5–25 copies per cell, minimum, for mRNA; Hough *et al.*, 1975; Galau *et al.*, 1977) in essentially all brain cells. In contrast, the lowest-abundance components of these RNAs are at such a low concentration that they could be present at a physiological concentration in only one out of several hundred cells. These data indicate that not all brain cells express the same genes. From such averaged calculations it is not possible to tell whether certain cell types express all brain genes and others virtually none, or whether a more equal distribution of sequence complexity among different cells exists.

From recent data of Kaplan *et al.* (1978) it is certain that such large and structurally complex brain regions as the cerebellum, cerebrum, and hypothalamus all have the same total poly(A)-RNA sequence complexity, which is also the same complexity as that of total brain. Thus, if many cell types contribute to the total complexity of brain RNA,

TABLE I

CHARACTERISTICS OF HnRNA AND POLY(A)-mRNA IN ADULT RAT BRAIN

Parameter	Abundance class			
	Frequent	Middle	Rare	Aggregate
HnRNA				
Nonrepetitive sequences hybridized (%)	1.0	4.5	11	16.5
Sequence complexity (nucleotides)	4.2×10^7	1.9×10^8	4.6×10^8	6.6×10^8
No. of HnRNA molecules in average cell	—	—	—	8.2×10^5
Average no. of molecules of each kind per brain	2.9×10^9	2.1×10^7	2.4×10^5	1.9×10^8
Average no. of molecules of each kind per cell	18	0.15	0.0015	1.2
Poly(A)-mRNA				
Nonrepetitive sequences hybridized (%)	0.5	1.5	1.2	3.2
Sequence complexity (nucleotides)	2.1×10^7	6.2×10^7	5.0×10^7	1.3×10^8
No. of mRNA molecules in average cell	—	—	—	6.45×10^4
Average no. of molecules of each kind per brain	8.9×10^8	7.8×10^6	3.2×10^5	—
Average no. of molecules of each kind per cell	5.5	0.05	0.002	0.086

they must be distributed in many or all brain regions. Data from Kaplan *et al* (1978), studying hippocampal total poly(A)-RNA, and Grouse *et al.* (1979a) studying visual cortex total RNA, raise the possibility that the complexity of RNA from certain small areas of brain may be less than the RNA complexity of total brain. This could result from the absence in these small amounts of tissue of certain cell types that express different sequences.

If it is assumed that the low-abundance RNA sequences referred to above, which are each present in only a small proportion of rodent brain cells, are distributed randomly among all brain cells, one calculates that there would be several hundred separate classes of cells containing about 50,000 cells each. Such classes of cells would be distinguished by the expression of 10–100 separate low-abundance HnRNAs and mRNAs which would not be present in any other cell class. This hypothetical situation would have certain consequences for the RNA complexity of different cells. First, a single brain cell class would have a lower complexity than total brain, since only part of the complex, low-abundance RNA would be expressed in any given class. Second, there would be extensive sequence overlap in different cell types as a result of the expression of similar (high-abundance) genes. From studies on poly(A)-mRNAs of a neuroblastoma clone (Schrier *et al.*, 1978) and a glioma cell line (Grouse *et al.*, 1979b) we have estimated that at least 50% of all mRNAs are expressed in both cell lines. Kaplan *et al.* (1978), in work on the total poly(A)-mRNA of rat neuroblastoma and glioma cell lines show an even greater amount, possibly as much as 90%, of overlapping of RNA sequences between the two cell lines. Moreover, as predicted, cell lines of neural origin have RNA complexities significantly less than that of total brain. Kaplan found the total poly(A)-RNA complexity of a neuroblastoma clone to contain RNA complementary to 8.7% of the nonrepetitive DNA. Total brain RNA reacted with 11.6% of the probe, and RNA from the C6 glioma reacted with 9.3%. Evidence that transcription in the C6 glial cell line might be different from transcription in glial cells *in vivo* is suggested by the complexity of total poly(A)-RNA from primary glial cultures (Kaplan *et al.*, 1978), which was 25% lower than that of the C6 RNA. It would be of interest if the RNA complexity of a single, physiological central nervous system cell type could be measured. Dorsal root ganglion cells might be the most accessible to study. Primary glial cell cultures (studied by Kaplan *et al.*, 1978) contain both astrocytes and oligodendroglia and may contain multiple classes of cells on the basis of gene expression, since gene expression in glial cells may be changed by the anatomical location of the cells.

Whatever the ideal single brain cell type may be, complexity measurements of the total poly(A)-RNA (Kaplan *et al.*, 1978) and poly(A)-mRNA (Schrier *et al.*, 1978; Grouse *et al.*, 1979a) of cell lines of neural origin suggest that neurally derived cell lines may have a greater RNA sequence complexity than either nonneural tissues or nonneural cell lines.

An alternative explanation of the low-abundance nuclear RNAs is that they are the result of "leaky" transcription and are not specifically regulated. For example, globin RNA sequences were found to be present in brain (Humphries *et al.*, 1976), a location where they would be supposed to be physiologically absent. However, more recent work (Knochel and Kohnert-Stavenhagen, 1977) suggests that the only globin RNA present in brain RNA arises from contamination of the brain by blood cells and that there is no evidence for expression of globin RNA in brain tissue itself.

V. Effects of Development and Aging on RNA Complexity

The effects of development on RNA complexity have been previously reviewed (Schultz and Church, 1974). The subject is, if anything, less clear than it was at that time. Schultz and Church (1974) described an increase in the expression of nonrepetitive DNA sequences in the total RNAs of mouse and rabbit embryos from about 1.8% of the nonrepetitive DNA at the blastocyst stage to 10% at birth. Brain total RNA data (Grouse *et al.*, 1972) also showed an increase in complexity during prenatal and early postnatal brain development. This impression was confirmed by Brown and Church (1972) and Cutler (1975). Moreover, Cutler also found evidence that total brain RNA complexity decreased with aging in the rat.

It is now realized that the interpretation of these earlier experiments is complicated by the fact that the complexity measurements were subsaturation and therefore were underestimates of the true complexity present. In experiments at saturation, Kaplan and his collaborators studied brain total poly(A)-RNA from rats of different ages (personal communication). Between 3 and 15 days of postnatal development they found a statistically significant 15% rise in the complexity of brain total poly(A)-RNA; however, they found no further change in complexity after 15 days of age. Similar results have also been found using brain polysomal poly(A)-mRNA from rats of different ages (B. Kaplan, personal communication). In addition to changes in the overall complexity of RNA during brain development, changes in the concentration of certain sequences have been observed. Experiments by Hahn and his co-workers, using cDNA prepared to brain poly(A)-mRNA from mice of different ages, have shown that certain RNA sequences present in high abundance immediately after birth are found at later stages of development to be present at greatly reduced concentrations (W. Hahn, personal communication). Thus, brain postnatal development appears to be associated with specific regulation of total poly(A)-RNA as well as poly(A)-mRNA sequences. Whether these changes are the result of transcriptional or post-transcriptional regulation is not certain. However, since data with nuclear and total brain RNA show increasing complexity with development, it is probable that transcriptional level changes occur. What is, nevertheless, most striking is the lack of dramatic RNA changes associated with the very profound alterations occurring in postnatal brain development. During active cell growth, cell migration, and synapse formation, only a 15% change in total poly(A)-RNA complexity was detected, with no changes occurring after 15 days of age.

VI. Sensory Experience and the Plasticity of Gene Expression

There are several lines of evidence indicating that neuronal development and differentiation occur concomitantly with, and possibly as a direct result of, selective gene expression. It is a different question whether experience modulates brain gene expression independently of development. Such an effect, dependent upon sensory experience, would be an example of neural plasticity. Plasticity has been defined as "adaptive modifications of neurons within the normal physiological range" or as "adjustments of the developing nervous system to changes in the internal or external milieu" (Jacobson, 1978). One could legitimately consider alterations or adaptations of brain transcription as a result of selective experience as a kind of neuronal plasticity, since such changes of the developing nervous system would certainly be responses to changes in the internal or external milieu.

We have investigated the effects of differential experience on gene expression in the developing brain using two experiential paradigms. The first paradigm was the environmental complexity model (Rozenzweig and Bennett, 1969). Weanling rats were placed in different environments (a "complex" environment had 12 rats together, with a maze, various stimulus objects, and frequent handling, and a "deprived" environment was one in which rats were raised alone with no handling or sensory stimulation). Total brain RNAs from rats in the two treatment conditions were then hybridized with rat nonrepetitive DNA to measure brain HnRNA sequence complexity. These results (Grouse et al., 1978a) agree with similar results previously published (Uphouse and Bonner, 1975) in showing (Table II) a greater RNA sequence complexity in the brain but not the liver RNA from the rats raised in the complex environment. However, it was not possible to tell whether the change occurred only in a particular brain region, or whether a specific part of the environmental enrichment caused the transcriptional change. Many of the other effects of environmental complexity on brain anatomy, weight, biochemistry, and pharmacology that have been studied appear to be all-or-none effects (Rosenzweig et al., 1969). That is, no single component of the enrichment produces the changes, only the combination. Whether this is true for complexity changes requires further investigation.

In order to examine the effects of a single sensory modality on gene expression, the second paradigm was tested in our laboratory. Pettigrew (1974) had shown that bilateral lid suture of newborn kittens disrupted the development of stimulus specificity in the kittens over a critical period during the first 3 months of life. These experiments and those of Wiesel and Hubel (1963), Hirsch and Spinelli (1971), and Blakemore and Cooper (1970) demonstrated experience-dependent developmental plasticity in the visual pathway. In this system, RNA complexity could be measured in the RNA of specific brain regions with and without lid suture. Results (Table II) of such experiments (Grouse et al., 1979a) have showed that (1) total RNA from the visual cortex of normally sighted kittens had a greater sequence complexity than RNA from the visual cortex of sutured animals, and (2) no changes in RNA complexity could be detected in brain areas other than the visual cortex. Thus, plasticity of gene expression in the developing mammalian brain was strongly suggested. This plasticity was stimulus-specific as well as brain region-specific. Table II summarizes the results of our work on the effects of environmental and genetic perturbations on RNA sequence complexity.

VII. Neuroblastoma Differentiation: A Model for Selective Gene Expression during Neural Development

Although the mouse neuroblastoma cell lines commonly used in neurobiological research cannot be considered "typical" neurons, they express characteristics of primary importance for neurobiologists: excitable membrane, regulation of neurotransmitter enzymes, neuronal morphology, and the ability to form competent synapses (Nelson et

TABLE II

Summary of Studies on RNA Sequence Diversity Responses to Environmental and Genetic Perturbations[a]

Paradigm	Species	Percentage of probe made double-stranded		Significance[b]	No. of gene equivalents changed[c]	Reference
		Control	Perturbed			
Environmental complexity	Rat					Grouse et al. (1978a)
Brain		10.6	8.2	<0.02	1.1×10^4	
Liver		5.3	5.3	N.S.	—	
Eyelid suture	Cat					Grouse et al. (1979a)
Visual cortex		11.8	8.7	<0.01	1.4×10^4	
Rest of cortex		13.8	13.1	N.S.	—	
Subcortical brain		13.0	12.9	N.S.	—	
Muscular dystrophy	Mouse					Grouse et al. (1978b)
Skeletal muscle		8.5	8.9	N.S.	—	
Spinal cord		5.1	6.2	N.S.	—	
Brain		12.0	12.4	N.S.	—	
Liver		5.2	4.9	N.S.	—	

[a] Total cellular RNA samples prepared from the indicated tissue were hybridized to tracers of homologous nonrepetitive DNA; extents of reaction were determined by S_1 nuclease digestion of unhybridized DNA.

[b] Calculated by the two-tailed t-test for independent samples; N.S., not significant.

[c] HnRNA transcript length of 4500 nucleotides is assumed.

al., 1976). Our goal was to obtain a tissue culture system that could be used to study these fundamental events of neuronal maturation and synaptogenesis at the molecular level. To this end we adapted the neuroblastoma NS20Y (a neuroblastoma clone shown to be capable of neuromuscular junction formation) to suspension culture where it was observed to dedifferentiate by biochemical and morphological criteria (Grouse *et al.,* 1980). The polysomal poly(A)-mRNA of these maximally dedifferentiated cells was used for comparison with the polysomal poly(A)-mRNA of the same cells following differentiation. Dibutyryl cyclic AMP was found to be the most effective means of bringing about this differentiation, which included neurite outgrowth, an increase in electrical excitability, a six-fold rise in choline acetyltransferase, the development of neuronal morphology, and competence in synapse formation.

To study the mRNA populations present in the differentiated and dedifferentiated cells, we isolated polyribosomal poly(A)-mRNA from NS20Y cells in each condition. This provided us with the mRNAs responsible for ongoing protein synthesis. We synthesized cDNA probes to each of the mRNA preparations. By hybridizing mRNA with the cDNA prepared from it, we determined the overall complexity and abundance classes of the mRNAs (Bishop *et al.,* 1974), as shown in Fig. 1. The differentiated cell poly(A)-mRNA has a total complexity of 1.30×10^7 nucleotides, while the mRNA from cells grown in suspension culture had a complexity of 1.34×10^7 nucleotides (approximately 8000 mRNAs) (Grouse *et al.,* 1980). Similar measurements had been previously made by Felsani *et al.* (1978) on poly(A)-mRNA of suspended neuroblastoma cells and on such cells differentiated by the absence of serum. When one mRNA population is used to drive the reaction of the cDNA copied from the other, the extent of overlapping of the two RNA populations can be assessed. Such heterologous reactions demonstrated that there were certain mRNA sequences present only in the differentiated cells; the presence of sequences unique to suspension cells could not be ruled out.

To quantitate and characterize the RNA sequences present in one cell condition but not in the other, specific cDNA probes to differentiation-specific mRNAs and to suspension culture-specific mRNAs were prepared by hybridization with heterologous mRNA to eliminate all cDNA sequences complementary to the heterologous mRNAs. These specific probes were then hybridized with homologous mRNA and, on the basis of the reaction, the number and concentration of species unique to a given treatment condition were determined.

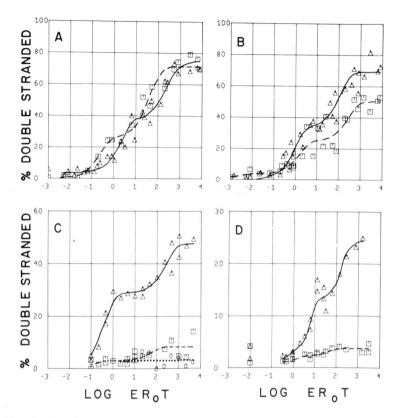

FIG. 1. Complementary DNA reactions with NS20Y poly(A)-mRNA. (A) Complementary DNA prepared from suspension cell poly(A)-mRNA was reacted with suspension cell poly(A)-mRNA (triangles, solid line) or with differentiated cell poly(A)-mRNA (squares, dashed line). (B) Complementary DNA prepared from differentiated cell poly(A)-mRNA was reacted with differentiated cell poly(A)-mRNA (triangles, solid line) or suspension cell poly(A)-mRNA (squares, dashed line). (C) A complementary DNA probe specific to sequences present only in the differentiated cell mRNA was reacted with differentiated cell poly(A)-mRNA (triangles, solid line), with suspension cell poly(A)-mRNA (squares, dashed line), or with yeast RNA (circles, dotted line). (D) A complementary DNA probe enriched in sequences present only in the suspension cell mRNA was reacted with suspension cell poly(A)-mRNA (triangles, solid line) or with differentiated cell poly(A)-mRNA (squares, dashed line).

Data from the hybridization experiments described above are presented in Fig. 1. At least three to four specific mRNAs were found to be present at a high concentration only in the differentiated cells, while at least another 360 mRNAs were present at a low concentration. There were also genes expressed in the undifferentiated cells but not in the differentiated cells. Three mRNAs at high concentration and 280 at low concentration could be detected, which were unique to the undifferentiated cells. This study provides evidence that neuro-blastoma differentiation involves the appearance of new, and the disappearance of old, mRNA sequences in polyribosomes. Felsani *et al.* (1978) found mRNA sequences unique to undifferentiated cells but found none specific to differentiated cells. Their differentiation paradigm (culture in serum-free medium) is quite different from ours [treatment with dibutyryl cAMP (dBcAMP)] and may account for the observed differences between our findings. We believe that the long dedifferentiation in suspension culture followed by dBcAMP treatment which we employed optimizes the increment of differentiation. Cells deprived of serum differentiate but do not survive for long periods of time following this treatment.

We are at present attempting to characterize the poly(A)-mRNA species present at high concentration exclusively in the differentiated cells by restriction endonuclease treatment and gel electrophoresis of the differentiation-specific cDNA.

VIII. Summary and Conclusions

Perhaps the most significant observation concerning transcription in the nervous system is the extremely high diversity of the genetic information transcribed in brain. In part this is probably related to the diversity of cell types in the brain. However, brain regions as different as cerebellum, cerebral cortex, and hypothalamus appear to express similarly high and essentially completely overlapping sequence complexities. Neuronal cell lines may also have a higher diversity of genetic information expressed than nonneuronal cell lines. These findings suggest that the major portion of the genetic complexity expressed in the nervous system is not concerned with specifying differences between neuronal cell types or regional differences in nervous system structure, but rather relates to other aspects of nervous system function.

Experiential factors seem to have an extremely powerful effect on transcriptional complexity. Either global perturbations such as the environmental enrichment–impoverishment paradigm or more specific sensory alterations (lid suture during the critical period for visual cortical development) produce changes of 20–30% in gene expression at the nuclear RNA level. As a working hypothesis, we propose then that the high sequence complexity of brain may reflect the enduring response of the brain to environmental conditions and therefore may be involved in a major way with the class of phenomena termed "plasticity."

For further progress in this area, however, we feel that more controllable systems must be exploited, and the differentiation paradigms available in neuronal cell lines seem attractive candidates in this regard. Differentiation-related sequences (either those appearing or those disappearing during differentiation) can be purified. Powerful methods for isolation and cloning of specific sequences are now available. When such purified probes are available, more complex systems may be approached with specific questions. Do the sequences that change with differentiation in neuroblastoma occur at all in brain mRNA or HnRNA? Are these sequences differentially expressed during development? During behavioral or experiential manipulations? What is the organization of these sequences in the genome? Does this organization change during development? Exciting analogies with the immune systems might be explored when specific probes for gene organization are available.

The conceptual framework for considering mechanisms of genetic structure and regulation is in an exciting state of flux. The new ideas currently developing may have great relevance to the understanding of brain function, and this application will probably be an area of great interest in the next few years.

REFERENCES

Bantle, J. A., and Hahn, W. E. (1976). Cell 8, 139–150.

Bester, A., Kennedy, D., and Heywood, S. (1975). *Proc. Natl. Acad. Sci. U.S.A.* 72, 1523–1527.

Bishop, J. O., Morton, J. G., Rosbash, M., and Richardson, M. (1974). *Nature (London)* 250, 199–204.

Blakemore, C., and Cooper, G. (1970). *Nature (London)* 228, 477–484.

Brown, I. R., and Church, R. B. (1972). *Dev. Biol.* 29, 73–84.

Chikaraishi, D. M., Deeb, S. S., and Sueoka, N. (1978). *Cell* 13, 111–120.

Cutler, R. G. (1975). *Exp. Gerontol.* 10, 37–60.

Davidson, E. H., Klein, W. H., and Britten, R. J. (1977). *Dev. Biol.* 55, 69–84.

Edmonds, M., and Caramela, M. G. (1969). *J. Biol. Chem.* 244, 1314–1324.

Felsani, A., Berthelot, F., Gors, F., and Broizat, B. (1978). *Eur. J. Biochem.* 92, 569–577.

Galau, G. A., Klein, W. H., Britten, R. J., and Davidson, E. H. (1977). *Arch. Biochem. Biophys.* 179, 584–599.

Garapin, A. C., Lepennec, J. P., Roskam, W., Perrin, F., Cami, B., Krust, A., Breathnach, R., Chambon, P., and Kourilsky, P. (1978). *Nature (London)* 273, 349–354.

Goldberg, S., Schwartz, H., and Darnell, J. E., Jr. (1977). *Proc. Natl. Acad. Sci. U.S.A.* 74, 4520–4523.

Greenberg, J. R. (1976). *Biochemistry* 15, 3516–3522.

Grouse, L. D., Chilton, M.-D., and McCarthy, B. J. (1972). *Biochemistry* 11, 798–805.

Grouse, L. D., Omenn, G., and McCarthy, B. (1973). *J. Neurochem.* 20, 1063–1073.

Grouse, L. D., Schrier, B. K., Bennett, E. L., Rosenzweig, M. R., and Nelson, P. G. (1978a). *J. Neurochem.* 30, 191–203.

Grouse, L. D., Nelson, P. G., Omenn, G. S., and Schrier, B. K. (1978b) *Exp. Neurol.* 59, 470–478.

Grouse, L. D., Schrier, B. K., and Nelson, P. G. (1979a). *Exp. Neurol.* 64, 354–359.

Grouse, L. D., Letendre, C. H., and Schrier, B. K. (1979b). *J. Neurochem.* 33, 583–585.

Grouse, L. D., Schrier, B. K., Letendre, C. H., Zubairi, Y., and Nelson, P. G. (1980). 255, 3871–3877.

Hahn, W. E., and Laird, C. D. (1971). *Science* 173, 158–161.

Harris, H. (1963). *Prog. Nucl. Acid Res.* 2, 19–29.

Hirsch, H. V. B., and Spinelli, D. N. (1971). *Exp. Brain Res.* 12, 509–515.

Hough, B. R., Smith, M. J., Britten, R. J., and Davidson, E. H. (1975). *Cell,* 5, 291–299.

Humphries, S., Windass, J., and Williamson, R. (1976). *Cell* 7, 267–277.

Irwin, D., Kumar, A., and Mait, R. A. (1975). *Cell* 4, 157–165.

Jacobson, M. (1978). *In* "Developmental Neurobiology," p. 89. Plenum, New York.

Kaplan, B. B., Schachter, B. S., Osterburg, H. H., deVellis, J. S., and Finch, C. E. (1978). *Biochemistry* 17, 5516–5524.

Kaufmann, R., Milcarek, C., Berissi, H., and Penman, S. (1977). *Proc. Natl. Acad. Sci. U.S.A.* 74, 4801–4805.

Knochel, W., and Kohnert-Stavenhagen, E. (1977). *Hoppe-Seyler Z. Physiol. Chem.* 358, 835–842.

Konkel, D. A., Tilghman, S. M., and Leder, P. (1978). *Cell* 15, 1125–1132.

Levy, B. W., and McCarthy, B. J. (1975). *Biochemistry* 14, 2440–2446.

McGrogan, M., and Raskas, H. J. (1978). *Proc. Natl. Acad. Sci. U.S.A.* 75, 625–629.

Maxwell, I., Lauth, M., and Hahn, W. (1978). *J. Cell Biol.* 79, 135a (Abstr.).

Nelson, P., Christian, C., and Nirenberg, M. (1976). *Proc. Natl. Acad. Sci. U.S.A.* 73, 123–127.

Nemer, M. (1975). *Cell* 6, 559–570.

Pettigrew, J. P. (1974). *J. Physiol.* 237, 49–74.

Rosenzweig, M. R., and Bennett, E. L. (1969). *Dev. Psychobiol.* 2, 87–95.

Rosenzweig, M. R., Bennett, E. L., Diamond, M. C., Wei, S.-Y., Slagle, R. W., and Saffran, E. (1969). *Brain Res.* 14, 427–445.

Ross, J., and Knecht, D. A. (1978). *J. Molec. Biol.* 119, 1–20.

Sakano, H., Rogers, J. H., Hiippi, K., Brack, C., Traunecker, A., Maki, R., Wall, R., and Tonegawa, S. (1979). *Nature (London)* 277, 627–633.

Sauerbier, W. (1976). *Adv. Radiat. Biol.* 6, 49–106.

Savage, M. J., Sala-Trepat, J. M., and Bonner, J. (1978). *Biochemistry* 17, 462–467.

Schrier, B. K., Zubairi, M. Y., Letendre, C. H., and Grouse, L. D. (1978). *Differentiation* 12, 23–30.

Schultz, G. A., and Church, R. B. (1974). *Curr. Top. Dev. Biol.* 8, 179–202.

Tilghman, S. M., Tiemier, D. C., Seidman, J. G., Peterlin, B. M., Sullivan, M., Maizel, J. V., and Leder, P. (1978). *Proc. Natl. Acad. Sci. U.S.A.* 75, 725–729.

Tonegawa, S., Maxam, A., Tizard, R., Bernard, O., and Gilbert W. (1978). *Proc. Natl. Acad. Sci. U.S.A.* 74, 3518–3522.

Uphouse, L., and Bonner, J. (1975). *Dev. Psychobiol.* 8, 171–176.

Van Ness, J., Maxwell, I., and Hahn, W. (1978). *J. Cell Biol.* 79, 341a (Abstr.).

Wiesel, T. N., and Hubel, D. (1963). *J. Neurophysiol.* 26, 1003–1017.

INDEX

CONTENTS OF PREVIOUS VOLUMES

405

Volume 14: Immunological Approaches to Embryonic Development and Differentiation, Part II

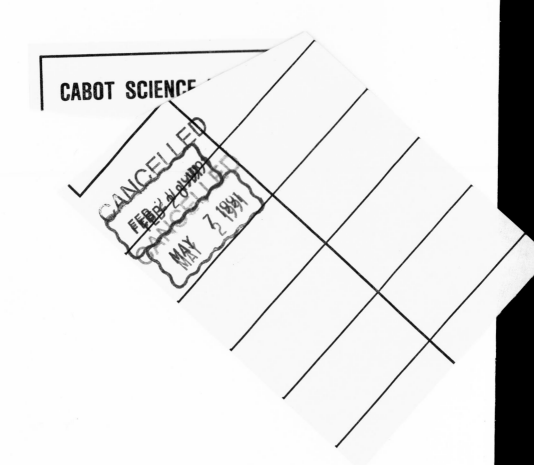